Jewish Environmentalism in the Post-Secular Age

Presenting Jewish environmentalism as a case study of contemporary post-secularism, this book exemplifies the responses of world religions to the global ecological crisis and situates Jewish environmental spirituality historically, socially, theologically, and politically.

Discussing key figures, texts, and organizations that have given Jewish environmentalism its distinctive character as "earthly spirituality," the book explains how science and technology functioned in the secularization of Judaism, on the one hand, as well as in the counter-process of sacralization of nature. It explores the various dimensions of Jewish environmentalism: official statements of Jewish denominations, eco-theologies, eco-hermeneutics, eco-feminism, and public theology; and also links Jewish environmentalism to main forces in modern Jewish history—Zionism and Americanization. Finally, the study explores the contribution of "non-Jewish Jews" to American environmentalism and explains how "radical Jews" of Leftist orientation became Spiritual Progressives who promoted the ecological wisdom of Judaism to address the social ills of American society as well as the global ecological crisis.

This book will be of great interest to students and scholars of Jewish studies, religious studies, secularism/post-secularism studies, religion and ecology, and religion and science.

Hava Tirosh-Samuelson (PhD Hebrew University of Jersualem, 1978) is Regents Professor of History and Irving and Miriam Lowe Professor of Modern Judaism at Arizona State University in Tempe, AZ. A Jewish intellectual historian, Tirosh-Samuelson writes on Jewish philosophy and mysticism, religion, science and technology, and religion and ecology. In addition to seventy-five essays and book chapters, she is the author of three monographs—the award-winning *Between Worlds: The Life and Thought of Rabbi David ben Judah Messer Leon* (1991); *Happiness in Premodern Judaism: Virtue, Knowledge and Well-Being* (2003), and *Religion and Environment: The Case of Judaism* (2020). She is also the editor of eight volumes including *Judaism and Ecology: Created World and Revealed Word* (2002) and *The Legacy of Hans Jonas: Judaism and the Phenomenon of Life* (2008). Tirosh-Samuelson is the Editor-in-Chief of the Library of Contemporary Jewish Philosophy (2012–2018), a set of 21 volumes.

Routledge Environmental Humanities

Series editors: Scott Slovic (University of Idaho, USA), Joni Adamson (Arizona State University, USA) and Yuki Masami (Aoyama Gakuin University, Japan)

The Routledge Environmental Humanities series is an original and inspiring venture recognising that today's world agricultural and water crises, ocean pollution and resource depletion, global warming from greenhouse gases, urban sprawl, overpopulation, food insecurity and environmental justice are all crises of culture.

The reality of understanding and finding adaptive solutions to our present and future environmental challenges has shifted the epicenter of environmental studies away from an exclusively scientific and technological framework to one that depends on the human-focused disciplines and ideas of the humanities and allied social sciences.

We thus welcome book proposals from all humanities and social sciences disciplines for an inclusive and interdisciplinary series. We favour manuscripts aimed at an international readership and written in a lively and accessible style. The readership comprises scholars and students from the humanities and social sciences and thoughtful readers concerned about the human dimensions of environmental change.

For more information about this series, please visit: www.routledge.com/Routledge-EnvironmentalHumanities/book-series/REH

Jewish Environmentalism in the Post-Secular Age

Earthly Spirituality

Hava Tirosh-Samuelson

Routledge
Taylor & Francis Group
LONDON AND NEW YORK

First published 2026
by Routledge
4 Park Square, Milton Park, Abingdon, Oxon OX14 4RN

and by Routledge
605 Third Avenue, New York, NY 10158

Routledge is an imprint of the Taylor & Francis Group, an informa business

For Product Safety Concerns and Information please contact our EU representative GPSR@taylorandfrancis.com. Taylor & Francis Verlag GmbH, Kaufingerstraße 24, 80331 München, Germany.

British Library Cataloguing in Publication Data
A catalogue record for this book is available from the British Library

ISBN: 978-0-8153-5773-5 (hbk)
ISBN: 978-1-041-22783-0 (pbk)
ISBN: 978-1-351-12394-5 (ebk)

DOI: 10.4324/9781351123945

Typeset in Times New Roman
by Taylor & Francis Books

In memory of Rabbi Arthur Waskow (1933–2025), who exemplified Jewish earthly spirituality.

Contents

Acknowledgments

This study integrates several intellectual disciplines: history, philosophy, religious studies, Jewish studies, religion and ecology, religion and science, secularism/post-secularism studies, and gender studies. Although interdisciplinarity has characterized my academic career, this book emerged out of the project "Beyond Secularization: Religion, Science, and Technology in the Post-Secular Age" (2019–2024) at Arizona State University. With external funding from the Templeton Religion Trust, the project was conducted with my colleagues J. Benjamin Hurlbut and the late Gaymon Bennett, who sadly passed away on February 2, 2024. I am grateful to my colleagues J. Benjamin Hurlbut, the late Gaymon Bennett, Craig Calhoun, Linell E. Cady, and Gregg Pascal Zachary for the intellectual camaraderie they have provided for over a decade. Their insights, knowledge, and probing exchanges have enabled me to grow as a thinker and a scholar. Our project included several Postdoctoral Fellows—Charles McCreary, Heather Melquist, and Schyler Marquez—and several ASU graduate students—Robert Fuller, Mario Oropesa Hernandez, Dan Affsprung, Anne Hammang, and Taylor R. Genovese. Our discussions during the project enabled me to clarify the interplay of religion, nature, and culture. The project was administered by the Center for the Study of Religion and Conflict at Arizona State University, and the work of its Assistant Director at the time, Carolyn Forbes, has made this large project go as smoothly and productively as possible. Special thanks are hereby expressed to the officers of the Templeton Religion Trust that funded the project, supporting the bi-weekly seminars, publications, public presentations, and research conferences. It was a truly enriching endeavor.

Learning and teaching are closely intertwined for me: I learn to teach, I teach what I have learned, and I learn as I teach. Teaching the undergraduate courses "Judaism and Ecology" and "Nature, Sustainability, and Religion" at ASU enabled me to experience the nexus of learning and teaching. Similarly, the mentoring of graduate students who study religion and ecology—the recently minted PhD, Dr. Robert Fuller, and the interaction with my current graduate students, Avalon Jade Theisen and David Krantz—has helped me to hone my approach to religion and ecology while helping them articulate their intellectual identity.

Oral instruction is different than written expression, in this project research, writing, and speaking were closely related. I am most grateful to two developmental editors, Jennifer Quincey and Regina Higgins, who showed me how to use language more economically, saying more with less, to Mike Isralewitz for copyediting the manuscript, and to Ina Gravitz for preparing the index. I am especially appreciative of my colleague Professor Joni Adamson, and her co-editors, Professors Scott Slovic and Yuki Masami, who supported the publication of this study in the Routledge series of Environmental Humanities.

While working on this project, I published a book and several articles on Jewish environmentalism. These publications have framed the views expressed in this study and they are cited as relevant. Additionally, I delivered public lectures about Jewish environmentalism in academic settings—Emory University, Humboldt University-Berlin, Bar-Ilan University, Hebrew University of Jerusalem, the University of Manchester, Case Western Reserve University—as well as in nonacademic settings, Jewish and interfaith, in Zurich, Switzerland; Riverside, Danville, and Palo Alto in California; and in Dallas, Texas. I was also interviewed for podcasts of the Association of Jewish Studies and the Voice of America, and a blog of the Times of Israel. I am convinced that public-facing scholarship is essential if environmentalism is to receive the attention it deserves from Jews and non-Jews. I hope that readers of this book will be persuaded that Judaism has responded to the ecological crisis in meaningful ways because caring for the Earth is a Jewish religious obligation.

Introduction: The Anthropocene

We are living through a massive ecological crisis that has been caused by cumulative human actions. Human activities since the industrial revolution have brought about the current ecological crisis evident in global warming, shifting weather patterns, extreme weather events, retreat of glaciers and melting of icecaps, rising sea levels, megadroughts and wide-spread desertification, devasting fires, mass extinction of species, loss of fisheries and forests, acidification of oceans, pollution of air and contamination of water and soil, and shifts in the range and prevalence of diseases (Henson 2014; Tirosh-Samuelson 2020, 11–19). Whether we refer to it as a "crisis," a "catastrophe," a "calamity," or an "emergency," it is quite clear that the widespread environmental degradation results from the disruption of "the fundamental relationship between carbon dioxide and climate, and more broadly … the interactions between Earth and life cycling carbon" (Knoll 2021, 216). This disruption if continued unabated will seriously compromise the inhabitability of the Earth (Wallace-Wells 2019). Because human activities have impacted all ecosystems and even the biosphere itself, scientists now speak about the dawn of a new geological era in which humanity has become a geological force. They named it the Anthropocene (Crutzen and Stoermer 2000; Crutzen and Steffen 2003; Ruddiman 2013), a name that captures the "dramatic influence that human societies are having on the physical world" (Davies 2016, 10).

Although the term 'Anthropocene' was coined about twenty-five years ago, to date, there is no consensus about the definition, temporal scope, meaning, or implications of the Anthropocene. How we understand, interpret, and evaluate the Anthropocene depends on the kind of narrative we construct to make sense of the known facts (Hamilton, Bonneuil, and Gemenne 2015; Angus 2016; Bonneuil and Fressoz, 2017; Ellis 2018). Christophe Bonneuil discussed four grand narratives that frame the Anthropocene discourse: the naturalist narrative, the post-nature narrative, the eco-catastrophic narrative, and the eco-Marxist narrative (Bonneuil 2015), all of which are secular narratives constructed by scientists for whom religion is not relevant to the ecological crisis. In contrast, this book and my previous publications join scholars who have argued that religion is very relevant to the ecological crisis

DOI: 10.4324/9781351123945-1

and that world religions can be part of the solution rather than part of the problem. Specifically, this book argues that Judaism offers a meaningful framework for facing the challenges of the Anthropocene and calls on Jews and non-Jews to become familiar with the ecological wisdom of the Jewish tradition that nourishes Jewish environmental spirituality. However, to appreciate the relevance of religion to the current ecological crisis, we need to acknowledge that religion is not merely a set of beliefs that one asserts in private, but a way of life that one practices in public. That acknowledgment marks the transition from modern secularism (1870–1950) to postmodern post-secularism (1980–present), which came at the very same time when the ecological crisis was fully recognized, and its causes became better understood. Since religious leaders, theologians, and ethicists have also responded to the ecological crisis, this book considers the religious responses to the eco-crisis as one manifestation of contemporary post-secularism, namely, the outlook that refuses to confine religion to the private sphere, insisting instead that religion has a role to play in shaping public discourse, policies, and education.

The Anthropocene discourse is riddled with ambiguities because the term 'Anthropocene' is employed both descriptively and prescriptively. As a *descriptive* term, the Anthropocene names a certain situation in which humans have impacted all ecosystems, thus "overcoming the great forces of nature" (Steffen, Crutzen, and McNeill 2007). Human activities have profoundly changed the atmosphere, water resources, oceans, soil, forests, and biodiversity so that the damage humans have inflicted is not irreversible and permanent. Due to human environmental impact, it is no longer possible to think about nature as "pristine" environment in which humans are not involved. With every increasing human impact on land, oceans, and the atmosphere, the Anthropocene signifies the breakdown of the binary dichotomy between "humanity" (namely, "society" and "culture") and "nature," and with the collapse of this conceptual dichotomy, the Anthropocene presumably inaugurates a "post-natural" or "post-wild" phase to human history (Purdy 2015). Accordingly, the wilderness does not exist at all, since all of it has been influenced by humans at some time. Bill McKibben names this condition "the end of nature," and he renamed our profoundly transformed planetary home "Eaarth" or "Earth 2" (McKibben 2011). According to McKibben, the damage humans have inflicted on the Earth is irreversible.

So, how should humans conduct themselves to survive, let alone thrive, in the Anthropocene? This question shifts the focus from the descriptive to the *prescriptive* aspect of the Anthropocene discourse. Crutzen defined the task facing humanity in *technological* terms, stating

> A daunting task lies ahead of scientists and engineers to guide society towards environmentally sustainable management during the era of the Anthropocene. This will require appropriate human behavior on all

> scales, and may well involve internationally accepted, large-scale geoengineering projects, for instance, to "optimize" climate.
>
> (Crutzen 2003, 23)

Crutzen's urgent call for engineering the climate responded to the failure of the world community to heed the scientists' warning about global warming (Crutzen 2006), but his endorsement of a technological fix manifests the paradoxes of the Anthropocene: Can human-made technology solve the human-made ecological crisis brought about by human technology? Can human-made technology save us from the calamity it has brought upon us?

Techno-optimists see the Anthropocene as an extraordinary opportunity for humanity to remake the world using innovative technologies to recreate Eden on Earth referred to variously as "ecomodernists," "neo-environmentalists," or "postmodernist greens" (Asafu-Adjaye 2015; Wuerthner, Crist, and Butler 2014). The people who signed "An Ecomodernist Manifesto" have maintained that humans should regard themselves as planetary gardeners who manage the planet rationally using new technologies and enlightened science (e.g., biotechnology, synthetic biology, nuclear power, and geoengineering), all for the benefit of humanity (e.g., Nordhaus and Shellenberger 2007; Marris 2011). For techno-optimists, humanization or domestication of the planet by means of technology is the best assurance that humans will flourish in the Anthropocene.

Ecomodernists see the Anthropocene as an "Engineered Age," an epoch in which humans have launched "the deliberate large-scale manipulation of the planetary environment to counteract anthropogenic climate change" (The Royal Society 2009). This has led promoters of geoengineering to see technology as the guarantor of human flourishing in the Anthropocene. More modest advocates of geoengineering regard these technological interventions merely remedial; they will only "buy time" enabling humans to facilitate "atmospheric restoration" (Keith 2000; 2013). By contrast, there are those who lament the forthcoming Anthropocene because it reflects the unwelcome mastery of Earth systems by humans. For them, the Anthropocene does not portend human flourishing by means of technology but rather the "requiem for a species." As Clive Hamilton argued in the book under that title,

> Recourse to climate engineering to counter human-induced warming is an unconscious attempt by one species to decouple the great process that links the composition of the atmosphere to the temperature of the Earth and the biotic system of the land and oceans. Instead of decoupling growth of the economy from the growth of carbon emission, the climate engineers want to decouple global warming from growth of carbon emission.
>
> (Hamilton 2015, 181)

Critics of geoengineering not only decry the environmental damage characteristic of the coming Anthropocene, they also counsel caution, claiming that the Anthropocene demands a modification of human behavior and curtailment of human hubris that has given us the environmental crisis in the first place. In between, there are the techno-pragmatists, who recognize that although geoengineering technologies are not unproblematic solutions to the problem of climate change and its environmental manifestations, geoengineering might be the best available option for the future of human species on a damaged planet. Whatever position one takes about climate engineering, it is evident that those issues are not merely techno-scientific, but rather laced with social, ethical, and political considerations (Gardiner 2011; 2016; 2021).

Technology, or more broadly technoscience, is a feature of contemporary life. We cannot reflect on human life in the present or the future without taking technoscience into consideration. Yet, technoscience is itself a product of culture that needs to be understood historically (Latour 2017). Modern technoscience was grounded in the assumption that "religion" and "science" are two distinct realms of human experience that are often in conflict with each other. But what if we challenge this assumption and recognize that religion and science have been historically, conceptually, and practically very close to each other, developing over time through cross-fertilization rather than through mutual negation?

The complex relationship between religion and technoscience in the context of the ecological crisis is most evident in the current debate about geoengineering. Although large-scale human impact on the environment is not new, geoengineering is qualitatively different because it is a "more comprehensive and invasive impact" and because "geoengineering projects seek to intentionally alter global climate systems" (Clingerman 2014). More than a mere technological fix of the eco-crisis, geoengineering is predicated on an implicit interpretation of the meaning of being human and the role of technoscience in our humanness and in human relationship with nature. Mike Hulme (2009) has insightfully noted that the vehement debates about climate change do not concern observable facts and physical realities but rather the interpretations of these facts, the meaning we ascribe to them, and the social, political, and cultural practices that flow from them. The debates on climate change and climate engineering, and by extension about the Anthropocene, are not about what this or that technology can do and should do but about the meaning of being human. These reflections are inseparable from our worldview and a host of non-scientific assumptions, beliefs, values, and norms.

Scholars in the field of religion and ecology have insisted that discussions about responses to climate change cannot ignore religion "since religion is fundamental to how human beings understand themselves and their place in the world" (Clingerman and O'Brien 2014, 29). Moreover, the various geoengineering proposals reflect an "implicit anthropology" about "the balance of human capability and fallibility" (Clingerman 2014, 16), so the presumably secular scientific discussion is laced with unacknowledged religious and ethical assumptions (Clingerman 2015). Indeed, the human species, whose conduct is responsible for the Anthropocene, is a *homo religiosus*, that is, a being that ponders the meaning of human existence in religious terms; a being who poses questions to which religion is the answer. Precisely because these debates are not just about what "is" (i.e., the facts of climate change) but rather about what "ought" to be (i.e., the meaning of these facts), the debates are suffused with "language taken from religion, theology, and morality" (Hulme 2009, 157; Cf. Hoffman 2015; Jenkins, Berry, and Kreiden 2018). Christopher Doran, a Christian environmentalist, has rightly noted that "environmental problems are not primarily scientific, political, or economic in nature; they are essentially moral and theological" (Doran 2017, 7), and this insight is endorsed by scholars in the field of religion and ecology and by religious environmentalists, including Jewish environmentalists. We cannot discuss human responses to the human-made ecological crisis or speculate about the future of humanity without taking religion into consideration.

Interdisciplinarity

Many participants in the Anthropocene discourse are secular scientists who assume that religion is either irrelevant to the scholarly (i.e., truthful) understanding of the world, or worse, that religious views necessarily conflict with scientific truths. The assumption that "religion" and "science" necessarily conflict with each other needs to be historicized, because the conflict thesis itself expressed the assumptions of the modern Enlightenment project that equated "science" not only with "reason" and "progress," but also with human "mastery of nature." The ecological crisis we currently experience was one of the tragic outcomes of modern constructions which separated humanity from nature and worshipped the unlimited powers of humans, encouraging them to exploit the resources of nature for the sole benefit of human advancement and happiness.

Allied to the modern ideology of "progress," modern technoscience not only venerated human powers to control nature, but it also presupposed the independence technoscience from religion or ethics. However, in the late 1960s, the academic discourse of science and religion emerged, challenging the binary dichotomy of religion and science and their presumably inevitable "conflict," instead urging us to appreciate the dynamic relations between science and religion historically, conceptually, and ethically (e.g., Brooke 1991; Ferngren 2002; Harrison 2017). When we recognize the interplay of religion and technoscience it becomes clear that we cannot think about the ecological crisis or about the Anthropocene without exploring the roles of both religion and technoscience, which makes our discussion necessarily interdisciplinary. Indeed, the discourses that inform this book—secularism/post-secularism studies, religion and science, religion and ecology, environmental ethics, religious studies, and gender studies—are all interdisciplinary inquiries, and so is the academic field of Jewish Studies, which integrates the disciplines of history, philosophy, religion, political science, sociology, literature, and the arts among others. This book approaches Judaism as a way of life that must be understood on its own terms but situated in the broad social, cultural, and historical context. By engaging the discourses of religion and science, religion and ecology, and secularism/post-secularism studies from the perspective of Judaism, this study shows how Judaism harbors deep insights relevant to the current ecological crisis.

Why and how does religion matter to the ecological crisis? Religion matters greatly to our thinking about the environment, because over 80 percent of people in the world construct reality in religious categories that express their culture (Grim and Tucker 2014, 28; Bratton 2021). Human beings understand themselves, their societies, and their daily life through sacred and symbolic rituals that point beyond themselves to an ultimate reality. Within religious worldviews human beings organize their life and find meaning, purpose, and hope as they face the unknown future. Religion provides the moral lens through which humans evaluate every aspect of life and decide what is good and bad, what is permitted and forbidden. Religious traditions express human existential and emotional needs, and frame that which we care most about, namely, our ultimate concern. Because religious traditions manifest a person's ultimate concern about ultimate reality, people are willing to sacrifice their life for it, and this commitment makes religion a force that mobilizes people to action. If we are to change human conduct to mitigate environmental degradation that will undermine human flourishing in the Anthropocene, we must concern ourselves with religious questions and use religious categories. In so doing, this study too has descriptive and prescriptive dimensions: it does not seek to remain disinterested and objective but rather calls on us to become engaged, concerned, and motivated to care for the Earth.

When the trope of the Anthropocene became widespread in the early 2000s, the academic fields of religion and science and religion and ecology were already fully developed, and a good number of scholars contributed to these inherently interdisciplinary discourses. This book too is interdisciplinary, integrating the above discourses of religion and science and religion and ecology along with history, philosophy, religious studies, Jewish studies, and gender studies. Complex, multifactorial, and multifaceted phenomena requires interdisciplinarity and this holds true for Jewish environmental spirituality as well.

I approach Jewish environmentalism as an intellectual historian. That means that I pay attention to historical processes, social institutions, influential individuals, key texts, and habits of mind. As an intellectual historian, I situate ideas in their sociocultural context, paying special attention to cross-cultural interaction between Jews and non-Jews, since Jewish culture always evolved through conversation with surrounding cultures. While I explicate ideas articulated in texts, I also pay attention to the life and public activities of the thinkers who articulated these ideas. Jewish environmental spirituality emerged out of specific life trajectories of certain people who brought these ideas to the attention of the public and who have tried to teach the public to think and behave in a certain way. By attending to the biographies of influential visionaries, I signal the importance of place, time, and embodiment in the interpretation of spirituality as well as the relevance of particularity, diversity, and plurality. However, since I am not a trained sociologist, ethnographer, or anthropologist, this study does not describe or analyze the lived experience of environmentally committed Jews, who call themselves EcoJews, although this ethnographic approach now dominates the field of religion and ecology (e.g., Baugh 2017; 2024; Wilkinson 2012). Instead, I focus on ideas, texts, processes, and thinkers rather than on what contemporary practitioners of Judaism feel about the ecological crisis or how they express environmental commitments. Although I focus on Jews and Judaism, I believe that this book could be of interest not only to Jews but to all people who care about the future of planet Earth. Given the centrality of the Bible in framing Jewish environmentalism, this study could be of interest to environmentalists of other religious traditions, especially Christians and Muslims. As a Jew by birth and by choice, I am convinced that Jewish environmental spirituality offers a meaningful response to our ecological crisis encouraging us to make the right choices that will impact the life of generations to come. That work must be done in collaboration with environmental thinkers and activists of other religious traditions.

The Plan of the Book

Chapter 1 lays out the theoretical frame of the book by clarifying key terms, including 'secular,' 'secularism,' 'post-secular,' 'post-secularism,' 'science,' 'religion,' and 'spirituality.' The chapter shows the historical development of these concepts and the intersection of two academic discourses that are usually kept apart: science and religion and religion and ecology. The chapter also explains what 'spirituality' means for various users and how it differs from 'religion.' The chapter argues that spirituality should not be viewed as an alternative to religion but as a dimension of religion so that different religious traditions develop distinctive spiritualities. Chapter 2 offers an overview of Jewish environmentalism in the United States, the locus of the largest and most influential Diaspora Jewry. Using broad brush strokes, the chapter identifies the leading visionaries of Jewish environmentalism, prominent environmental organizations, types of activities, and persistent challenges. Since Judaism is a religious tradition, Jewish environmentalism necessarily uses religious categories, even when it is articulated by Jews who underwent processes of secularization. Jewish environmentalists care about the natural world for religious reasons, and they invest "secular" (i.e., "down-to-earth") activities with religious significance. Chapter 3 focuses on

Jewish theologians who reflected on the relationship between God, humanity, and nature and whose thought inspired the environmental discourse. The chapter shows that eco-theology is not the prerogative of one strand of contemporary Judaism but rather that it cuts across all strands of Judaism (i.e., Reform, Conservative, Reconstructionist, Orthodox, and Jewish Renewal). Therefore, there is no contradiction between environmentalism and Judaism, and there are many ways to be a Jewish environmentalist.

Environmentalism is concerned about the well-being of the Earth and all its inhabitants, but biological species are internally differentiated by sex and gender. Chapter 4 explores the impact of feminism on "earth-based Judaism" and the relevance of gender. Feminism and environmentalism emerged side by side in the early 1970s, but their impact on society and the responses to them have differed. In the 1970s and 1980s both feminism and environmentalism transformed public awareness and brought about many tangible achievements. Since then, both movements experienced a backlash, generating criticism and resistance from their opponents who consider feminism and environmentalism as part of the progressive political agenda to which they oppose. Chapter 4 discusses feminist variants of Earth-based Judaism that share deep concern for embodiment, physicality, sensuality, and sexuality. The chapter highlights the contribution of Jewish women to American environmentalism and contribution of Jewish theologians, activists, and educators to Earth-based Judaism. Jewish environmentalism is not necessarily feminist, but feminism did contribute to Jewish environmentalism.

One geographical territory—the Land of Israel—plays a central role in Judaism. Regarded as the Holy Land, it is pivotal to Jewish collective identity, religious practices, social habits, literary imagination, hopes, and aspirations. Chapter 5 focuses on the dialectic of holiness in Jewish attitudes toward the Land of Israel from antiquity to the present. Modern Zionism, which was launched by secularized Jews in the late 19th century, insisted that all Jews should resettle in the Land of Israel to ensure the future of the Jewish People, because Europe is inherently anti-Semitic and hence inhospitable to Jews. A revolution within Judaism, Zionism sought to create the New Jew, one who would be rooted in the natural world rather than in the study of sacred texts. But Zionism could revolutionize modern Jewish life only because Zionism was rooted in the very utopian and messianic impulses of the Judaic textual tradition. Zionism was both a revolution against traditional Judaism as well as its continuation through transformation.

The depth of this complex paradox was fully exposed since the Six-Day War of 1967 in which Israel won over several Arab states, gaining control of new territories in the West Bank, the Golan Heights, and the Sinai Peninsula. For some, the conquered territories, especially the West Bank, were the core of the biblical Holy Land and therefore have religious significance that cannot be compromised. For others, the occupied territories have no sacred status, and they should only be viewed as the territorial basis for a future sovereign Palestinian state. The internal Jewish debate about the future of these territories has an environmental aspect as Jewish settlers in the West Bank engage in environmental vandalism or prevent Palestinians from access to natural resources. Israel has a robust environmental movement, and there is some collaboration and cooperation with Palestinian and Arab environmentalists that show how environmentalism can be a path for peace. Without covering the environmental movement in Israel in detail, the chapter suggests that because the Land of Israel is believed to be a Holy Land, Jewish environmentalism is inseparable from political theology, on the one hand, and political ecology, on the other hand.

Tragically, today there is a growing rift between Diaspora Jewry and Israel, precisely because of the controversies about the policies of the State of Israel. Chapter 6 and Chapter 7 focus on the American Jewish experience, tracing not how that rift with Israel emerged, but why and how Jewish environmentalism in the United States tends to focus on social justice rather than on the sacrality of land. Chapter 6 discusses the contribution of secularized Jews to American environmentalism, before and after WWII. The chapter shows how the secularization of Jewish religious values, such as responsibility for the common good or commitment to justice, accounts for their distinctive environmental work and how that work was inspired by faith in the power of science. Chapter 7 focuses on the late 20th century and early 21st century, during which Jews became more aware of the limits of science and more critical of secularism and scientism because of their negative social ramifications. Calling themselves Spiritual Progressives, Left-oriented Jews re-embraced and transformed forms of traditional Judaism, articulating faith-based environmental spirituality which they promoted as solutions to America's social ills and humanity's existential challenges. Promoting the religious value of *Tikkun Olam*, the Spiritual Progressives sought to resacralize not only the physical world of nature but every aspect of social life, including politics and public discourse. Ironically, the Spiritual Progressives who came out of the secular Left turned Jewish environmental spirituality into post-secular public theology that could be endorsed by traditional Jews.

The Conclusion spells out the contribution of Jewish environmental spirituality to the academic field of religion and ecology, to religious environmental activism, and to contemporary Jewish life. Jewish earthly spirituality illustrates that in the post-secular age a binary dichotomy between religiosity and secularity is untenable and that the relationship between them should be understood as a dialectical process. To wit, the post-secular world emerged out of the processes of secularization while being critical of them. In Judaism and particularly in Jewish environmentalism, the "secular" and the "religious," the "earthly" and the "spiritual" dimensions are inherently intertwined. Therefore, there is no contradiction between being a religiously committed Jew and caring deeply about the well-being of the Earth, and there are many ways to express the ethics of care and responsibility for the Earth and all its inhabitants. Jewish earthly spirituality is a powerful, critical, value-based response to the technoscientific "engineered age" of the Anthropocene.

References

Angus, Ian. 2016. *Facing the Anthropocene: Fossil Capitalism and the Crisis of Earth Systems.* New York: Monthly Review Press.

Asafu-Adjaye, John. 2015. "*An Ecomodernist Manifesto.*" Available at: http://www.ecomodernism.org/s/An-Ecomodernist-Manifesto.pdf, 6.

Baugh, Amanda J. 2017. *God and the Green Divide: Religious Environmentalism in Black and White.* Oakland, CA: University of California Press.

Baugh, Amanda J. 2024. *Falling in Love with Nature: The Values of Latinx Catholic Environmentalism.* New York: New York University Press.

Bonneuil, Christophe. 2015. "The Geological Turn: Narratives of the Anthropocene." In Clive Hamilton, Christophe Bonneuil, and Françoise Gemenne (eds.), *The Anthropocene and the Global Environmental Crisis*, 17–31. London and New York: Routledge.

Bonneuil, Christophe and Jean-Baptiste Fressoz. 2017. *The Shock of the Anthropocene*, transl. David Fernbach. London: Verso.

Bratton, Susan P. 2021. *Religion and Environment: An Introduction.* Abingdon, UK: Taylor and Francis.

Brooke, John Hedley. 1991. *Science and Religion: Some Historical Perspectives.* Cambridge: Cambridge University Press.

Clingerman, Forrest. 2014. "Geoengineering, Theology and the Meaning of Being Human." *Zygon: Journal of Religion and Science* 49 (1): 6–21. Available at: doi:10.1111.zygon.12072.

Clingerman, Forrest. 2015. "Redeeming the Climate: Investigating a Theological Model of Geoengineering." In *Technofutures, Nature and the Sacred: Transdisciplinary Perspectives*, ed. Celia Deane-Drummond, Sigurd Bergmann, and Bronislaw Szerszynski, 175–192. London and New York: Routledge.

Clingerman, Forrest and Kevin J. O'Brien. 2014. "Playing God: Why Religion Belongs in the Climate Engineering Debate." *Bulletin of the Atomic Scientists* 70 (3): 27–39.

Crutzen, Paul J. 2003. "The Geology of Mankind." *Nature* 415: 23.

Crutzen, Paul J. 2006. "Albedo Enhancement by Stratospheric Sulfur Injection: A Contribution to Resolve a Policy Dilemma?" *Climate Change* 77: 211–220.

Crutzen, Paul J. and Eugene F. Stoermer. 2000. "The Anthropocene." *IGBP Global Change Newsletter* 41: 17–18.

Crutzen Paul J. and Will Steffen. 2003. "How Long Have We Been in the Anthropocene Era?" *Climate Change* 61: 252–257.

Davies, Jeremy. 2016. *The Birth of the Anthropocene.* University of California Press.

Doran, Christopher. 2017. *Hope in the Age of Climate Change: Creation Care This Side of the Resurrection.* Eugene, OR: Cascade Books.

Ferngren, Gary A. 2017 [2002]. *Science and Religion: A Historical Introduction*, 2nd ed. Baltimore: John Hopkins University Press.

Ellis, Erle C. 2018. *The Anthropocene: A Very Short Introduction.* Oxford: Oxford University Press.

Gardiner, Stephen M. 2011. *A Perfect Moral Storm.* Oxford: Oxford University Press.

Gardiner, Stephen M. 2016. *Debating Climate Ethics.* Oxford: Oxford University Press.

Gardiner, Stephen M. 2021. *The Ethics of "Geoengineering" the Global Climate: Justice, Legitimacy and Governance.* New York and London: Routledge.

Grim, John and Mary Evelyn Tucker. 2014. *Religion and Ecology*. Washington, DC: Island Press.
Hamilton, Clive. 2015. *Requiem for a Species: Why We Resist the Truth about Climate Change*. London and New York: Routledge.
Hamilton, Clive A., Christophe Bonneuil and Francoise Gemenne (eds.). 2015. *The Anthropocene and the Global Environmental Crisis: Rethinking Modernity in a New Epoch*. London and New York: Routledge.
Harrison, Peter A. 2017. *The Territories of Science and Religion*. Chicago: University of Chicago Press.
Henson, Robert. 2014. *The Thinking Person's Guide to Climate Change*, rev. edition. American Meteorological Society.
Hoffman, Andrew. 2015. *How Culture Shapes the Climate Change Debate*. Stanford, CA: Stanford University Press.
Hulme, Mike. 2009. *Why We Disagree about Climate Change: Understanding Controversy, Inaction and Opportunity*. Cambridge: Cambridge University Press.
Jenkins, Willis, Evan Berry, and Luke B. Kreiden. 2018. "Religion and Climate Change," *Annual Review of Environment and Resources* 43: 9–15.
Keith, David W. 2000. "Geoengineering the Climate: History and Prospect." *Annual Review of Energy and Environment* 25: 245–284.
Keith, David W. 2013. *A Case for Climate Engineering*. Cambridge, MA: MIT Press.
Knoll, Andrew K. 2021. *A Brief History of Earth*. New York and Boston: Mariner Books.
Latour, Bruno. 2017. *Facing Gaia: Eight Lectures on the New Climatic Regime*, trans. Catherine Porter. London: Polity Press.
Marris, Emma. 2011. *Rambunctious Garden: Saving Nature in the Post-Wild World*. London: Bloomsbury.
McKibben, Bill. 2011. *Eaarth: Making a Life on a Tough New Planet*. New York: Random House.
Nordhaus, Ted and Michael Shellenberger. 2007. *Break Through: From the Death of Environmentalism to the Politics of Possibility*. Boston and New York: Houghton Mifflin.
Purdy, Jedediah. 2015. *After Nature: A Politics for the Anthropocene*. Cambridge, MA: Harvard University Press.
Ruddiman, William F. 2013. "The Anthropocene." *Annual Review of Earth Planetary Science* 41: 45–68.
Steffen, W., Paul J. Crutzen, and Robert McNeil. 2007. "The Anthropocene: Are Humans Now Overwhelming the Great Forces of Nature?" *AMBIO, A Journal of Human Environment* 36: 614–621.
The Royal Society, 2009. *Geoengineering the Climate: Science, Governance and Uncertainty*. London: The Royal Society.
Tirosh-Samuelson, Hava. 2020. *Religion and Environment: The Case of Judaism*. Kitchener, Ontario: Pandora Press.
Wallace-Wells, David. 2019. *The Uninhabitable Earth: A Story for the Future*. London: Allen Lane.
Wilkinson, Katharine K. 2012. *Between God and Green: How Evangelicals Are Cultivating a Middle Ground on Climate Change*. New York: Oxford University Press.
Wuerthner, George, Eileen Crist, and Tom Butler (eds.). 2014. *Keeping the Wild: Against the Domestication of Earth*. Washington, DC: Island Press.

1 Post-Secularism

Religion, Science, and Spirituality

Terminological Ambiguity

Conceptions of nature in Jewish environmental spirituality shed light on the complexity of the post-secular world. Jewish environmentalism both exemplifies larger trends of contemporary post-secularism while also challenging formations of the post-secular from a Judaic perspective. We begin by noting the inherent ambiguity of the terms in our disposal, while setting the stage for the main argument of the book that Jewish environmental spirituality manifests the trends of post-secularism. All the terms used in this study—'secular,' 'secularity,' 'secularism,' 'secularization,' 'Secularization Theory,' and 'post-secularism'—have been interpreted in various and conflicting ways and all have very dense history. I will not be able to do justice to this history but instead flag the need to be attentive to the historical context that gave rise to such ambiguous terminology. The ambiguity arises first and foremost from the fact that 'secular' is a dyadic term that has been used to mean 'nonreligious' or 'a-religious.' Recognizing the relational nature of the term 'secular' does not help much because the terms 'religion' and 'religious' are no less ambiguous and loaded with contested meanings that cannot be separated from the historical context in which they developed.

To clarify some of the ambiguity it helps to keep in mind that the word 'secular' comes from the Latin term *saeculum*, which originally denoted a person's potential lifetime, or equivalently, the complete renewal of a human population (like our term 'generation'). *Saeculum* was a temporal category of roughly 100 years. Temporal categories became more complex with the Christianization of the ancient world, as the *saeculum* was understood to mean profane time in contrast to sacred time. The differentiation between profane time (*hol*) and sacred time (*kodesh*) was Judaic in origin. It referred to the differentiation between the six days of the week, during which there are productive and creative activities, as opposed to the Sabbath, the seventh day, during which no creative activity is allowed. Originally a Jewish sect, Christianity took over the Judaic differentiation between profane time and sacred time, but since Christianity spread in the Greco-Roman world, it translated this distinction into Greek philosophical terms and aligned it with Platonic metaphysical dualism.

DOI: 10.4324/9781351123945-2

Sacred time was now understood to mean that which pertains to the eternal, the immutable, and the unknowable reality experienced by the soul after the death of the body, as opposed to "profane time," which pertains to material, physical, sensuous, bodily, mutable existence prior to death.

With the Christianization of the ancient world, the secular was no longer a temporal category but pertained more broadly to spatial activities in the world such as farming, commerce, and trade, in contrast to religious activities such as prayer, fasting, and Scripture meditation. Whereas the former were necessary for human physical survival the latter were relevant to the soul's salvation in the afterlife (Lim 2009). How does the world of nature fit into this schema? The answer is that the term 'nature' is also ambiguous. On the one hand, in Platonic metaphysics, the physical, material objects we perceive through our bodily senses are not considered real because they are only images of non-material essences (i.e., the Ideas) that exist eternally but non-spatially. In this regard the natural world is "profane" and ontologically inferior to the Ideas that exist eternally. But on the other hand, Platonic dualism made it possible for Christians to invest the natural world with symbolic meaning that points beyond itself to the other-worldly realm of eternal, non-physical, or spiritual meaning. The natural world can thus be interpreted to have both secular and religious meanings.

The relationship between the secular and religious realms was therefore inherently ambiguous. Since the work of daily life was independent of belief in God, religious people were engaged in "secular" (i.e., this-worldly) activities. It was difficult to differentiate secular and religious activities in individuals' lives. More importantly, the way in which the Roman world was Christianized gave political meaning to the distinction between the secular and the religious. The Roman state was a secular (i.e., a-religious) institution, engaged in temporal affairs, but when the Roman Emperors converted to Christianity in the 4th century and made Christianity the official religion of the Roman Empire, the boundary between the secular and the religious domains was blurred (Taylor 2011).

In medieval Christendom the religious and the secular were indelibly intertwined as much as they were engaged in perpetual conflict. The Church, which represented the "kingdom of heaven" on Earth, claimed total control over the spiritual destiny of humans, but did not give up its aspiration to have a say in economic, social, and political dimensions of life. Conversely, the secular princes who governed their subjects' "earthly life" were themselves religious and justified their worldly activities by appealing to their deeply held religious convictions. Because medieval Christianity endorsed Augustine's differentiation between the "City of God" and the "City of Man" as separate domains with distinctive authority and responsibility as well as distinctive human orientations, the story of the Christian West consists of their constant negotiation. Medieval society was a Christian polity, and the duality of the *Saeculum* turned explicitly political and institutional. In medieval Christendom, Christian kings and emperors were considered "secular rulers" whose civic authority pertained only to temporal affairs, whereas popes, bishops, and clergymen were ecclesiastical authorities whose authority was derived from divinely revealed scriptures and whose rituals made salvation of individual Christians possible. The religion and secular spheres of medieval society were inextricably entangled so that all worldly affairs (e.g., politics, economic, law, education, science, and medicine) were configured with a religious worldview and pertained to salvation.

Renaissance humanism in the 15th and 16th centuries challenged the medieval worldview by recovering writings of ancient Greek and Roman pagan (non-Christian) authors. Renaissance humanists rebelled against the medieval Church, especially Scholasticism, advancing a new appreciation of the inherent goodness of earthly life and rejecting the Church's control over the spiritual destiny of humans. Renaissance humanism recognized the inherent value of humans because they are created in the "image of God" and their ability to interact with God without the Church's mediation. Inspired by humanism, the Protestant Reformation called for extricating various aspects of life from the control of the Church, declaring God alone (rather than the institutions of the Church) as the salvific power, and enabling each believer to have direct unmediated relationship with God through reading of scripture. Renaissance humanism, the Protestant Reformation's critique of the Catholic Church, and the bitter religious wars between Protestants and Catholics in the 16th and early 17th centuries all paved the way for the profound changes in Western culture we call 'secularism,' namely an ideology critical of religious institutions, beliefs, and practices (Asad 2003; Copson 2019; Berlinerblau 2021).

From the mid-17th century on, a new secularity has taken hold in the West according to which religion has no authority or social power. In the modern, secular world, large swaths of human activities (e.g., economics, politics, government, law, and education) were declared to have no relationship to religion and religion could have no power or authority over these domains. The most radical version of modern secularity was articulated in the French Revolution and its ideal of *laïcité*, according to which the public arena must remain devoid of any religious symbol, practice, gesture, or meaning. Religion could exist in the modern secular world only if practiced in the domestic, private sphere. The historical processes that transformed Western society in terms of its organization and practices is referred to as 'secularization,' and the sociological theory that seeks to explain and predict the conditions of secularity in each country is referred to as the "Secularization Thesis."

Post-Secularism as a Critique of the Secularization Thesis

The Standard Model

The Secularization Thesis claimed to explain a universal trajectory: all human societies that move from the pre-modern to the modern must undergo the process of secularization. Failure to do so meant either failed modernization or failed secularization. According to the Secularization Thesis, the secular was a necessary feature of the modern: to be modern meant necessarily to be secular and to be secular indicated attainment of modernity. The Secularization Thesis presented itself as an objective, impartial, scientific theory, but in truth it was highly partial and even normative because it privileged "the secular" over "the religious." Whereas the secular stood for progress, enlightenment, and rationality, all highly esteemed ideals and goals of "modern" people, the religious was regarded as superstitious, backward, or primitive. Secularization was the process that created the neutral public space in which communication was based on universal reason rather than on tradition or religious beliefs. Public civic institutions had to be secular in the sense of being a-religious. The Secularization Thesis claimed to be a universal theory that explains and predicts how all societies regardless of historical specificity transition from "premodern" existence to modernity. Presenting itself as a global explanatory theory, the Secularization Thesis was decidedly Eurocentric and inseparable from the experience of imperialism and colonialism in which (Christian) Europeans dismissed the religious (non-Christian) societies they came to control in the 19th century.

In the standard model of the Secularization Thesis, all modern societies witness the inevitable disappearance of belief in supernatural powers, a decline in religious authority, decrease in religious attendance and practice, and the diminishing influence of religious values, leaders, and institutions over individual behavior, social institutions, and public discourse (Swatos and Christiano 1999). Accordingly, the loss of religious faith erodes habitual religious practices, eviscerating observances of religious rituals and ceremonies. To the extent that religious beliefs are still relevant, they remain a private matter, a set of beliefs held by individuals that have little or no bearing on social relations or the functioning of the public sphere. As religion becomes privatized and large swaths of public life free themselves from the control of religious institutions, doctrines, and people, modernization necessarily undermines religion.

Critique of the Secularization Thesis

The Secularization Thesis became the regnant paradigm of sociology, an academic discipline that came into existence with the birth of the modern. But already in 1965 David Martin, a British sociologist, critiqued the concept of secularization as a "hotch-potch of ideas, some of them contradictory" (Martin 2005, 19; referring to Martin 1965), arguing that because modernity and sociology came into existence about the same time, Enlightenment assumptions inflected the discipline of sociology. Martin expanded his critique and revised the Secularization Thesis (Martin 1978), and his critique was endorsed by Jeffrey K. Hadden in a Presidential address of the Southern Sociological Society in 1986. Reviewing the history of sociology, Hadden asserted that the "idea of secularization was itself sacralized" by sociologists of religion: their "belief in secularization has been sustained by a deep and abiding antagonism to religious belief and various expressions of organized religion" (Hadden 1987, 588). Hadden showed how the Secularization Thesis emerged to become a "doctrine rather than a theory," how it reflected the secularist biases of sociologists, and why the Secularization Thesis was flawed as a research tool. Others endorsed his analysis (Dobbelaere 1987).

The main disconfirmation of the Secularization Thesis came from the United States, a modern, industrialized, and urbanized nation that showed no evidence for the decline or decay of religion. As the first Western nation to separate Church and State, the United States gave rise to a highly pluralistic society, in which "religion is no more than an option, yet it is an option which a very considerable proportion of the population chooses" (Macleod 1992, 64). Was the United States an exception to the rule of secular modernity, or rather was secularized Europe the exception? The comparative analysis of Europe and the United States required sociologists to revise the Secularization Thesis to account for the differences between these two modern, industrialized societies, and to acknowledge different types of secularization in Europe (Ziebertz and Ulrich 2009). The more sociologists pondered the differences between "secular Europe" and "religious America," the more it became clear that the standard model of the Secularization Thesis is inadequate.

When comparative social science was applied to the developing world, the limitations of the Secularization Thesis became even more pronounced. Worldwide, religious commitment was by no means on the decline, religious beliefs and practices were not limited to the private sphere, and religion played a significant role in politics, law, and education. In the second half of the 20th century, religion reasserted itself in the public sphere, contrary to the predictions and analysis of the Secularization Thesis: the Shiite Revolution in Iran, the Moral Majority and Evangelical Protestantism in the United States, the spread of Pentecostalism in Latin America and in Africa, the restoration of Confucianism and Daoism in China, charismatic New Religious Movements in Europe, United States, and Asia, the spread of Buddhism from Taiwan to the United States, the mobilization of faith-based organizations to address social problems, the Islamic revival of belief and practice in North Africa and South-East Asia, the rise of political Islam in the Middle East, Europe, and Asia, and the emergence of ethnic-religious nationalism in Eastern Europe. Clearly the Secularization Thesis failed to predict or account for these developments.

Refining the Secularization Thesis

Throughout the 1990s, sociologists were embroiled in a debate about the Secularization Thesis. Bryan R. Wilson, who launched the Secularization Thesis in England, observed that "secularization, after all, is a many-sided phenomenon affecting different aspects of the social and cultural structures of society in diverse ways and in difference sequence" (1992, 206). The debate about the Secularization Thesis, he correctly observed, had suffered from confusion between 'secularism,' referring to "an ideology that seeks to minimize, if not to eliminate, the role of religion in social affairs," and 'secularization' as "the description of a process of social change in which religion loses social significance" and "for which empirical evidence can be advanced" (ibid, 209). José Casanova (1994; 2008; 2010; 2011) made great strides toward clarification of the Secularization Thesis when he showed that it consists of three distinct sub-themes: (1) the decline of religious belief; (2) the differentiation of religious and nonreligious spheres; and (3) the privatization of religious commitment. Casanova argued that only the notion of differentiation is plausible and that secularization (i.e., differentiation) sets the stage for the emergence of public religions (i.e., de-secularization). He identified the counter-process of de-privatization to signify those indicators that religions are increasing in public relevance. Religion refused to be privatized and marginalized, and it articulated many of the concerns of civil society: environmental degradation, abortion, immigration, racial prejudice, state repression, and economic exploitation.

Calls for the renunciation of the Secularization Thesis were now voiced by leading sociologists, notably Rodney Stark (1999) and Peter Berger, who had promoted the Secularization Thesis but now declared that "modernity does not necessarily bring about secularization" (Berger, Davie, and Fokas 2008, 12). Renouncing the singular model of secular modernity, Berger adopted instead Shmuel Eisenstadt's notion of "multiple modernities" according to which "modernity and Westernization are not identical; Western patterns of modernity are not the only 'authentic modernities,' though they enjoy historical precedence and continued to be a basic reference point for others" (Eisenstadt 2000, 2–3). Concurring with Eisenstadt, Berger summed up his revised views saying that

> modernity does not necessarily lead to secularization; where it does, this development cannot be assumed but must be explained. Modernity necessarily leads to pluralism. This indeed presents faith with a significant challenge, but it is different from the challenge of secularization. (Berger 2014, 20)

Casanova too adopted the concept of "multiple modernities" as a "more adequate conceptualization and pragmatic vision of modern global trends than either secular cosmopolitanism or the clash of civilization" (Casanova 2008, 106).

By the beginning of the 21st century, it became clear that "what is needed is not a simpleminded theory of inevitable religious decline, but a theory to explain variation" (Stark and Finke 2000, 33). Pippa Norris and Ronald Inglehart responded to the challenge and articulated a revised theory of secularization that seeks to explain "most of the variants in religious practices found in the world" (Norris and Inglehart 2011 [2004], 13). Although their theory did not become the reigning paradigm of the social sciences, by the beginning of the 21st century it became clear that the terms 'post-secular,' 'post-secularism,' and 'post-secularity' emerged as part of a new framework to account for the resurgence of religion in the public sphere, the remarkable vitality of religion, and new creative forms of spirituality which flatly contradicted the Secularization Thesis. But what exactly does the 'post-secular' mean? In truth, the term 'post-secular' is multivalent because of the imprecision of the terms 'secular,' 'secularism,' 'secularity,' and 'secularization' as well as the inherent ambiguity of the term 'post.' The meaning of 'post-secular' depends on the disciplinary context in which it is used.

The 'Post-Secular' World

In the post-secular world, the relationship between the secular and the religious is reconfigured not as adversarial and antagonistic but as complementary and mutually supportive. The term 'post-secular' appeared first in 1982 in Richard John Neuhaus's critique of the rigid interpretation of the separation of Church and State in the United States (Neuhaus 1982). Though he would eventually become the most prominent conservative Roman Catholic in America, Neuhaus was originally ordained as a Lutheran minister and was a liberal activist involved in the civil rights struggle and in the anti-Vietnam war effort. At the time a close friend of Peter Berger, Neuhaus was familiar with the debates on secularization and was most attentive to the rise of the Moral Majority and the Christian Right. Neuhaus did not analyze the 'post-secular,' nor did he link it to postmodernism, since that term just began to make its mark on American higher education and in public discourse; but as postmodernism spread in the 1980s the 'post-secular' gradually began to circulate in different academic discourses.

The association between the post-secular and the post-modern was first stated in 1990 by the Italian sociologist Filippo Barbano, who critiqued the Weberian theory of secularization and post-Weberian Sociology of Religion (Pierucci 2000, 130). Contrary to the linkage of secularization with modernization, the 'post-secular' depicted a postmodern society in which countless phenomena of de-secularization take place. In the United States, however, postmodernism thrived not among social scientists, whose empirical scientific research was governed by rational choice theory, but among literary theorists. In 1995 the literary critic John McClure used the term 'post-secular' to capture a "counter-construction of the postmodern" in the writings of theorists such as Zygmunt Bauman and Terry Eagleton and the "artistic engagement with non-secular constructions of reality" in the works of authors such as Thomas Pynchon, Don DeLillo, Ishmael Reed, Toni Morrison, Louise Erdrich, and others, all of whom illustrate "the untidy resurgence of magic, sacred, pre-modern, and non-western construction of reality" (McLure 1995, 148). In postmodernist literary discourses, the 'post-secular' signifies new spiritualities and new sacred discourses promoted by marginalized communities such as feminists and eco-theologians who were "attempting to blur the distinction between magic and economics, magic and science" (ibid., 158). A decade later, McLure drew a "map of the broader post-secular movements with which fiction is engaged," including

> explosive growth of fundamentalist and pneumatic forms of organized religions practice, 'New Age' experiments in alternative spiritualities, and the turn toward religion in certain philosophical circles—all reflect a strong but selective disenchantment with secular values and modes of being and a determination to invent alternatives.
>
> (McClure 2007, 7)

Whether fiction imitates, mirrors, or constructs reality, the category of the 'post-secular' depicted contemporary social reality as well as the literature it generates.

The postmodernist "disenchantment with secular values" is even more pronounced in contemporary philosophy (more precisely, Continental Philosophy), although the nexus of the postmodern and the post-secular was somewhat different than in literary studies. In philosophy the 'post-secular' refers to the religious turn in phenomenology (e.g., Jacques Derrida, Emmanuel Levinas, and Jean-Luc Marion), a Christian brand of deconstruction (e.g., John Caputo), the feminist appeal to Mariology (i.e., Luce Irigaray), and the discourse of political theology (e.g., Giorgio Agamben), all of which illustrate the theologization of philosophy. Importantly, the trope of the 'post-secular' made little impact on Anglo-American analytic philosophy, because it continues to perpetuate the modernist assumptions of Logical Positivism and the Enlightenment assumptions that undergirded it. By contrast, the trope of the 'post-secular' is central in the thought of Christian, socially minded theologians (e.g., John Milbank, Graham Ward, and Catherine Pickstock) who are most critical of the social sciences, because they are products of modernist secularism, and in the works of philosophers who emphasize ethics and consider the conditions of possibility of ethics in pluralist societies (e.g., William Connolly, Philip Blond, and Judith Butler). How ethics and values, especially religious values, should shape public education has been the concern of public theologians (e.g., Hogue 2010; Graham 2013) as well as of urban planners and civic engineers who grapple with the physical presence of religious institutions in public spaces. Finally, in the arts, the 'post-secular' describes writings or artistic performances that have spiritual dimension or context (King 2005). In all these discourses the 'post-secular' means something specific, but all these usages acknowledge the social relevance of religion and articulate a self-reflexive and critical attitude toward secular modernity (McLennan 2010).

Social scientists and political theorists adopted the term 'post-secular' to signal religion's renewed visibility in contemporary culture and politics. The term gained currency after 9/11 in the speech that Jürgen Habermas delivered when he received the Peace Prize of the German Publishers and Booksellers Association on October 15, 2001. In a later speech delivered on the Holberg prize in 2005 and subsequently expanded into a longer article (Habermas and Ratzinger 2006), Habermas insisted that secular citizens must learn to live in a post-secular society. In doing so, they will follow the example of religious citizens, who have already come to terms with the ethical expectations of democratic citizenship in the sense that they have adopted appropriate epistemic attitudes toward their secular environment.

For Habermas a post-secular society is epistemically attuned to the continued existence or religious communities and it displays "an affinity to post-metaphysical thinking," which abstains from big ontological statements on the constitution of the whole of beings; but this does not mean reducing our knowledge to the sum total of statements that at each time represents the respects of "state of science" (Habermas, 2006, 16; Cf. Habermas 2008). Post-metaphysical thinkers and post-secular individuals are agnostic but willing to learn from religion and the post-secular consciousness involves a loss of the hitherto firm secularist conviction that religion will eventually disappear in the continuing process of modernization, and it is also a normative notion: it embodies the de-legitimation of early political-philosophical conviction (namely, the liberal, secular nation-state). Habermas's analysis of the rising prominence of religion and his theoretization of post-secular society have generated a vast literature among sociologists and social theorists (Harrington 2007; Warner, VanAntwerpen, and Calhoun 2010; Molendijk, Beaumont, and Jedan 2010; Calhoun, Mendieta, and VanAntwerpen, 2013), but Habermas by no means exhausted how the post-secular is understood or what post-secular society and culture are all about.

Other dimensions such as globalization, late-capitalism, post-colonialism, consumerism, media and digital art, environmentalism, urbanism, and technoscience are no less relevant to understanding the post-secular society than the theoretical parameters he articulated (Naynäs et al. 2011). For some observers of contemporary culture, the 'post-secular' signifies a new faith-based political approach to public life, consumption, and global issues in late modernity, the revival of New Age spiritualities, or "the reemergence of the sacred in the development of urban space and community development" (Beaumont 2010, 9). In short, 'post-secular' is a fluid, multivalent construct that expresses the globalized Zeitgeist in which the assumptions of Western, secular modernity are questioned, critiqued, revised, and/or replaced. It is important to bring to mind Casanova's observation that

> becoming post-secular does not mean necessarily becoming religious again, but questioning our stadial consciousness, de-stabilizing if not our secular immanent frame, at least, the possibilities of transcendence within the immanent frame, and being open, receptive, at least curious to all the manifold forms of being religiously human.
>
> (Casanova 2010, 278)

The post-secular age is a self-reflexive age.

Casanova's reference to the "immanent frame," is indebted, of course, to Charles Taylor's influential contribution to the debate about secularization. Although Taylor has deliberately avoided the term 'post-secular,' many regard him as a post-secular thinker because he argued that it is possible and desirable for a secular (or secularized) person to seek the transcendent within "the immanent frame," the "default setting" for many people in the contemporary world. Criticizing the Secularization Thesis, Taylor has rejected the notion that in modernizing societies there is inevitable decline in belief or practice, and he refuses to see the secular age as a story of "subtraction." Contrary to Weber, for whom modernization is the result of the loss of belief in magic and the supernatural, Taylor sought to show that religious beliefs still have merit and religious experience still has grandeur and depth because "it shows itself to be before we start trying to explain it away" (Gordon 2008, 653). In the secular age, according to Taylor, "belief in God is no longer axiomatic. There are alternatives" (Taylor 2007, 3), but even as religion became a social contingency rather than an integral part of what it meant to be human, it is still possible for human beings to long for the transcendence and to seek to experience its depth and awe within the "immanent frame." His magisterial work explains how the secular consciousness emerged as a new, modern "social imaginary" in which the human being was solidified into what Taylor called the "buffered self," namely a self who is no longer vulnerable to the transcendent or the demonic because it is isolated in mind. His analysis exposes the strengths and weaknesses of living in a disenchanted world, of living "as if God does not exist."

The "immanent frame" is Taylor's shorthand term for interpreting the cosmos without reference to a nonhuman or transcendent source of meaning. In the disenchanted, secularized world, however, the longing for transcendence did not disappear. As Peter Gordon put it, "the immanent frame is haunted by the feeling of the transcendence it has rejected. It leaves behind a trace of malaise and the sense that something is missing even if one is not entirely sure what that something is" (Gordon 2008, 663–64). In the modern world we all live within the "immanent frame," "a constructed social space that frames our lives entirely within a natural (rather than supernatural) order" (Smith 2014, 141), but some live with the desire for something beyond, and some live as if the immanent frame is closed. The "immanent frame" allows for both options but does not demand any one of them. Taylor is critical of the denial of transcendence because it is "bound to lead to a crumbling and eventual breakdown of all moral standards. First, secular humanism, and then eventually its pieties and values come under challenge. And in the end nihilism" (Taylor 2007, 638). As Taylor sees it, "human beings have an ineradicable bent to respond to something beyond life. Denying this stifles" (ibid.).

In short, for the past two decades, the terms 'post-secular,' 'post-secularism,' and 'post-secularity' have come to signify the contemporary resurgence of religion, in contrast to the Secularization Thesis (or Secularization Theory) that predicted the disappearance of religion in modern societies (Swatos and Christiano, 1999; Gorski and Altinordu 2008; Torpey 2010; Gorski, Kim, Torpey, and VanAntwerpen 2012). However, the precise meaning of 'post-secular' varies depending on the disciplinary context in which it is used, and there is no consensus on what it means to live in a post-secular age. Nonetheless, the trope of the 'post-secular' is useful because it captures the hybridity of secular and religious elements that can be witnessed in numerous cultural forms. In the post-secular age, religion is by no means private but takes its place in the center of public discourse in the public sphere. Religion, however, is not limited to institutional forms of inherited traditions, but encompasses a range of new spiritual pursuits in variety of forms. These forms are inseparable from the consumerist culture of late capitalism as exercised by individuals as well as societies, and these forms no longer conform to distinctive traditions of East and West, but merge and fuse elements from various religious and spiritual traditions. In the highly pluralistic, consumerist, and globalized world, all sorts of hybrids emerge that allow people to experiment, explore, and express their spiritual impulses as they seek to re-enchant the world that modern secularism had (allegedly) disenchanted. Nature or "the world" is re-enchanted when it is understood to be the realm of spiritual forces or vitality that modern science cannot explain or quantify.

Unlike sociologists of religion, historians of science have already challenged the Secularization Thesis when they dismantled yet another powerful modernist master-narrative: the so-called Conflict Thesis between "science" and "religion." (Tirosh-Samuelson 2017, 23–31). The Conflict Thesis emerged in the 1860s primarily due to relentless efforts of Thomas Henry Huxley and his colleagues in the X-Club to give natural history a new, secular, "scientific" status and to make secular science the center for cultural life in Victorian England. Coming at the expense of amateurs, clergymen, and women, the professionalization of science deployed the rhetoric of conflict between "science" and "religion" to achieve political gains, namely, to overthrow the hegemony of the English Church. It was a political struggle to professionalize science by dissociating it from natural history and natural philosophy, both of which were dominated by the clergy. Modern science, especially the new disciplines of biology and geology, was to be independent of clerical influence "while at the same time legitimizing a new set of non-ecclesiastical authorities" (Harrison 2010, 27). The professionalization of science and its separation from ecclesiastical control clearly fit into the differentiation aspect of the Secularization Thesis but it hardly fits historical reality. Remarkably, despite mounting historical evidence against the Conflict Thesis, the perception that science and religion are antagonistic forces not only persisted into the 1990s and 2000s but became even more acute (Hardin, Numbers, and Binzley 2018). Why? The answer lies in the resurgence of religion in the public sphere during the 1980s, which signified the emergence of the post-secular age.

By the turn of the 20th century, when Max Weber, the putative "father" of the Secularization Thesis, developed his sociological theories, the Conflict Thesis was accepted by academics and the public in the United States and in Europe. In truth, Weber said relatively little about modern science per-se and used the term 'secularization' (*Säkularisation*) rather sparingly, primarily in the context of law (Pierucci 2000). More often than 'secularization,' Weber spoke about the "de-magification of the world" (*Entzaubeung der Welt*, commonly but less accurately translated as the "disenchantment of the world"). When Weber coined the phrase in "Science as a Vocation" (Weber 2004 [1917]), he had in mind what the Romantic poet Friedrich Schiller called the "de-deification of nature" (*Entgötterung der Natur*), a term that accurately captured how modern science rejected God in explaining the workings of nature. For Weber, "de-magification" expressed the rationalization of the world, but it is important to note that he did not interpret that process as a conflict between "science" and "religion," but rather as a development from within a long religious process that had begun with ancient Judaism and continued in Christianity, culminating in Puritanism. In the last edition of the *Protestant Ethic* from 1919–1920 (originally published in 1904–1905) Weber stated:

> The great historical religious process of disenchantment of the world, which had begun with the prophets of ancient Judaism, and, in conjunction with Hellenistic scientific thought, had repudiated all magical means to salvation as superstition and sacrilege, came here to its fulfillment.
>
> (Weber 1920, 105)

The ancient insistence on divine transcendence in ancient Judaism created the possibility for the rationalization of human conduct and for the secularization of the world, but the process of rationalization was larger than the disenchantment of the world, and, conversely, the disenchantment of the world has a much longer history than secularization.

Post-Secular Spirituality and the Ecological Crisis

New Age Spirituality and New Religious Movements

In the post-secular age religion refuses to be privatized or marginalized, demanding instead to be at the center of the public sphere and to frame public discourse on a range of issues. Religion, however, is not limited to institutional forms of inherited traditions but encompasses a range of new spiritual pursuits in literature, film, music, art, performance, technoscience, and even work. The post-secular age is marked by "de-traditionalization" resulting in the emergence of New Age Religions or New Religious Movements (Heelas 1996; Urban 2013). The term "New Age" refers to the "Age of Aquarius" an astrological term popularized by the hippies of the 1960s and 1970s. As a cultural category, "New Age" refers to nonconventional or alternative spiritual movements that include "a diverse array of alternative spiritual practices, such as channeling, crystals, astrology, aromatherapy and various ideas drawn from Eastern religions" (Urban 2013, 5). New Age religions tend to be "individualistic, focused on the optimum physical, psychological and spiritual development of the individual practitioner who is free to pick and choose from a wide array of spiritual options" (ibid). As a result, all sorts of spiritual movements have emerged as spiritual seekers experiment, explore, and express their spiritual impulses to re-enchant the world that modern secularism had disenchanted. Today, practitioners of spirituality distance themselves from their New-Age antecedents of the 1960s (Crockford 2021), but this view is historically inaccurate.

Heelas and Woodhead (2005) explain how the process of de-traditionalization occurs through two complimentary moves: on the one hand, there is a move inward to focus on subjective life of the Self so that individuals' intense experiences come at the expense of conformity to external authority and traditional institutions. On the other hand, "the subjective turn is bound up with the development of a wide array of provisions and activities" (Heelas and Woodhead 2005, 79) in the name of a holistic approach to life that encompasses body, mind, and spirit. Thus the "personal, relational life now has less to do with belonging to a specific, ordered community, and more to do with developing an array of interactions which serve to cater for different aspects of subjective life concerns" (ibid., 80). Moving inward enables people the freedom to find their own path rather than following a prescribed one, but the paths they find are tested in activities that best work for their experience of life. People do not look for the sacred that transcends the particularities of their life or that depends on a belief system, but rather on numerous activities that enhance their subjective well-being here and now. Such "secular" activities can include yoga, reiki, aromatherapy, meditation, dance, or Tai-Chi, among numerous others. This massive turn inward results in the sacralization or spiritualization of secular life. Recently, Carolyn Chen has shown that even "work has become a spiritual practice that inspires religious fervor" (Chen 2022, 4).

'Spirituality' is the term that expresses the co-existence of sacralization and secularization characteristic of the post-secular age. Admittedly vague, spirituality is used to express a range of values, attitudes, preferences, aspirations, and commitments. Recognizing the "fuzziness, indistinctness and, and multiplicity of definitions," Courtney Bender observes that spirituality "is entangled in social life, in history, and in our academic and nonacademic imaginations" (Bender 2010, 5). Therefore, Bender insists, "it is necessary to engage spirituality, historically, institutionally, and imaginatively without pulling it completely together into a single thing" (ibid., 6). In agreement with Bender's approach, this study positions spirituality in the conceptual space between the "secular" and the "religious" and blends elements from both: it is "secular" in the sense that spirituality is experienced in the earthly world rather than the afterlife, but spirituality is "religious" in the sense that it concerns ultimate meaning and ultimate concern. For that reason, all religious traditions encompass a spiritual dimension so that we can identify spiritual strands within world religions, be they Christian, Islamic, or Buddhist and others (e.g., Holder 2005 [2011]; Perrin 2024). But 'spirituality' may also refer to the realm of personal subjectivity, namely, the interior life of human thoughts, imagination, desire, ideals, values, aspirations, and hopes.

Roger S. Gottlieb, a leading Jewish environmental philosopher, offers this definition: "Spirituality is an understanding of how life should be lived and an attempt to live that way" (Gottlieb 2013, 5). As such, spirituality reflects "a movement away from what we come to see as unreal, painful, disappointing, trivial, or meaningless and toward the ultimate, true, vital, real, or sacred" (ibid., 7–8). In his approach, spirituality is an overarching vision that articulates a framework for how we should live our life. Put differently, spirituality is about *meaning making* through which we experience *belonging, identity, and purpose*, without which it is difficult to be fully human. Indeed, Sallie McFague, a prominent feminist Christian eco-theologian, employs the Scottish Churches Council's 1977 definition of spirituality, according to which spirituality is "an exploration into what is involved in becoming human." In this statement "becoming human" is understood as "an attempt to grow in sensitivity to self, to others, to nonhuman creation, and to God who is within and beyond this totality." Thus, spirituality is grounded in relationships, and it involves developing "awareness of, feeling for and sympathy with others" (McFague 1997, 10). Spirituality is necessarily social rather than self-centered. For both Gottlieb and McFague being spiritual is a crucial component of what it means to be human, and spirituality is necessarily relational. This awareness characterizes the so-called "participatory turn" in religious studies that affirms "the immanence of the sacred" along with "a plea for the re-sacralization of everyday life" (Ferrer and Sherman 2007, 12). Unlike those who insist on a radical separation between "spirituality" and "religion" (e.g., Harris 2014), or those who identify "spirituality" with "otherworldliness," in the so-called "participatory turn" spirituality itself is embodied, manifested in perceptions, sensations, and actions. Jewish environmental spirituality is a clear case of embodied spirituality characteristic of the "participatory turn," since human relationship with God is manifested in actions toward land, soil, vegetation, animals, and other human beings.

Spirituality in Historical Perspective

Those who define themselves as "spiritual but not religious," use the term 'spirituality' in contradistinction to 'religion.' Historically this has not been the case. All world religions have spiritual dimensions; they all see life as spiritual journeys, and they devise rituals to express human spiritual quests. Nonetheless, today people invoke 'spirituality' to express a certain discontent with religious traditions and their ability or inability to address the human quest for meaning. Those who promote spirituality express a discomfort with the rigid modernist divide between the "secular" and the "religious," and a desire to re-enchant the world but without falling into the trappings of organized religions. In the religious history of the United States, the term 'spirituality' covers a wide array of magical and occult practices of unchurched religiosity in colonial America including divination, fortune-telling, astrology, witchcraft, and even folk medicine, which competed with organized Christianity (Fuller 2001, 14–15). In the 19th century, spirituality referred to the religious enthusiasm of the Great Awakening movement and the spread of the mystical philosophy of Swedenborgianism (based on the teaching of Emanuel Swedenborg, 1688–1771) and Transcendentalism (the metaphysical philosophy of Ralph Emerson, who held that every human being has the capacity to become receptive to the inflow of spiritual energy). At the turn of the 20th century the term 'spiritualism' referred to a blend of esoteric philosophies "transmitted through a privileged instructor or the cultivation of special states of mystical awareness" (ibid., 52). These pursuits have informed contemporary spirituality in America which has moved beyond Western culture to absorb indigenous and Eastern forms of spirituality.

The upheavals of the 1960s and early 1970s were crucial to the emergence of New Religious Movements (NRM), or "cults" as they are popularly known (Wilson and Cresswell, 2002; Daschke and Ashcraft 2005, 1–21). The anti-establishment Zeitgeist of the 1960s was directed against many causes, including nuclear armament and testing, the US War in Vietnam, the oppression and marginalization of women, the lack of civil rights for African Americans and other racial and ethnic minorities, the stodgy universities, and more. But what is the connection between the New Spirituality and nature? The social and political activism of the 1960s expressed a certain yearning for responsibility toward the physical world and toward the future. Put differently, the spiritual impulse of the 1960s fused social and environmental responsibilities. Thus, there were experiments with "back to the land" movements, simple communal living, and resistance and repudiation of the perceived artificial and hypocritical values of the ruling social order. Young spiritual seekers explored new opportunities to expand their spiritual horizons, and they were attracted to beliefs, especially of Eastern provenance, which previously were considered "exotic," but which now have been de-exotified and incorporated into Western paradigms.

As Western philosophies welcomed and absorbed their Eastern counterparts, a New Spirituality, also referred to as New Age Spirituality, became the marker of a conceptual shift in the West. The emerging New Spirituality rejected any form of religion which locates authority in a source that transcends the individual (e.g., God, scriptures, a particular community, rituals, sacraments, or priesthood). Instead, the New Spirituality located authority in the heart, feelings, intuition or individuals' experiences. Translated into metaphysical language, the New Spirituality focused on immanence rather than on transcendence since it viewed every day or phenomenal reality as manifestation of a deep and unifying Spirit or a Life Force. Such this-worldly, monistic stance maintains that "All is One" and that it is through the phenomenal world (natural and human) that we gain access to "the One." The New Spirituality of the 1960s and 1970s was optimistic in its outlook, linked to evolutionary science, and socially progressive, maintaining that a New Age of unity, peace, and spiritual enlightenment is dawning (Partridge 2004, 106).

The 1960s culture of seekers was highly individualistic, inward-looking, and based on personal choice. The Self is understood to be divine or a "particle" of the divine; there is no higher authority than the Self and personal experience is the final arbiter of truth. In the market of ideas, individuals are encouraged to peruse and select from a bewildering array of spiritual options, since they are, like all people in late capitalism, first and foremost consumers. While eclecticism has enabled the individual to experiment with diverse forms derived from different traditions, the New Spirituality of the 1960s emphasized holism. It rejected the reductionism of the modern scientific worldview, which rigidly separated body and soul or mind and matter, and it criticized modern medicine's treatment of people as a collection of parts rather than as whole persons. The theme of connectedness and interrelatedness pervades the New Age worldview. If previously health and happiness were understood to be the by-products of religious life, now health and happiness were the goal or the promise of spirituality.

Sociologists of religions have shown that although the New Spirituality was critical of secular modernity, in fact the New Spirituality was not an "antidote to secularization" but rather "the style and form of religion well-suited to the secular world" (Partridge 2004, 43). As an alternative to traditional religions, New Religious Movements provided eclectic, individualized forms of self-expression for disenchanted Westerners who explored various traditions, regarded the Self as inherently divine, and invested the Self with final authority. Precisely because they were profoundly individualistic, the 1960s devotees of counterculture did not manage to transform social institutions and they did not create their own alternative institutions (e.g., schools, nurseries, colleges, resettlement houses, women's shelters, or projects of urban renewal). The New Spirituality was more a sensibility or a posture than an ideology that called for social transformation. In fact, the New Spirituality was inseparable from Capitalism as yoga, the macrobiotic diet, Tai Chi, and aromatherapy became big successful industries that enriched their promoters (Lau 2000). And yet the New Spirituality positioned itself in contrast to and as superior to traditional religions. Whereas religion was narrow-minded, spirituality was open-minded; whereas religion was considered shallow, spirituality was deep; whereas religion was rigid, setting boundaries and creating limitations, spirituality was fluid, breaking and crossing boundaries; whereas religion was about death and the afterlife, spirituality was life-enhancing in this world; whereas religion was about the past, spirituality was about the present. In short, "spirituality" had what "religion" presumably lacked.

New Age spirituality took off in the 1960s as part of the countercultural movement, leading an increasing number of people to define themselves as "spiritual but not religious" or having "no religion." The Baby Boomers were fascinated by the spiritual traditions of the East, using them to critique whatever they found wrong or missing in the West. In 1972, 5 percent of the people surveyed in the General Social Survey were defined as "nones" and in 1975 the number went up to 7 percent but during the 1990s the number of "nones" began to rise significantly. In the 1990s, those self-identifying as "nones" more than doubled, going from 14.3 million to 29.4 million. By the end of the first decade of the 2000s at least 46 million, or more than one-fifth of all Americans, had no religious affiliation. The "nones" are now considered the fastest-growing "religious group" in America, outnumbering those affiliated with mainline forms of Protestantism. From 2008 to 2012, the percentage of the "nones" rose again by 3.2 percent, and especially "young people are now more religiously unaffiliated than earlier generations were at the same age" (Mercandante 2014, 1–2; Cf., Chen 2022, 4).

The self-defined "spiritual-but-not-religious" people exhibit diffuse, diverse, and even idiosyncratic beliefs. They usually do not belong to religious institutions, may not participate in traditional religious rituals, and do not abide by denominational doctrines, but they are deeply preoccupied with finding meaning in life and ponder the same questions to which religious traditions provide an answer: where we come from, why we are here, what happens when we die. Dissatisfied with religious institutions, the seeker of spirituality is interested in personal religious experience, places greater emphasis on the quality of inner life, and seeks holistic, life-affirming worldviews, often culling them from diverse religious and intellectual traditions. Eclecticism itself is a feature of contemporary spirituality.

Adherents of New Age religions sought to find "God within" rather than the "God without." Rejecting supernaturalism as well as metaphysical dualism, New Age religions aspire to experience transcendence within the "immanent frame" by experiencing "life-itself," "life force," or "energy" (Heelas 2008, 27) within one's own interiority. For the seeker of spirituality, the truths of life cannot be summarized in propositional or doctrinal terms, nor can they be expressed merely in sacred texts or rituals; instead, they should be experienced subjectively through love, harmony, joy, inner peace, vitality, self-confidence, positive emotions, and warm, supportive relations. Seekers of spirituality can define their goals in numerous ways but together they convey yearning for wholeness and wholesomeness in a profoundly broken world, a world that is out of balance due to stress, bad habits, addiction, isolation, negative emotions, and despair. In the highly pluralistic, consumerist, and globalized world, all sorts of hybrids emerge that allow people to experiment, explore, and express their spiritual impulses as they seek to re-enchant the world that modern secularism and modern science had disenchanted.

Contemporary spirituality is distinctly this-worldly: instead of seeking to unite with transcendent other-worldly deities, seekers of spirituality seek enlightenment in the sense of release from ego-attachments to achieve spiritual growth, and they engage in activities such as Yoga, Reiki, health and fitness, meditation, and art that assure subjective well-being. Because they are this-worldly focused, such activities have been characterized as "secular spiritualities" (Hanegraaff 1999, 152) but their relationship to the socio-cultural world is rather complex. On the one hand, New Age forms of spirituality are critical of the world's social ills and seek to dissociate themselves from it by focusing on subjective well-being. But on the other hand, New Age religions are an integral part of consumerist culture of late capitalism, enhanced by the communication revolution, transportation revolution, and globalization (Carrette and King, 2005). With gurus who promote spirituality selling their "spiritual merchandise" (e.g., books, tapes, clothes, retreats, memberships, seminars, counseling and healing services, and more), spirituality is a big business—and spiritual seekers are its producers and consumers.

Spirituality and the Re-enchantment of Nature

The quest for spirituality became more acute since the mid-20th century because of humanity's existential crisis. The mass killings of two world wars, the industrialization of death in the Holocaust, the threat of nuclear annihilation, and most relevant to this book, the ecological crisis all made clear that the modernist meta-narrative has collapsed, and a new understanding of the human condition is in order (Lyotard 1984 [1979]). Postmodernism captured that point when it critiqued the assumptions and presumed verities of the Enlightenment that yielded modern science, the ideology of progress, the nation-state and its perpetual wars, and the colonialism and imperialism that have subjugated people to exploit the natural riches of their societies. Modernity itself was now implicated in the world's dis-ease and brokenness. The postmodern condition entailed comprehensive rethinking of metaphysics, ethics, and politics.

Modern science, the main accomplishment of the Enlightenment itself, became subject for critique. In the 20th century, two world wars, Auschwitz, and Hiroshima exposed science's dark side and called into question the heroic image of science as a driver of modern progress and human betterment. In the early 1960s scientists became very concerned with the power of nuclear arms to annihilate all forms of life. Short of total annihilation, scientists noticed the negative impact of technoscience on all forms of life. Murray Bookchin (1962), Rachel Carson (1962), and Barry Commoner

(1966; 1971), as we shall discuss in Chapter 6, warned about the destructive power of pesticides and the threat that synthetic chemicals pose to all forms of life, including humans. In the following decades it became clear that humanity is now faced with massive environmental degradation and indeed the collapse of many ecosystems as the result of human behavior. Nature itself has become dis-eased, out of balance, as its fecundity has been diminished or depleted.

But who or what was responsible for the ecological crisis? Was the eco-crisis due to the modern "disenchantment of nature" brought about by modern science that regarded nature as inert matter devoid of intrinsic moral worth, or was the eco-crisis due to the Judeo-Christian religion that gave humanity the mandate "to have dominion" over the natural world and exploit natural resources for its sole benefit? Or perhaps the eco-crisis came about because of a unique confluence of modern science and one religion: Christianity?

That modern science—or more accurately, modern scientism—was the culprit of our eco-crisis was articulated most vociferously by religious philosophers (Christian, Muslim, and Jewish) who viewed the world as God's creation and who rejected philosophic naturalism and materialism. Scientism reduces nature to its material components, failing to appreciate the created world's magnificent beauty and awesome mystery. The premodern, theocentric worldview expressed awe and wonder at God's creation, which leads to respect, reverence, concern, and care for creation. In contrast, modern science's human-centeredness has demystified, objectified, mechanized, and commodified nature, viewing it merely for its instrumental value. For modern technoscience, nature is no more than a resource whose exploitation is morally justified because it facilitates human progress, liberation, and flourishing. For Alister McGrath, a chemist who became a historian of Christianity and constructive theologian, the solution to the ecological crisis will come with the re-enchantment of nature through the integration of religion and science (McGrath 2002; 2003). Modern science is not scientism (as atheist popularizers such as Richard Dawkins have led us to believe), since no less than religion, science, and scientists have deep appreciation for the wonder of nature, its inherent grandeur and beauty. A deeper and more accurate understanding of both science and the history of religions can bridge the harmful perception that science and religion are necessarily in conflict with each other and show how science itself can lead us to heal the wounds inflicted on the physical world. A very similar critique of scientism from a Muslim perspective has been argued by Sayed Hossein Nasr (1997 [1968]).

By contrast, Lynn White Jr., a historian of medieval technology, placed the blame for the eco-crisis at the door of Christianity, "the most anthropocentric religion" (White 1967) and especially the Bible which instructed humanity "to fill the earth and master it and rule the fish of the sea, the birds of the sky, and all the living things that creep on earth" (Gen. 1:28). Lynn

White charged that the belief that humans were created in the "image of God" gave humanity, especially in the technologically advanced West, the justification to exploit the natural world. Like McGrath, White also held that the cause of the eco-crisis is ideational or mental: "What people do about their ecology depends on what they think about themselves in relations to things around them. Human ecology is deeply conditioned by beliefs about our nature and destiny" (White 1967, 1205). White was fully aware that the eco-crisis is multifactorial, but he highlighted the fact that modern science emerged in the Christian West and both science and Christianity articulated a dualistic, hierarchical, and utilitarian worldview that resulted in the exploitation of nature. Christian metaphysical dualism made possible the desacralization, commodification, and exploitation of nature that yielded the eco-crisis. Yet, Christian spiritual traditions, especially Franciscan spirituality, also offered resources for a nonexploitative and caring attitude toward the physical world. Christianity (or religion more generally) need not be only the cause of the crisis; it can also be part of the solution.

White's critique of Christianity was a turning point in public awareness since it gave rise to a new discourse variously known as "religion and ecology" (Gottlieb 2006a; Bauman, Bohannon, and O'Brien 2023 [2011]; Grim and Tucker 2014; Jenkins, Tucker, and Grim 2016; Hart 2017), "religion and nature" (Taylor 2005), "religious environmentalism" (Gottlieb 2006; Berry 2011), or "ecological spirituality" (Sponsel 2012; Vaughan-Lee 2016 [2013]). Taking root within the discipline of religious studies rather than within the discourse of religion and science, the new field emerged that generated numerous monographs, articles, reference books, and conferences as scholars, theologians, and religious believers engaged in the "retrieval, reevaluation and reconstruction" of their respective traditions. Exploring, critiquing, and reinterpreting the relationship between humanity and nature within the contours of specific religious traditions, contributions to the discourse have not only studied religious conceptions of nature comparatively but also joined forces in interfaith activism to address environmental degradation and ecological collapse. Mary Evelyn Tucker and John Grim, the leaders of this international, interfaith conversation, regard the new discourse not only as an academic field but as a "movement" and even a "force" that seeks to affect a "quiet revolution" (Tucker and Grim, 2016). The academic discourse could inspire a social movement because it insisted on the connection between theory and practice, between how we think about nature and what we do to protect its vitality.

However we name this academic discourse, it is quite clear that faith-based environmentalism sought to reconfigure the relationship between humanity and the natural world. The "re-enchantment of nature" has become the shorthand for that new understanding, but what that means differs greatly. Environmental spirituality, or eco-spirituality, cuts across all religious traditions and denominations and breaks down the barriers between academic scholars and non-academic activists (Wheeler 2022). Those who care about the future of the Earth recognize that humanity cannot address the ecological crisis without reference to religious beliefs, values, rituals, and norms, since they shape human actions, policies, and politics. Faith communities who recognized the severity of the eco-crisis have linked theology and ethics to articulate new ecologically sensitive rituals, educational programs, and communal activities that address the social, economic, cultural, and spiritual dimensions of the eco-crisis. Today, ecologically minded faith communities re-read their sacred texts, revised their liturgies, re-read their histories, restructure their places of worship, make public pronouncements, and engage in social and political action all to protect and heal the wounded Earth. Environmental spirituality seeks to close the gap between "spirit" and "nature" that has dominated Western culture, offering instead an *embodied spirituality* (Hessel-Robinson and McNamara 2011).

The spiritual and ecological impulses of the 1960s converged to yield the re-enchantment of nature which modern science had disenchanted. Max Weber, as noted above, analyzed the "disenchantment of the world" as a linear process of "progress" towards moral perfection or towards cumulative technological rationalism. Weber did not welcome the process, but he correctly identified its driving force: the rationalization and bureaucratization that removed from the natural world any reference to nonmaterial forces be they spirits, faeries, gods, or God. Yet, when did the process of de-magification of nature start? For Weber the answer was in antiquity when prophetic monotheism replaced pagan polytheism. In the polytheistic worldview, capricious natural forces were deified and demanded human obedience, but nature could also be manipulated by magic, divination, and incantations. By contrast, monotheism posited a transcendent God who is not part of nature and who could not be controlled or manipulated by humans or by other nondivine forces. The monotheistic worshipper was concerned with obeying only one God and had one single system of religious ethics. In the Judaic monotheistic worldview, the natural world was the creation of a good and loving God, but it was not itself divine. According to Weber, then, the seeds of disenchantment of nature were sown by Jewish monotheism, which was transmitted to the West through Christianity. As we shall see in this book, the relationship between Judaic monotheism and paganism is more complex than Weber presented, but the relevant point here is that the New Spirituality of the 1960s that sought to re-enchant the world—in part so as to address the ecological crisis—manifested itself among Jews no less than in the society at large.

Jewish Environmental Spirituality as a Case Study

This book explores the Jewish responses to the eco-crisis in some detail. The focus on Judaism is no coincidence since I am a Jew by birth and by choice and a scholar of Judaism by academic career. As an intellectual historian, I treat all religions, including Judaism, as cultural expressions of specific historical conditions as they changed over time. In several publications I have offered historical overviews of Judaic conceptions of nature and how this intellectual history frames contemporary Jewish environmentalism. In this book, I examine Jewish environmental spirituality more closely considering the dynamics of the post-secular age outlined above. I argue that Jewish environmental spirituality has much in common with other religious environmental movements and other forms of environmental or ecological spirituality, but also that Judaism calls into question basic concepts taken for granted in Western (i.e., Christian or post-Christian) environmental discourse. Most importantly Jewish environmentalism calls into question the religious-secular divide, the dichotomy of the term 'sacred' and 'profane,' the distinction between the 'physical' and 'spiritual' aspects of human beings, the relationship of 'mind' and 'body,' and the interaction between the 'natural' and the 'supernatural' realms. Understanding Jewish environmentalism on its own terms will therefore enable us to appreciate the importance of difference as well as the merit of crossing boundaries (be they cultural, religious, ethnic, or racial).

As an exercise in Jewish intellectual history, itself an interdisciplinary practice, the book argues that Jewish environmentalism calls on all Jews to invest their Jewish identity and practice of Judaism with commitment to the natural environment whose integrity is now under duress. Earth-based Judaism enables environmentally concerned Jews to worship God not through bookish Torah study and performance of prescribed commandments but through environmental activism, reconnecting to the rhythms of nature, and treating the nonhuman as a moral Other that imposes on humans, and especially on Jews, infinite responsibility. At the same time, attention to environmental issues enables Jews who are affiliated with Reform, Conservative, Reconstructionist, and Orthodox forms of Judaism to show that secularism is not necessary for the modernization, social integration, and acculturation of Jews. If by "secular" we mean "the non-Jewish" world, then all Jews belong to and operate within the "secular" world even as they express their commitment to Judaism in "post-secular" forms. Jewish environmentalism is even more complex because of the place of the Land of Israel in Jewish consciousness, especially in the modern period. Zionists have argued for the centrality of the Land of Israel in environmental spirituality, whereas post-Zionists or anti-Zionists minimize the importance of the Land of Israel and promote diasporic environmental spirituality. The examination of Jewish environmental spirituality thus offers a comparative dimension to the study of religious environmentalism in the post-secular age.

References

Asad, Talal. 2003. *Formations of the Secular: Christianity, Islam, Modernity.* Palo Alto: Stanford University Press.

Bauman, Whitney A., Richard Bohannon, and Kevin J. O'Brien (eds.). 2023 [2011]. *Grounding Religion: A Field Guide to the Study of Religion and Ecology*, 3rd edition. London: Routledge.

Beaumont, Justin. 2010. “Transcending the Particular in Postsecular Cities.” In *Exploring the Postsecular: The Religious, the Political and the Urban*, ed. Arie L. Molendijk, Justin Beaumont, and Christopher Jedan, 3–38. Leiden and Boston: Brill.

Bender, Courtney. 2010. *The New Metaphysicals: Spirituality and the American Religious Imagination*. Chicago: University of Chicago Press.

Berger, Peter. 2014. *The Many Altars of Modernity: Toward a Paradigm for Religion in a Pluralist Age*. Boston and Berlin: De Gruyter.

Berger, Peter L., Grace Davie, and Effie Fokas. 2008. *Religious America, Secular Europe?: A Theme and Variation*. Farnham Surrey, England and Burlington, VT: Ashgate Publishing Company.

Berlinerblau, Jacques. 2021. *Secularism: The Basics*. London: Routledge. Taylor & Francis Group.

Berry, Evan. 2011. *Devoted to Nature: The Religious Roots of American Environmentalism*. Oakland, CA: University of California Press.

Bookchin, Murray (under the pseudonym, “Lewis Herber”). 1962. *Our Synthetic Environment*. New York: Knopf.

Calhoun, Craig, Eduardo Mendieta, and Jonathan VanAntwerpen (eds.). 2013. *Habermas on Religion*. Cambridge, UK: Polity.

Carson, Rachel. 1962. *Silent Spring*. New York: Houghton Mifflin.

Carrette, Jeremy and Richard King. 2005. *Selling Spirituality: The Silent Takeover of Religion*. London and New York: Routledge.

Casanova, José. 1994. *Public Religions in the Modern World*. Chicago: University of Chicago Press.

Casanova, José. 2008. “Public Religion Revisited.” In *Religion: Beyond the Concept*, ed. Hent de Vries, 101–119. New York: Fordham University Press.

Casanova, José. 2010. “A Secular Dawn or Twilight.” In *Varieties of Secularism in a Secular Age*, ed. Michael Warner, Jonathan Van Antwerpen, and Craig Calhoun, 265–281. Cambridge, MA and London: Harvard University Press.

Casanova, José. 2011. “The Secular, Secularizations, Secularisms.” In *Rethinking Secularism*, ed.

Craig Calhoun, Mark Juergensmeyer, and Jonathan VanAntwerpen, 54–74. Oxford and New York: Oxford University Press.

Chen, Carolyn. 2022. *Work, Pray, Code: When Work Becomes Religion in Silicon Valley*. Princeton and Oxford: Princeton University Press.

Commoner, Barry. 1966. *Science and Survival*. New York: Viking.

Commoner, Barry. 1971. *The Closing Circle: Nature, Man and Technology*. New York: Knopf.

Copson, Andrew. 2019. *Secularism: A Very Short Introduction*. Oxford: Oxford University Press.

Crockford, Susannah. 2021. *Ripples of the Universe: Spirituality in Sedona, Arizona*. Chicago: University of Chicago Press.

Daschke, Derek and Michael Ashcraft (eds.). 2005. *New Religious Movements: A Documentary Reader*. New York: New York University Press.

Dobbelaere, Karel. 1987.“Some Trends in European Sociology of Religion: The Secularization Debate.” *Sociological Analysis* 48 (2): 107–137.

Eisenstadt, Shmuel N. 2000. “Multiple Modernities.” *Daedalus*, 129 (1): 1–29.

Ferrer, Jorge N. and Jacob H. Sherman (eds.). 2007. *The Participatory Turn: Spirituality, Mysticism, Religious Studies*. Albany, NY: SUNY Press.

Fuller, Robert C. 2001. *Spiritual but Not Religious: Understanding Unchurched America*. Oxford and New York: Oxford University Press.

Gordon, Peter E. 2008. "The Place of the Sacred in the Absence of God: Charles Taylor's A Secular Age." *Journal of the History of Ideas* 69 (4): 647–673.

Gorski, Philip S. and Ates Altinordu. 2008. "After Secularization?" *Annual Review of Sociology* 34: 55–85.

Gorski, Philip S, David Kyuman Kim, John Torpey, and Jonathan VanAntwerpen (eds.). 2012. *The Post-Secular in Question: Religion in Contemporary Society*. New York: Social Science Research Council and New York University Press.

Gottlieb, Roger S. 2006. *A Greener Faith: Religious Environmentalism and Our Planet's Future*. New York and Oxford: Oxford University Press.

Gottlieb, Roger S. 2013. *Spirituality: What It Is and Why It Matters*. Oxford and New York: Oxford University Press.

Gottlieb, Roger S. (ed.). 2006a. *The Oxford Handbook of Religion and Ecology*. Oxford: Oxford University Press.

Graham, Elaine. 2013. *Between a Rock and a Hard Place: Public Theology in a Post-Secular Age*. London: SCM Press.

Grim and Tucker. 2014. *Ecology and Religion*. Washington, DC: Island Press.

Habermas, Jürgen. 2006. "Religion in the Public Sphere." *European Journal of Philosophy* 14 (1): 1–25.

Habermas, Jürgen. 2008. "Notes on a Post-Secular Society." *New Perspective Quarterly*, 17–29.

Habermas, Jürgen and Joseph Ratzinger. 2006. *The Dialectics of Secularization: On Reason and Religion*. San Francisco: Ignatius Press.

Hadden, Jeffrey K. 1987. "Toward Desacralizing Secularization Theory." *Social Forces* 65: 587–611.

Hanegraaff, Wouter J. 1999. "New Age Spiritualities as Secular Religion: A Historian's Perspective." *Social Compass* 46 (2): 145–160.

Hardin, Jeff, Ronald L. Numbers, and Ronald A. Binzley. 2018. *The Warfare between Science and Religion*. Baltimore: The Johns Hopkins University Press.

Harris, Sam. 2014. *Waking Up: A Guide to Spirituality Without Religion*. New York, London, Toronto, Sydney, and New Delhi: Simon & Schuster Paperbacks.

Harrison, Peter. 2010. "'Science' and 'Religion': Constructing the Boundaries." In *Science and Religion: New Historical Perspectives*, ed. Thomas Dixon, Geoffrey Cantor, and Stephen Pumphrey, 23–49. Cambridge, UK: Cambridge University Press.

Harrington, Austin. 2007. "Habermas and the 'Post-Secular' Society." *European Journal of Social Theory* 10 (4): 543–560.

Hart, John (ed.). 2017. *The Wiley-Blackwell Companion of Religion and Ecology*. Malden, MA, and Oxford: Wiley-Blackwell.

Heelas, Paul. 1996. *The New Age Movement: The Celebration of the Self and the Sacralization of Modernity*. Oxford: Blackwell.

Heelas, Paul. 2008. *Spiritualities of Life: New Age Romanticism and Consumptive Capitalism*. Oxford: Blackwell.

Heelas, Paul and Linda Woodhead. 2005. *The Spiritual Revolution: Why Religion is Giving Way to Spirituality*. Oxford: Blackwell Publishing.

Hessel-Robinson, Timothy, and Ray Maria McNamara (ed.). 2011. *Spirit and Nature: The Study of Christian Spirituality in a Time of Ecological Urgency*. Eugene, OR: Pickwick Publications.

Hogue, Michael S. 2010. "After the Secular: Toward a Pragmatic Public Theology." *Journal of the American Academic of Religion* 78 (2): 346–374.
Holder, Arthur (ed.). 2005 [2011]. *The Blackwell Companion to Christian Spirituality.* Chidester, West Sussex: Wiley-Blackwell.
Jenkins, Willis, Mary Evelyn Tucker, and John Grim (eds.). 2016. *The Routledge Companion of Religion and Ecology.* London and New York: Routledge.
King, Mike. 2005. "Art and the Postsecular." *Journal of Visual Art Practice* 4: 3–17.
Lau, Kimberly J. 2000. *New Age Capitalism: Making Money East of Eden.* Philadelphia: University of Pennsylvania Press.
Lim, Richard. 2009. "Christianization, Secularization, and the Transformation of Public Life." In Philip Rousseau (ed.). *A Companion to Late Antiquity*, 497–511. Oxford: Oxford University Press.
Lyotard, Jean-Francoise. 1984 [1979]. *The Postmodern Condition: A Report on Knowledge.* Minneapolis: University of Minnesota Press.
Macleod, Hugh. 1992. "Secular Cities? Berlin, London, and New York in the Later Nineteenth and Early Twentieth Centuries." In *Religion and Modernization*, ed. Steve Bruce, 59–89. Oxford: Clarendon Press.
Martin, David. 1965. "Toward Eliminating the Concept of Secularisation." In *Penguin Survey of the Social Sciences*, ed. Julius Gould. London: Penguin.
Martin, David. 1978. *A General Theory of Secularization.* Oxford: Blackwell.
Martin, David. 2005. *On Secularization: Toward a Revised General Theory.* Aldershot England and Burlington, VT: Ashgate Publishing Company.
McGrath, Alister. 2002. *The Re-enchantment of Nature: Science, Religion and the Human Sense of Wonder.* London, Sydney, and Auckland: Hodder & Stoughton.
McGrath, Alister. 2003. *The Re-enchantment of Nature: The Denial of Religion and the Ecological Crisis.* London, Sydney, and Auckland: Hodder & Stoughton.
McFague, Sallie. *Super, Natural Christians.* Minneapolis: Fortress Press, 1997.
McLennan, Gregor. 2010. "The Postsecular Turn." *Theory, Culture & Society* 27 (4): 3–20.
McClure, John A. 1995. "Postmodern/Post-Secular: Contemporary Fiction and Spirituality." *Modern Fiction Studies* 41 (1): 141–163.
McClure, John A. 2007. *Partial Faiths: Postsecular Fiction in the Age of Pynchon and Morrison.* Athens, GA: University of Georgia Press.
Mercandante, Linda A. 2014. *Belief without Borders: Inside the Minds of the Spiritual but Not Religious.* Oxford and New York: Oxford University Press.
Molendijk, Arie L., Justin Beaumont, and Christoph Jedan. 2010. *Exploring the Postsecular: The Religion, the Political and the Urban.* Leiden and Boston: Brill.
Nasr, Sayed Hossein. 1997 [1968]. *Man and Nature: The Spiritual Crisis in Modern Man.* Santa Barbara: ABC International Group.
Neuhaus, Richard John. 1982. "Educational Diversity in Post-Secular America." 77 (3): 309–320.
Norris, Pippa and Ronald Inglehart. 2011 [2004]. *Sacred and Secular: Religion and Politics Worldwide.* Cambridge, UK: Cambridge University Press.
Partridge, Christopher (ed.). 2004. *New Religions: A Guide. New Religious Movements, Sects and Alternative Spiritualities.* New York: Oxford University Press.
Perrin, David B. 2024. *Studying Christian Spirituality.* New York and London: Routledge.
Pierucci, Antonio Flavio. 2000. "Secularization in Max Weber: On Current Usefulness of Re-Accessing that Old Meaning." *Brazilian Review of Social Sciences*, special issue, no. 1: 129–158.

Smith, James K.A. 2014. *How (Not) to be Secular: Reading Charles Taylor.* Grand Rapids: Eerdmans.

Sponsel, Leslie E. 2012. *Ecological Spirituality: A Quiet Revolution.* Santa Barbara: Praeger.

Stark, Rodney. 1999. "Secularization, R.I.P." *Sociology of Religion* 60 (3): 249–273.

Stark, Rodney and Roger Finke. 2000. *Acts of Faith: Explaining the Human Side.* Berkeley: University of California Press.

Swatos, William H. and Kevin J. Christiano. 1999. "Secularization Theory: The Course of a Concept." *Sociology of Religion* 60 (3): 209–228.

Taylor, Bron R. (ed.). 2005. *The Encyclopedia of Religion and Nature.* London: Continuum.

Taylor, Charles. 2007. *A Secular Age.* 2007. Cambridge, MA: Harvard University Press.

Taylor, Charles. 2011. "Western Secularity." In *Rethinking Secularism*, ed. Craig Calhounet al., 31–53. Oxford: Oxford University Press.

Tirosh-Samuelson, Hava. 2017. "Religion, Science and Technology in the Post-Secular Age: The Case of Trans/Posthumanism." *Philosophy, Theology and the Sciences* 4 (1): 7–45.

Torpey, John. 2010. "A (Post-) Secular Age? Religion and the Two Exceptionalisms." *Social Research* 77 (1): 269–296.

Tucker, Mary Evelyn and John Grim. 2016. "The Movement of Religion and Ecology." In *Routledge Handbook of Religion and Ecology*, ed. Willis Jenkins, Mary Evelyn Tucker and John Grim, 3–12.

Urban, Hugh B. 2013. *New Age, Neopagan and New Religious Movements: Alternative Spirituality in Contemporary America.* Oakland, CA: University of California Press.

Vaughan-Lee, Llewellyn (ed.). 2016 [2013]. *Spiritual Ecology: The Cry of the Earth*, 2nd ed. Point Reyes, CA: The Golden Sufi Center.

Warner, Michael, Jonathan van Antwerpen, and Craig Calhoun (eds.). 2010. *Varieties of Secularism in a Secular Age.* Cambridge, MA: Harvard University Press.

Weber, Max. 2004 [1917]. "*Science as a Vocation.*" In *The Vocation Lectures*, ed. David Owen and Tracy B. Strong, trans. Rodney Livingstone. Indianapolis: Hackett.

Weber, Max. 1920. "Protesnticsche Ethic und der Geist des Kapitalismus." In *Gessamelte Aufsatze zur Relgionssociologie.* Tübingen: Mohr. Originally published in *Archiv für Sozialwissenschaft und Sozialpolitic* 1904–1905.

Wheeler, Rachel. 2022. *Ecospirituality: An Introduction.* Minneapolis: Fortress Press.

White, Lynn Jr. 1967. "The Historical Roots of Our Ecologic Crisis." *Science* 167: 1204–1207.

Wilson, Bryan R. 1992. "Reflections on Many Sided Controversy." In *Religion and Modernisation: Sociologists and Historians Debate the Secularization Thesis*, ed. Steve Bruce, 195–210. Oxford Clarendon Press.

Wilson, Bryan and Jamie Cresswell. 2002. *New Religious Movements: Challenge and Response.* London: Routledge.

Ziebertz, Hans-Georg and Ulrich Riegel. 2009. "Europe: A Post-secular Society? International Report." *International Journal of Practical Theology* 13: 293–308.

2 Jewish Environmentalism

An Overview

Judaism Faces Modernity

The concern for the well-being of the environment is an attitude, an orientation, or a mindset that leads to certain behaviors. Environmentalism is the behavioral manifestation of that attitude, and it can express itself in numerous activities, including articulation of principles, public protests and marches, planting trees, urban gardening, organic farming, letter writing to legislators, environmental education, and policy deliberation and legislation. Religious environmentalism refers to these activities when done by religious people or when environmental activities are justified by appeal to religious beliefs, texts, and practices (Gottlieb 2010; Jenkins and Chappel 2011; Kearns 2011). Jewish environmentalism is the expression of environmental concerns, attitudes, and actions of Jews, but the phrase is ambiguous, and the ambiguity goes to the heart of the relationship between the "religious" and the "secular" spheres. If a born Jew engages in any environmental activity, does that make the activity an expression of "Jewish environmentalism"? That question is open for debate. The problem of definition arises from the fact that Judaism is not simply a "religion," if by that we mean a set of beliefs or dogmas; rather, Judaism is a way of life that encompasses all aspects of life of a particular group of people, the Jews. Since Judaism changed over time in response to historical circumstances, generalizing about Judaism is problematic if not futile. Jews determine what Judaism means to them and how they practice it, if at all.

For most of their history the Jews lived as a distinctive religious minority governed by special laws: their own authoritative legal tradition believed to be divinely revealed and the laws that non-Jewish rulers imposed on the diasporic community. The Enlightenment and the rise of the modern nation-state in the 18th century challenged the very core of Jewish corporate existence. If the Jews were to receive civil rights as individuals and become citizens in the country of residence, they would have to surrender their corporate existence. In the modern period, Judaism was recognized as a "religion" that could be practiced in the private sphere (Batnitzky 2011), but the Jewish community (*kehilah*) could no longer exist as a legal entity with semi-political status. The rabbis would lose their legal authority, and the Jewish

DOI: 10.4324/9781351123945-3

community would no longer be able to enforce Jewish law or excommunicate its members. In the modern nation-state, Jews would become "Germans of the Mosaic Faith" or "Frenchmen of the Mosaic Faith" but Judaism as a faith community would be neither a legal nor political entity. The modern nation-state expected Judaism to become a private affair so that one could be "a Jew at home and a human being in public."

Granting civil rights to Jews was no simple matter for Christian Europe. For centuries Jews have been marginalized, segregated, persecuted, and despised, even though they also enjoyed the formal protection of religious and secular authorities. Christian anti-Jewish literature associated Jews with certain animals to accentuate the subhuman nature of the Jew, the ultimate Other, indeed the Anti-Christ (Geller 2018). Granting civil rights to Jews required Christian Europe to begin to see Jews as humans and as such grant them civil rights as a matter of logical consistency. That was one of the accomplishments of the French Revolution, which granted civil rights to Jews in 1790. By 1871 the Jews of western and central Europe were legally emancipated and could integrate into European society and culture, although they were by no means accepted as social equals. Indeed, modern anti-Semitism began as a political movement in the 1870s signaling the backlash against Jewish emancipation. In the Russian Empire, where the Jews lived confined to the Pale of Settlement, the Jews were not formally emancipated until the Bolshevik Revolution of 1917. Until then a mixture of policies intended to Russianize the Jews were imposed on them although the Jewish community continued to function as a legal entity.

These processes were not merely a matter of politics. Rather, emancipation transformed the very meaning of being Jewish. If Jews are now citizens of nation-states, why should they remain Jewish? And if they choose to remain Jewish, how should they express their Jewishness? Jews responded to emancipation's challenges in myriad ways: In France and German-speaking Central Europe, Jews largely welcomed emancipation and enthusiastically pursued the benefits of the emancipation in economics, residential mobility, education, civil service, and the liberal professions. Moreover, those who welcomed emancipation, social integration, and acculturation were willing to modernize Judaism by changing the language and style of traditional rituals to make the practice of Judaism more aligned with the intellectual outlook and social mores of Christian Europe. By contrast, many traditional Jews (especially in Eastern Europe) were critical of the emancipation and Jewish integration into European majority culture. These Ultra-Orthodox Jews resisted religious reforms on the ground that Jewish divinely revealed Law (the Torah) is eternal; Judaism should not be evaluated or changed to fit human contingent realities. Between the poles of Reform Judaism and Ultra-Orthodoxy lay many intermediate variants, such as the Positive Historical School (later to become Conservative Judaism) and Neo-Orthodoxy (later to be known as Modern Orthodoxy), which sought to balance loyalty to the Jewish tradition with selective adaptation to modernity.

Precisely when the emancipation was achieved in Western Europe in 1870, Jewish secularism and Jewish nationalism were on the rise in Eastern Europe. Both were an outcome of the Jews' peculiar political conditions in the Russian Empire. The Enlightenment did spread in the Russian Empire, imported by the Tsarina, Catherine the Great. In the divisions of the Polish Lithuanian Commonwealth in 1795, Russia inherited a large population of Jews but instead of extending them civil rights, Catherine the Great restricted Jews to the Pale of Settlement, where their economic life was strictly governed by discriminatory policies. During the 19th century, the Tsars Nicholas I and Alexander II desired to Russianize the Jews through forced conscription to the army and through obligatory education in state-sponsored schools, a policy which Jewish proponents of the Enlightenment (*maskilim*) actively supported. While some Jews were able to move to the cities and enter the university and the liberal professions with the intention of integrating into Russian culture (Nathans 2002), the masses of Jews in the Russian Empire continued to live traditional life governed by rabbinic authorities, whether the traditional rabbis or their Hasidic alternatives. Jewish promoters of the Enlightenment subjected the beliefs and practices of traditional Judaism to scathing critique, but that did not yield to religious reforms as it did in western and central Europe. Instead, the critique of traditional Judaism resulted in the secularization of Jews and Judaism.

The most universal expression of Jewish secularism was when Jews adopted Marxism, Socialism, and Communism as alternative ideologies to Judaism. Since these were not religious ideologies, endorsing them did not require formal "conversion," but in effect Jews who joined revolutionary movements were not so different from religious converts. However, there were also many Jews who adopted revolutionary ideologies but did not wish to erase their Jewish identity. Rather, they sought to reinterpret Judaism anew in secular, materialist terms (Stern 2018). The Jewish secularists were thus deeply concerned with the material well-being of the Jewish people and looked to modern science rather than to canonic texts, rituals, or dogmas to understand the conditions of the Jews, since it is the physical world itself that "connected modern Jews to each other" (Stern 2018, 7). The rise of Jewish secularism meant attention to aspects of peoplehood, land, and embodiment and to the particular social ills that afflicted Jewish society. The Jewish secularists offered various solutions, but historically the most successful was Jewish nationalism, namely Zionism. Variations of the Zionist idea evolved from the 1860s onward, reflecting the development of European nationalism and the reality of the Zionist project, but the key idea was that Jews must end their exile from the Land of Israel and return to their ancestral home. In the old/new land, new economic, political, and social conditions would be created to enable Jews to live justly without oppression and discrimination. Between the universalist (i.e., Marxism, Socialism, and Communism) and the particularist (i.e., Zionist) solutions of the Jewish problem, there were intermediate positions that advocated autonomism (i.e., the position of Simon Dubnow) or cultural nationalism (i.e., the position of Jewish Socialists).

In short, at the beginning of the 20th century there was no consensus among Jews about how to be Jewish or what it means to be Jewish. All forms of modern Judaism responded to the challenges of modernity by offering various blends of "secularity" and "religiosity." Precisely because in Europe the Jews were a distinct religious and ethnic minority, processes of secularization and modernization were expressed differently in the Jewish case. To be "modern" and "secular" did not pertain merely to religious beliefs, to synagogue attendance, or to observance of rituals, but also to attire, food, labor, embodiment, sexuality, languages, education, law, politics, art, and time—in short, to all aspects of life and culture (Finer 2010). Theories of modernization and secularization were developed regarding majority Christian society, and they simply miss most of the nuances of secularization and modernization among Jews. Jewish attitudes toward the physical environment, space and place, and toward land (especially the Land of Israel) are examples of what secularization theories fail to capture in relation to Jews. This is the focus of our book.

The breakdown of legal and political barriers meant that Jews increasingly integrated into non-Jewish society and culture although Jewish Otherness remained. Anti-Semitism, which was not translated from religious to secular pseudo-scientific categories, perpetuated Jewish Otherness as a matter of biology, since Jews are presumably a separate and inferior "race." The Nazis would seek to eliminate the Jewish race to bring a "Final Solution" to the "Jewish Question" that preoccupied Europe throughout the 19th century. The Holocaust did not annihilate the Jewish people or erase Judaism, but the differentiation between Judaism as a "religion" (or a "faith") and Judaism as a "people," a "nation" or an "ethnicity" would have a profound impact on the complexity of Jewish existence in the 20th century. The secular, ethnic definition of Judaism meant that one could be a Jew by birth but hold all sorts of views and beliefs or adopt various spiritual practices to express one's Jewishness. Moreover, the modern secularization of Jews did not erase the traces of the Jewish religious tradition. Even though secular Jews did not abide by the strictures of traditional Judaism, Jewish values, norms, sensibilities, and habits were part of their cultural identity, consciously or not. The secularization of Judaism resulted in cultural diversity, but all forms of secular Jewish culture were this-worldly, or earthly, and their "Jewishness" was informed by the very sources of Judaism from which secular Jews disengaged or categorically rejected.

A case in point is Murray Bookchin (1921–2006), the founder of Social Ecology, one of the main strands of environmental thought. Bookchin was the son of Russian Jewish immigrants in New York who was a member of Communist youth organizations and joined the Socialist Workers Party in the 1930s (Biehl 2015). Working in factories, he became a labor organizer, but after the General Motors strike in 1945–1946, he broke with Marxism-Leninism as a revolutionary ideology. After a short-lived association with Trotzkyism, Bookchin began to develop nonauthoritarian, libertarian

Communism with deep ecological concerns (Biehl 1997). We will have much more to say about Bookchin's Social Ecology and Deep Ecology in Chapter 6, but the relevant point here is this: Bookchin was a Jew by birth, but in what way were his environmental theories and activism "Jewish"? I take an expansive approach to this question. The story of Jewish environmentalism encompasses Jews of all stripes—observant, non-observant, secularized, secularist, and "spiritual but not religious"—whose views must be understood on their own terms and in their proper sociocultural context. The story of Jewish environmentalism also encompasses intellectual, social, political, and cultural dimensions of Jewish life in the Diaspora as well as in the Land of Israel and the relationship between them. Furthermore, the story of Jewish environmentalism includes Jews as well as non-Jews, and it engages Judaism along with other religious and spiritual traditions, be they Christianity, paganism, and shamanism. The story of Jewish environmentalism is therefore inherently pluralistic because Judaism lacks centralized authority to determine what it means to be Jewish or how to be Jewish. In the modern period, being Jewish is ultimately a subjective decision of individuals who define their Jewishness on their own terms as they see fit, and therefore, modern Judaism cannot be reduced to any one way of being Jewish or to any one category such as "religion," "faith," "nationality," "ethnicity," "peoplehood," or "culture." This chapter provides an overview of Jewish environmentalism in the United States with a focus on the interplay of faith, scholarship, and activism (Cf. Seidenberg 2005; Tirosh-Samuelson 2023).

The Emergence of Jewish Environmentalism

Although Jews like Murray Bookchin and Barry Commoner, who will be discussed in Chapter 6, warned about a pending eco-crisis already in the 1960s, the Jewish environmental movement emerged only in the 1970s. The impetus for it was the short essay by Lynn White Jr. (1967) we cited in the previous chapter, who charged that the responsibility for the ecological crisis lies at the doorstep of Christianity, the "most anthropocentric religion." He found the Bible guilty of the ecological crisis because humanity was given license to dominate the natural world (Gen. 1:28), sanctioning human conduct that has eventually brought about the ecological crisis. Responding to this charge, Jewish theologians and religious leaders disputed White's reading of the Bible arguing that the Bible does not give license to dominate the natural world but rather regulates human interaction with it. Rejecting White's accusations, rabbis, theologians, and Judaica scholars (e.g., Helfand 1971; Gordis 1971; 1986; Lamm 1971; Dresner and Sherwin 1978; Rakover 1982; Cohen 1989) argued that he had focused on Genesis to the exclusion of the rest of the Bible and that he had utterly ignored rabbinic Judaism, the normative interpretation of Bible, or the post-biblical developments of Judaism. They also noted the richness and complexity of the biblical portrayal of nature and the role that nature played in the covenantal relationship between

God and Israel. The desired relationship between humanity and the natural world is not exploitative; rather, humans must care for the natural world of which they are part. Published in Jewish journals whose primary readership was Jewish, these early essays explained how the belief in God as Creator of the world leads to protection of that which properly belongs to God; they explicated biblical and rabbinic norms and values about the environment; they articulated the ethics of responsibility toward the Earth and its inhabitants; and they highlighted the degree to which Judaism is not a glorification or deification of nature, the position characteristic of paganism, but rather a critique of amoral nature (Schwarzschild 1984 [2001]; Wyschogrod 1991 [2001]).

Environmentalism in the United States

To understand the emergence of Jewish environmentalism, we need to say a few words about environmentalism in America. In the United States, environmentalism began with the Conservation movement, which was concerned with fisheries and wildlife management, water, soil conservation, and sustainable forestry. Conservation became a national issue in America during the Progressive Era (1890–1920s), advocating the shift of responsibility from local to state and federal agencies. President Theodore Roosevelt, an ardent conservationist, gave the issue high visibility by establishing the US Forest Service, creating the National Parks system along with bird reserves, game preserves, and National Forests, and designating millions of acres as protected public land. A different approach to the environment was articulated by John Muir, the founder of the Sierra Club (1892), who wanted nature preserved for the sake of its beauty (S. Fox 1981; Worster 2008 [2011]). Muir reflected the Protestant (particularly its Calvinist strand) approach to nature, according to which nature is the pure ongoing creation of God, whose innocence was a foil to human corruption (Stoll 2015, 137–71). He saw the environment as a holistic organism and opposed the exploitation of natural resources, especially water, for the sake of urban and industrial development. The conservation of nature as an ideal informed the policies of Franklin Delano Roosevelt, whose New Deal programs during the 1930s sent millions of people into rural areas to work on conservationist projects.

A milestone in the development of the environmental movement was Aldo Leopold's *Sand County Almanac* (1949). His short essay "The Land Ethic," "provided an evocative and profound effort to articulate ethical guidelines for human interactions with nature" (Taylor 2005, 598). Leopold's land ethics insisted that "soils, waters, plants, and animals, or collectively the land" (ibid.) should count as the moral community to which humans belong. As members of such a large ecological community, humans can no longer act as conquerors and exploiters of nature, but rather as responsible citizens. Leopold linked ethics, politics, and environmentalism, but he made a nod to religion when he evoked the Mosaic Decalogue (i.e., the Ten Commandments) as an example of a set of moral standards that help define rights and wrongs in human relations. He also cites the Golden Rule and democracy as foundations that inform our societal code of conduct.

Already before Leopold's untimely death in 1948, Rachel Carson began to publish magazine articles in which she exposed the dangers of radioactive materials, pesticides, and herbicides that followed the economic boom of America in the post–WWII years. Like Muir, she was a Presbyterian who had a deep reverence for nature and an appreciation for the connectedness of all living things. In her influential *Silent Spring* (1962), Carson argued that industrial society decimated the avian population and threatened the health of many organisms, including humans. Carson's environmental warning was deeply rooted in her religious convictions, defending vulnerable nature from self-seeking, thoughtless modern industrial society (Sideris and Moore 2008). Carson is just one example of the general point relevant to us: American environmentalism manifested the deep religiosity of its founders. When Lynn White accused Christianity of responsibility for the environmental crisis, he compelled all Americans, but especially Christians and Jews, to reflect about the relationship between religion and the environment.

Since the establishment of Earth Day on April 22, 1970, religious practitioners have become increasingly engaged with the eco-crisis in the United States and worldwide. Through the 1980s and the 1990s various religious denominations articulated official positions about human relations to the natural environment.

Christian ecological spirituality is earthly and this-worldly. In this regards it affirms its closeness to Judaism. By the dawn of the 21st century, ecological Christianity was fully in place, led by theologians such as Roderick Nash, Sallie McFague, Rosemary R. Reuther, Dietrich Hessel, John Hart, Larry L. Rasmussen, Willis Jenkins, and numerous others. Through eco-theology, eco-hermeneutics, ecologically oriented rituals, environmental ethics, and environmental education, Christianity has engaged the social, economic, cultural and spiritual dimensions of our eco-crisis. The engagement involved also political activism, which was inspired by the secular environmental movement but developed independently of it (Ellingson, 2018). Faith-based environmental organizations hold that humanity cannot address the eco-crisis without reference to deeply held religious beliefs, values, rituals, and norms, since they shape human actions, politics, and policies and therefore religious environmentalism is "secular," in the sense that it is concerned with the well-being of the Earth and all its inhabitants, including human beings. Earthly in its scope and orientation, the justification for changing humanity's relationship with nature was "religious," because it appealed to canonical texts believed to be divinely revealed.

Environmentalism transformed the academy as well. In 1971 environmental ethics emerged as a distinctive academic sub-discipline within the discipline of philosophy with the publication of the journal *Environmental Ethics*. Departing from over two millennia of moral philosophy, environmental philosophers and ethicists began to extend moral considerability to nonhuman entities and identify humans' ethical obligations toward biotic and abiotic collectives. The academic discourse of environmental ethics, however, largely ignored the environmental significance of religious beliefs, texts, or rituals. These were to be the purview of a different academic sub-discipline variously referred to as "religion and ecology," "religion and nature," or "religious environmentalism." Within the discipline of religious studies this discourse engaged in the retrieval, reevaluation, and reconstruction of religious traditions in the light of the eco-crisis, and new journals were launched to examine a range of related subjects. The field was comparative and global in scope, engaging all world religions, including Judaism.

The Greening of Organized Judaism

The academic discourse of religion and ecology stimulated Jewish environmentalism as Jewish scholars were invited to participate in edited anthologies and conferences (e.g., Hargrove 1986; Foltz 2003). Under the leadership of Mary Evelyn Tucker and John Grim, the Harvard Divinity School sponsored a series of ten conferences on world religions and the environment (1996–1998), which resulted in ten edited volumes, including one on Judaism and ecology (Tirosh-Samuelson 2002). The Jewish response to the ecological crisis began on an apologetic note, but through the 1970s and 1980s all strands of modern Judaism began to mine the sources of Judaism for their ecological wisdom and environmental sensibilities, and Jewish environmental activism began to take shape both within Jewish institutions and in interfaith settings. By 2000, a new Jewish environmental awareness emerged: the eco-crisis should be of concern to Jews, and Judaism must address it based on its own religious sources. Judaism should be seen as part of the solution and not just part of the problem of the ecological crisis.

Reform Judaism

The "greening" of Judaism in the second half of the 20th century took place within the academy and outside it, and rabbis were crucial to both efforts. Reform Judaism, the largest strand of modern Judaism in America, took the lead in responding to the environmental crisis within the organized Jewish community in the United States. Already in November 1965, even before Lynn White Jr. published his influential paper, the Union of American Hebrew Congregations (UAHC, known as the Union for Reform Judaism [URJ] since 2003) noted that "America, in thoughtlessly abusing its natural resources has disregarded the Biblical injunction to conserve God's creation for the good of all" (UAHC 1965). In 1969, the General Assembly of the UAHC resolved "to urge that appropriate measures be taken by local state and national governments to remove or ameliorate the growing threats of environmental pollution and to afford protection to the environment"

(UAHC 1969). In resolutions on environmental pollution (1969), energy policy (1977; 1979) and the dangers of industrial toxins (1983) the leadership of Reform Judaism recognized the connection between environmental degradation and energy policies and accorded moral standing to the natural environment. In accord with the theology of Reform Judaism, the resolutions made their case for conservation by appealing to the ideal of justice, the ethics of responsibility, and the obligation to save life (*pikuach nefesh*).

In 1977 the Central Conference of American Rabbis (CCAR), representing the professional leadership of the movement, issued a resolution that called for "swift development of an effective national energy policy in consonance with the needs of an environment which will nurture life on earth" (CCAR 1977). In the following decades, Reform Judaism issued numerous resolutions on environmental matters and initiated many programs and projects to implement its environmental commitments. Currently, the official "Position of the Reform Movement on the Environment" states that

> as heirs to a tradition of stewardship that goes back to Genesis and teaches us to be partners in the ongoing work of Creation, we cannot accept the escalating destruction of our environment and its effect on human health and livelihood. It is our sacred duty to alleviate environmental degradation and the human suffering it causes instead of despoiling our air, land and water.
>
> (RAC 2023)

In 1991 the URJ passed a comprehensive resolution on the environment, calling on the American and Canadian governments to protect endangered species; protect and create new wilderness areas; ensure clean air and water; and prevent environmental injustice to poor and minority community. The same year the URJ passed the "New North America Energy Strategy: A Resolution" supporting energy efficiency and conservation, development of non-nuclear energy alternatives, and opposing expansion of offshore oil drilling. In 2009 the Reform Movement passed a resolution on "Climate Change and Energy" that ties together many of these concerns in a comprehensive statement on the need for all congregations and institutions of the Reform Movement to take action to protect the environment through local actions and advocacy. In 2017 the Reform Movement passed the "Resolution on Addressing the Impacts of Climate Change," reiterating its support for fighting climate change and pledging to work on supporting adaptation measures and relief efforts for those impacted by natural disasters exacerbated by climate change. Clearly Reform Judaism has been very attentive to the environmental crisis, but translating environmental concern into environmental action depends on the commitment of individual rabbis or congregational leaders.

Since 1998, climate change has emerged as the "face" of the ecological crisis. The CCAR recognized climate change in terms of religious responsibility to future generations, stating that "minimizing climate change requires us to learn how to live within the ecological limits of the Earth so that we will not compromise the ecological or economic security of those who come after us" (CCAR 1998). Along with responsibility to future generations, the resolution spoke of the "integrity of creation" and the "equitable distribution of responsibility." The resolution demanded that we recognize "our nation's responsibility to reduce global carbon emissions" and urged "institutions within the Jewish community to conduct energy audits of private homes and communal facilities, including synagogues, schools, community centers and commercial buildings" (ibid.). Since the Brundtland Report of 1987, "sustainable development" became the organizing principle for meeting human development goals while also sustaining the ability of natural systems to provide the natural resources and ecosystem services on which the economy and society depend. Yet how to translate the goal of sustainability into action has generated spirited debate, since some consider development to be inherently unsustainable while others are disappointed by the lack of progress on what has been achieved to date. If in the 1990s the goal of sustainability was the banner of Jewish environmentalism, since the 2000s, the focus has shifted to addressing climate change.

Conservative Judaism

No less attentive to the dangers of environmental collapse is Conservative Judaism, which positions itself between Reform Judaism and Modern Orthodoxy. Unlike Reform Judaism, which challenged the authority of rabbinic Judaism and called for the modernization of the Jewish tradition, Conservative Judaism declared its commitment to Jewish law (Halakhah) and responded to the challenges of American life by looking at all aspects of American society and culture through the prism of Jewish canonic texts. And unlike Modern Orthodoxy, Conservative Judaism emphasizes the historical development of Judaism through the collective decision-making of the Jewish community. During the 1980s and 1990s, Conservative rabbis trained at the Jewish Theological Seminary in New York (JTS) were engaged in anchoring Jewish environmentalism in the literary sources of the Jewish tradition. In 1991, a special issue of *Conservative Judaism* 44 (no. 1) featured the work of eco-hermeneutics by Conservative rabbis, including Rabbi Evertt Gendler (d. 2022), an early advocate and practitioner of organic farming and vegetarianism; Rabbi Ismar Schorsch, Chancellor emeritus of JTS (1986–2006); and Rabbi Lawrence Troster (d. 2019), who promoted environmentalism and who were involved in the creation of the Coalition of Environment and Jewish Life (COEJL) and in interfaith environmental activism.

During the 1990s the official bodies of the Conservative Movement, the Rabbinical Assembly (RA), the United Synagogue of Conservative Judaism (USCJ), and the Committee of Law and Standards (CLS), issued many resolutions on environmental matters. For example, in 1991 the Rabbinical

Assembly called for growing, producing, and manufacturing food and food products in a way that protects the Earth, using consumables such as paper products, cleaning material, etc., that are both recycled and recyclable, and investing in companies that practice environmentally friendly practices. The congregations of the Conservative Movement were called to recycle and reduce waste in both its institutions and homes of members and constituents, and congregations were encouraged to conduct environmental audits, to integrate Jewish environmental education into the educational curriculum, and to avoid wasting food in Jewish celebrations (1994). When the United States refused to sign the Kyoto Agreement of 1997 and climate change had emerged as the "face" of the ecological crisis, the Conservative Movement, like Reform Judaism, issued many resolutions that called for legislation to reduce greenhouse gas emissions, including financial incentives such as a carbon tax, enforce high regulation standards, regulate harmful chemicals and substances, and aggressively support pollution prevention measures. The Conservative Movement spelled specific measures. For example, in 2008 the Rabbinical Assembly called on the institutions of the Conservative Movement to sign the "Green Covenant: A Jewish Pledge toward Carbon Neutrality" to reduce their carbon footprint and to call on the governments of our countries to reduce carbon emissions by 50 percent (RA 2008).

Attentiveness to climate change also characterized the resolutions of the Conservative Movement. In resolutions from 1998, 1999, 2003, and 2013, the Rabbinical Assembly called upon officials to also urged its members to advocate for energy audit for their own institutions and to promote the EPA Energy Standards for all future building projects. Members were encouraged to become educated about the value of recycling, green transportation, options such as public transit and high-mileage vehicles, and energy efficiency practices in the home. Whereas the recommended measurements and policies concerned "down-to-earth" (or "secular") activities, the justification was decidedly religious. Thus, the resolution of the Rabbinical Assembly on energy and environment (2001) stated that humanity must serve as "faithful stewards of Creation that are morally compelled to choose safe, clean, and sustainable sources of energy which will protect and preserve the world" (RA 2001). The appeal to the doctrine of creation cuts across all forms of Jewish environmentalism, expressing a shared religious worldview, as will become clearer in Chapter 3.

Reconstructing Judaism

Created by Mordecai M. Kaplan (d. 1983), the movement currently named Reconstructing Judaism, has been most attentive to the ecological crisis. Already in the 1980s the movement issued a series of resolutions about environmental matters based on the religious premise that "The Earth is the Lord's and the fullness thereof" (Ps. 24:1) and that "our tradition obligates every Jew to work for *Tikkun Olam*, the repair of the world." Recognizing the "devastating despoliation of our environment, directly or indirectly contributed to by the vast majority of the human inhabitants of the Earth," which increasingly "threatens the health, and, indeed, the very existence of animal and plant life on the Earth, including human life," the Federation of Reconstructionist Congregations and Havurot (FRCH) has supported legislation that deals with wide-ranging environmental problems. In a series of resolutions since the 1980s, the movement concerned itself with a plethora of environmental issues including air pollution; the greenhouse effect; acid rain; ozone smog and ozone depletion; agricultural runoffs; discharge limitations; wetland protection; ocean protection; hazardous and hazardous wastes; wildlife and natural resources; insecticide, herbicides, and fungicides; energy consumption; habitat restoration and loss prevention; and resource waste reduction. Asserting that "education and action go hand in hand," the Reconstructionist movement has viewed environmentalism as one of its values because it expresses "caring for our earthly home" and has encouraged its congregants and institutions act in environmentally correct ways. Reconstructionist rabbis, such as Rabbi Fred Scherlinder Dobb and Rabbi Jeff Sultar, have promoted environmental activism as a form of Jewish spirituality (RJ 2001/2002) and have inspired the movement to take a stand on global climate change (RRA 2008). In 2017 the movement condemned the withdrawal of the first Trump administration from the Paris Accord (2015) and declared its support of a renewable energy bill that divests from Fossil fuels (RJ 2017). Eco-Judaism, or the integration of ecology and Judaism, has become a distinctive commitment of Reconstructing Judaism not only because of the legacy of Mordecai Kaplan but also because of the Jewish Renewal Movement with which it is closely aligned and to which we shall turn below.

Modern Orthodoxy

Attentiveness to environmentalism did not belong exclusively to the liberal and progressive forms of contemporary Judaism. Modern Orthodoxy too was not oblivious to the environmental crisis, although it did not respond to it with the same vigor. Already in the early 1970s, Rabbi Norman Lamm, the President of Yeshivah University (1976–2003), insisted that "'ecology' is now a universally major concern: unless men change their ways immediately, we are bound to suffer irrevocably from our thoughtless abuse and exploitation of nature" (Lamm 1974, 213). He made it clear that

> in the Jewish view man has certain duties and responsibilities toward the natural environment. It was given to him in trust not outright, and therefore he must answer to God for his conduct toward it. "The Earth is the Lord's" (Ps. 24:1).
>
> (ibid.; Cf. Lamm 1971)

Similarly, several Modern Orthodox Jews championed Jewish environmentalism and veganism in response to the ecological crisis. For example, in 1975 Jonathan Wolf, an Orthodox Jewish social activist, founded the Jewish Vegetarians of North America (JVNA), which changed its name to Jewish Veg in 2015, and he was the first to integrate the kabbalistic practice of a Tu Bi'shvat Seder with explicit references to environmental issues. Richard H. Schwartz (b. 1935) is another example of an Orthodox Jew who persistently argued that traditional Judaism articulates a meaningful response to the eco-crisis because it supports vegetarianism, environmental protection, compassionate treatment of animals, conservation of natural resources, and world peace (Schwartz 1982 [2001]; 1984 [2002]).

The leadership of Modern Orthodoxy adopted that approach in 1988, when the Rabbinic Council of America (RCA) issued a resolution on energy and attended to the seeming conflict between environmentalism and the language of Genesis 1:28. In 1991 the RCA's statement suggests that humans have only a "limited dominion" over the Earth, which means "Stewardship over God's creation, to tend and to protect the world for future generations" (RCA 1991). Although the RCA did not engage environmentalism as rigorously as the other Jewish denominations, in 1999 the RCA issued a resolution on energy policy and in 2006 there was a call to action to reduce oil consumption and American dependence on oil and another resolution on the importance of preserving the environment.

Commitment to environmental action has been voiced by prominent Modern Orthodox rabbis—especially the late Rabbi Jonathan Sacks (d. 2020) in the UK and Rabbi Irving (Yitz) Greenberg in the United States (b. 1928). We will have more to say about Rabbi Sacks in Chapter 7, but here we should also mention Rabbi Shmuli Yanklowitz, a disciple of Rabbi Greenberg (Yanklowitz 2018). An Orthodox environmental activist and public theologian, Yanklowitz founded environmental organizations—Shamayim Va'Aretz Institute, later renamed Shamayim: Jewish Animal Advocacy, and Uri L'Tzedek: Orthodox Social Justice—that are guided by Torah values in their attempts to combat suffering and oppression of people, animals, and the environment. In these initiatives and in his several books, Yanklowitz has illustrated how Torah-based social justice coheres with veganism, environmental activism, and Jewish learning (Yanklowitz 2012; 2015; Labendz and Yanklowitz 2019).

Environmental activism by and among Orthodox Jews is not limited to men. In 2004 Evonne Marzouk, a Philadelphia-born Orthodox woman, who was inspired by Rabbi Arthur Waskow, founded Canfei Nesharim: Sustainable Living Inspired by Torah. This organization generated Judaic and scientific resources for environmental education in Orthodox communities, schools, and organizations, spreading the message that the Jewish tradition teaches deep wisdom about protecting the physical world created by God (Neril and Marzouk 2013). In 2019 Kanfei Nesharim merged with GrowTorah: while Canfei Nesharim runs a Jewish camp that emphasizes Torah learning and community, GrowTorah manages experiential and environmental Torah education programs. If Torah commands Jews to protect the integrity of the divinely created world, Judaism and environmentalism are fully compatible.

That message was articulated not only in the Diaspora but also in Israel. For example, Manfred Gerstenfeld (b. 1937), an Austrian born Modern Orthodox environmental scientist and political theorist argued that the values and precepts of traditional Judaism support and promote sustainability (Gerstenfeld 1998; 2001). Critical of the association between environmentalism and neo-paganism (Gerstenfeld 1998a), Gerstenfeld is one of the few Jewish religious authors who is well informed about environmental and ecological sciences. Today Torah-based environmentalism is promoted by Jewish Eco-Seminars, a unique collaboration between Orthodox and Evangelical communities to disseminate faith-based environmentalism that promotes creation care (Jewish Eco-Seminars 2025).

How to make environmentalism more widespread among Orthodox Jews has been the focus of Professor Tanhum Yoreh, a Modern Orthodox Jew who is also a scholar of environmental studies and a sociologist of religion. His studies of the history of the environmental precept *Bal Tashchit* (Do Not Destroy), have shown that pro-environment activism in the Orthodox community exists despite barriers that hampers its spread among Orthodox Jews (Yoreh 2010; 2018; 2019; 2019a). In Israel too, Modern Orthodoxy has become more attentive to environmentalism as is evident in the establishment of Ma`ayan Center for Jewish Philosophy and Sustainability at the Department of Jewish philosophy at Bar-Ilan University.

Jewish Environmental Spirituality: Inspirational Visionaries

The Jewish quest for spiritual renewal must be understood in the context of the countercultural upheavals of the late 1960s and early 1970s: the Civil Rights movement, the anti–Vietnam War movement, the movement for nuclear disarmament, the students' movement, and the women's movement, all of which challenged the established order and called for rethinking philosophical assumptions and social conventions. In all these movements Jews were deeply involved as visionaries, leaders, or rank-and-file members. Disaffected (or disenchanted) Jews not only critiqued America's social ills or demanded a change in its underlying social structures—they also critiqued the ethos of the organized Jewish community, the social mores of American Jews, and the limitations of the institutional forms of American Judaism. Whether the critique was directed at the preoccupation with social mobility and financial success, the overemphasis on Zionism and the State of Israel, or the flatness of synagogue worship, Jewish spiritual seekers insisted that "spirituality" was sorely missing in the organized life of American Jewry. Where could Jews find an answer to the spiritual malaise of the day? The spiritual quest led Jews both outside and within Judaism.

Many Jewish spiritual seekers found a responsive chord for their spiritual quest in Eastern traditions, especially Buddhism. As part of the global trend

Eastward characteristic of New Age religions, Jewish Baby Boomers flocked to Ashrams in India or Buddhist monasteries in Nepal or Japan, practicing meditation, chanting, and mindfulness. Bringing the wisdom of the "mystical East" to the United States, some Jews (e.g., Ram Dass [aka Richard Alpert], Joseph Goldstein, Jack Kornfield, Daniel Goleman, Sharon Salzberg, Sylvia Burstein, Jon Kabat-Zinn, and many others) became leaders in American Buddhism as they interpreted Buddhist and Hindu teachings to address the psychological and spiritual concerns of the day (e.g., anxiety, addiction, discontent, boredom, etc.). To be a Jewish Buddhist (that is, a "JuBu") one did not have to deny one's Jewishness; in fact, it was the very encounter with Buddhism that would allow for one's Jewish identity to be articulated with greater conviction (Kamenetz 1994). Since Jewishness was a matter of birth and ethnicity, practicing meditation or mindfulness that brought inner peace and serenity did not negate one's Jewishness. It could only enrich it. Spiritual practices, techniques, and sensibilities of other traditions (e.g., Buddhism, Hinduism, or Shamanism) could be easily "Judaized" by absorbing them into traditional Jewish life and the result was a deliberately eclectic or syncretistic Judaism. Crossing cultural boundaries was a matter of choice; eclecticism was a hallmark of Jewish environmental spirituality (e.g., Winkler 1998 [2004]; 2003).

Simultaneously an opposite trend took hold as tens of thousands of Jews were gravitating toward traditional Jewish observance, Orthodox, Ultra-Orthodox (Haredi), or Hasidic. Known as "Ba`aley Teshuvah" (literally, "masters of repentance") these returnees to traditional life could be secular Jews who had no religious upbringing or no prior formal affiliation with the organized Jewish community, or they could be Jews who grew up in liberal and progressive forms of Judaism (e.g., Reform, Conservative, or Reconstructionist) but considered them to be lacking spiritual depth (Kaplan 2009, 285–90). Those who gravitated to traditional Jewish life found it to be intellectually, socially, and emotionally richer than the life they had before. The Ba`aley Teshuvah movement reflected disenchantment with organized Jewish life, both in Israel and in the Diaspora and a rejection of the social values of consumerist society or a search for a psychologically more meaningful life.

For some, the return to traditional Judaism was inspired by the victory of Israel in the Six-Day War of June 1967, which unleashed messianic fervor of sorts. Orthodox theologians interpreted this remarkable military success in messianic terms, as direct divine intervention in history and the beginning of redemption after the destruction of the Holocaust.

We will explore these developments in Chapter 5, but for now it is important to recognize how the influx of newcomers to Israel in the early 1970s reflected the spiritual awakening of the late 1960s and early 1970s, especially in American Jewry. At the center of that spiritual awakening stood several visionaries: Rabbi Mordecai M. Kaplan (1881–1983), Rabbi Abraham Joshua Heschel (1907–1972), Rabbi Zalman Schachter-Shalomi (1924–2014), Rabbi Arthur Waskow (1933–2025), and Rabbi Arthur Green (b. 1941), whose ideas and actions framed Jewish ecological spirituality. If "environmental spirituality" concerns the well-being of the physical environment, "ecological spirituality" is a more abstract concept that concerns the metaphysical framework within which God, humanity, and other beings relate to each other. Below and in the following chapters we tell the biographies of key individuals in some detail because they inspired the growth of Jewish environmentalism and because they created institutions that facilitated Jewish environmental activism. These visionaries embody and promote Jewish earthly spirituality, namely, they express their awareness of the divine presence in the world and experience the sacredness of life by means of "down-to-earth" actions that sanctify the "secular" world.

Abraham Joshua Heschel

Abraham Joshua Heschel inspired a new attitude toward the natural world; one based on wonder and "radical amazement." Born in Warsaw, Poland, in 1907, Heschel was privately tutored in Talmudic studies and was recognized as a prodigy from a young age (Kaplan and Dresner, 1998). But unlike other youngsters of his Hasidic community in Warsaw, he was attracted to secular studies and was determined to pursue it. Mastering Polish and German with private tutors, Heschel traveled to Vilna to be educated in a Jewish Gymnasium whose diploma would enable him to enter a German university. After Vilna, the young Heschel moved to Berlin, where once again he broke cultural conventions by studying philosophy at the Friedrich Wilhelm Universität, the renowned University of Berlin, as well as studying at the liberal Hochschule für die Wissenschaft des Judentums, the rabbinic seminary of Reform Judaism.

In 1933, soon after Heschel received his PhD and rabbinic ordination, Heschel witnessed Hitler's rise to power and the gradual exclusion of Jews from the public sphere, the professions, and educational institutions. Under the leadership of Martin Buber and the Fries Jüdisches Lehrhaus in Frankfurt-am-Main, German Jews experienced a "Jewish renaissance," an outburst of creativity through adult education. Heschel joined the project, teaching, inspiring, and uplifting German Jews, especially after Buber migrated to Palestine in 1938 and Heschel remained to run the Lehrhaus. As a Polish Jew, however, Heschel was forcibly deported to Poland in November 1939, where he taught at the Warsaw Institute of Jewish Studies for several months while tirelessly trying to arrange his escape from Nazi-occupied Europe. With Buber's help, and especially with the help of Julian Morgenstern, the President of Hebrew Union College, the Reform seminary in Cincinnati, Heschel was finally able to flee Poland, first to London and from there to the United States in 1940.

In the United States he emerged as "one of the leading voices of conscience, interreligious dialogue, and spiritual awakening in American society" (Marmur 2016, 7). After five years at the Hebrew Union College (1940–

1945), he joined the faculty of the Jewish Theological Seminary in New York, the rabbinic seminary of the Conservative Movement, where he taught from 1946 until his death in 1972. During those most productive years, Heschel had one primary commitment: to awaken the spiritual consciousness of American Jewry. Although Heschel was a deeply acculturated Jew who was at home in Western philosophy and science, he was alarmed by the profound secularization of American Jewry and what he regarded as the negative impact of modernity and the Enlightenment. Precisely because he was so well rooted in the secular humanities, Heschel understood the predicament of the modern person who is locked in the "iron cage," as Weber called it, of objectivity, rationality, and expediency.

Heschel was fully aware that modernity for Jews signaled a rupture with the Jewish tradition but believed that "the tools for responding to that rupture were present within the Jewish tradition itself" (Marmur 2016, 18). Heschel offered a new interpretation of Judaism intended to spiritually renew not only Jews but all modern people. He saw himself as incarnating the Jewish spirit that spoke a universal message through the myriad sources of the Jewish tradition. Heschel called American Jews to return to the sources of Judaism and showed them how Jewish piety and academic learning could enrich each other. In so doing he greatly contributed to the revival of Jewish Studies in America in the 1970s, championed by the Jewish Theological Seminary of America in New York, Heschel's academic home.

Heschel articulated Judaism's spiritual message as follows: existence is a mystery to which we respond in wonder and radical amazement (Held 2013). Heschel's task was to awaken in all modern people the spiritual capacity "that sees through rather than gazes at objects" (Marmur 2016, 159). This consists of not only seeking self-transcendence through sacred deeds (i.e., the commandments of Judaism) but also understanding that God Himself is engaged in self-transcendence. As Heschel put it, "God is unwilling to be alone, and man cannot forever remain impervious to what He longs to show" (Heschel 1951, 91). For Heschel spirituality meant that God needs humans to improve the world, and humans must respond to God; humans are called by God to act. In Heschel's philosophical anthropology the human self is not an "isolated entity confined in itself," or as Charles Taylor called it, the "buffered self," but rather a relational self whose essence is to engage in self-transcendence through action. Heschel's spirituality did not consist of a "leap of faith" or a leap of thought but a "leap of action." In Heschel's words,

> A Jew is asked to take a leap of action rather than a leap of thought: to surpass his needs, to do more than he understands in order to understand more than he does … Through ecstasy of deeds, he learns to be certain of the presence of God.
>
> (Heschel 1996, 137–38)

Heschel's mission was to revive humanity's spiritual sensitivity, beginning with the Jews. He taught Jews and non-Jews to experience reverence and awe for the natural world. In so doing he rebuked the tendency of the modern age to ignore the sublimity of nature and to view nature simply as a useful resource that industry exploits. Instead, Heschel called on all of us to "stand still and behold," imploring us to "behold not only in order to explain, to fit what we see into our notions, behold in order to stand face to face with the beauty and grandeur of the universe" (Heschel 1965, 118; Cf. Merkle 1978; M. Fox 2014, 7). This posture of wonder and appreciation is the very opposite of expediency, exploitation, and manipulation that brought about the ecological crisis. Wonder and radical amazement led Heschel's disciples to engage in environmental activism especially through education. By developing the qualities (or sensibilities) of uniqueness, expectations, choice, creativity, spirituality, and ethical living, the individual can act within the social context of the family, the group, the civic order, the nation, and finally humanity.

To cultivate the qualities necessary for a social and civic life Heschel got involved in a wide range of social issues: Jewish–Christian dialogue, the plight of Soviet Jewry, the Civil Rights movement, the anti–Vietnam War protest, and many more. Heschel inspired an entire generation of American Jews to express their social activism through the medium of Jewish traditional sources and to do so within Jewish educational institutions. Not surprisingly, when American Jews committed to environmentalism migrated to Israel in the early 1970s—Jeremy Benstein and Eilon Schwartz—they named their institution, the Heschel Center for Environmental Learning and Leadership. In the model of public intellectual exemplified by Heschel, ecological spirituality and social activism went hand in hand although Heschel himself was not an environmental activist. His numerous disciples perpetuated this blend of spirituality and activism as is best illustrated in Michael Lerner (d. 2024), the founder of *Tikkun* magazine, the intellectual home of "spiritual progressives" in America, who will be discussed in Chapter 7.

Zalman Schachter-Shalomi

Heschel was not the only Jewish thinker in the post-Holocaust years engaged with the revival of Judaism. Even more revolutionary and transformative was Rabbi Zalman Schachter (who in the 1980s called himself Schachter-Shalomi), another Polish-born Jew associated with Hasidism who managed to flee Nazi-occupied Europe and settle in the United States. Schachter was born in 1924 in Zolkiew, a town in southeastern Poland (now in Ukraine), but that very year his family moved to Vienna so that the father could avoid conscription to the Polish army. The father, a Hasid associated with the Belz strand of Hasidism, encouraged integrating religious and secular studies, so Schachter was educated in a Zionist high school as well in an Orthodox Yeshivah. The 1938 invasion of Austria by Nazi Germany caused the family to flee briefly to Antwerp, Belgium, where the young Schachter encountered and joined the Chabad branch of Hasidism and adopted its worldview. Belgium fell to the Nazi regime soon after, and the family fled southward to France, which soon also fell to the Nazis. Under the Vichy regime the family was caught in the roundup of Jews and ended up in a Marseilles holding camp. Remarkably, the family secured exit visas and arrived in the United States, settling in New York. There the young Schachter enrolled in a Chabad-Lubavitch Yeshiva and in 1947 was ordained as a rabbi. Hasidism would play a major role in the revival of Orthodoxy in America in the Post-Holocaust years as much as it exerted a profound impact on Jewish religious environmentalism.

After his rabbinic ordination, Schachter and his close friend Rabbi Shlomo Carlebach were sent by the sixth Lubavitch Rebbe, Joseph Isaac Schneerson (1880–1950) to promote Chabad teaching to nonaffiliated, secular Jews (Ariel 2003). Schachter was sent to Brandeis University, the newly established Jewish university in Newton, Massachusetts, and Carlebach was sent to the University of California-Berkeley, the stronghold of Jewish academic secularism on the West Coast. As Chabad emissaries, Schachter and Carlebach incorporated stories, music, and folklore to reintroduce Americanized, secular Jews to traditional Judaism. To make a living, Schachter served as a congregational rabbi and as an educational Hillel director and staff member in Ramah summer camps. Yet intellectually Schachter was restless, seeking deeper and broader intellectual engagement. He enrolled at Boston University and in 1956 graduated with an MA in psychology of religion, focusing on pastoral counseling. This degree facilitated his move from a congregational rabbi to an academic, but the first position he obtained at the University of Manitoba in Winnipeg, Canada, actually combined the two functions as rabbi of Hillel. In the 1950s and 1960s, before the discipline of Jewish studies was established in secular universities in North America, the Hillel Foundation was the primary context for American youth to engage in the academic study of Judaism on campuses.

In Winnipeg, Schachter began to carve out his own spiritual path, rooted in Hasidism but open to all spiritual traditions as well as to psychology and popular science. Schachter's relationship with the Chabad movement, however, soured in 1962 because he was too immersed in contemporary secular culture, including experimenting with psychedelic drugs with Timothy Leary (Schachter-Shalomi and Hoffman 2012, 141–54). Under the seventh Rebbe, Menachem Mendel Schneerson, Chabad became more stringent and less tolerant of secularization. Schachter was forced to leave the Chabad organization but remained a Hasidic practitioner–teacher–counselor–spiritual guide throughout his life. Perpetuating the Hasidic style of teaching through tales, anecdotes, and homilies (e.g., Schachter-Shalomi and Miles-Yépez 2003), Schachter functioned as the Hasidic Rebbe of non-Hasidic Jews, who referred to him as "Reb Zalman." Indeed, Schachter deeply believed that the Hasidic Rebbe was the correct model of a "spiritual leader" who could renew contemporary Judaism (ibid., 11). As an "independent" rabbi, Schachter would privately ordain his own disciples to whom he imparted his unique interpretation of Judaism which addressed all aspects of modern life, including the environmental crisis. As he was experimenting with new forms of Jewish life, Schachter also helped to establish the Department and Clinic of Pastoral Psychology of United College (later the University of Winnipeg) and earned a Doctorate of Hebrew Letters (DHL) from the Hebrew Union College, the Reform Seminary in Cincinnati, in 1968. Crossing boundaries (denominational, religious, or cultural) characterized Schachter's career and intellectual posture; this same eclecticism was the hallmark of the Jewish Renewal Movement.

In Canada, Schachter began to experiment with new structures for Jewish life. He founded the liturgical brotherhood Order of B'nai Or (Sons of Light). This community did not last long, but it would be later revived in Philadelphia, albeit with more realistic expectations. Schachter also tried to start up a "seminary without walls" modeled after the counterculture's free universities of the late 1960s but that too failed. Despite the organizational failures, Schachter's spiritual quest made him attractive to other spiritual seekers who were dissatisfied with traditional Jewish education and worship. Schachter's profound interest in other spiritual traditions (especially Buddhism, Hinduism, and Sufism) made his Hasidism inherently syncretistic and opened Jews to New Religious Movements that sprouted at the time (Ariel 2011).

In 1968 Schachter was on a Sabbatical from the University of Manitoba to study as a Post-Doctoral Fellow at Brandeis University. The year 1968, of course, was an extraordinary and turbulent year worldwide due to the countercultural students' rebellions, political upheavals in Eastern Europe, and the civic unrest in the United States. Jews were deeply involved in the countercultural movement, and they directed their disenchantment inward in a scathing critique of the organized Jewish community. From the 1920s to 1960s American Jews, most of whom were migrants from Eastern Europe, actively pursued Americanization. That meant abandoning the observance of Jewish religious rituals and adopting American social mores but without converting to Christianity.

After WWII, American Jews moved to suburbs where they built massive synagogues and a very impressive philanthropic system that supported many educational institutions and civic organizations. However, in the late 1960s the young counterculture critics found organized Jewish life to be overly materialistic, spiritually empty, joyless, intolerant, lacking democracy, and overly focused on Zionism and the State of Israel or on success in American consumerist society. In response, they created small, egalitarian, participatory fellowships in which they could express their Jewishness through new rituals and social forms (Prell 1989). The Havurah movement was to create an alternative, "spiritual" Judaism that could overcome the shortcomings of established forms of Judaism.

Arthur Green

One of the Havurah movement's first expressions was Havurat Shalom (Fellowship of Peace), founded by Arthur Green in 1968 in Sommerville, Massachusetts. If Heschel and Schachter moved from traditional Judaism toward the secular world, Green's was the opposite trajectory. He grew up in a secular Yiddish-speaking family in New York, raised by an atheist father (his mother died when he was eleven) who ensured his son's Jewishness by sending him to supplementary Hebrew School and the Ramah summer camps of the Conservative Movement. In the Ramah summer camps, however, Green was exposed to the depth of traditional Judaism that led him to clash with his secularist father. Fortunately, Green completed high school early at sixteen and he left home to start undergraduate studies at Brandeis University in 1957. In those years, this nascent Jewish institution of higher learning was exceptionally vibrant, attracting enormously talented faculty and students and serving as the intellectual and artistic hub of Jewish culture in America. At Brandeis, Green studied philosophy, literature, and psychology and was particularly taken by Existentialist philosophers (e.g., Sartre and Camus) and by the philosophy of Nietzsche (Green 2010, 1–3; Mayse 2015, 3). There he also encountered European Jewish intellectuals (e.g., Alexander Altmann and Nahum Glatzer) who introduced him to the academic study of Judaism with a focus on Jewish philosophy and mysticism and to the comparative study of religion. Although Glatzer and Altmann differed in their approach to Jewish studies, both showed Green the spiritual depth of Jewish textual scholarship.

After earning his BA in 1960, Green studied for a year at the Hebrew University in Jerusalem, where he met the world-renowned scholar of Jewish mysticism, Gershom Scholem. Like him, Green would employ scholarship of Jewish mysticism to effect a change in modern Judaism, albeit it was for divergent purposes. Scholem saw scholarship on Jewish mysticism as part of the Zionist enterprise of nation-building, whereas for Arthur Green scholarship of Jewish mysticism was to be used for spiritual renewal of Diaspora Jews. To sharpen his scholarly and interpretative skills, Green enrolled at the Jewish Theological Seminary (JTS) in New York, where Abraham Joshua Heschel became his mentor. Critical of the highly cerebral, emotionally disengaged and pedantic academic study of Judaism that thrived at JTS, Green was attracted to Heschel because his scholarship was invested with spiritual meaning and engaged social and political concerns. Although Green "refused to become Heschel's *hasid*" (Mayse 2015, 5), "Heschel became a role model of a scholar who was an academic, but also public intellectual, communal activist, and a deeply religious person engaged [in] reading the classical sources to create a Jewish theology for the twentieth century" (ibid, 6). Ordained as a Conservative Rabbi at JTS, Green returned to Brandeis in 1967 to embark on doctoral studies in Jewish mysticism under Alexander Altmann's supervision. Considering his criticism of JTS, Green also wanted to establish a new kind of Jewish rabbinic seminary, one that combined rigorous study, meaningful worship, and political activism. He would accomplish that dream by founding Havurat Shalom Community Seminary in Sommerville, Massachusetts, in 1968.

Havurat Shalom was to appeal to "New Jews" who wanted to merge spirituality, learning, and deep Jewish commitment (Sleeper and Mintz 1971). Green assembled an impressive advisory committee of prominent professors, writers, rabbis, and journalists and persuaded friends and acquaintances to join. As noted above, in 1968 Schachter was in Boston as a postdoctoral fellow at Brandeis University, and Green recruited him to serve on the faculty of the new seminary. At Havurat Shalom, Schachter taught courses on prayer techniques, introduction to Talmud, Jewish mysticism, and the Jewish prayer book, and his spiritual style was most attractive to Green. For Green, the five years he spent with Havurat Shalom as founder and sometimes leader were most transformative, intellectually, spiritually, and devotionally, and the interaction with Schachter-Shalomi was crucial to this transformation. Although Havurat Shalom gave up on its attempt to function as a seminary that ordains rabbis, in part in response to internal criticism that the seminary was no more than a ploy to avoid the draft, the organization became the model for other fellowships of Jews who wished to revitalize their Jewish commitment by combining worship with textual study and spiritual devotion.

Out of the experience of Havurat Shalom and his own academic studies of Hasidism came Green's commitment to Neo-Hasidism as the framework for the renewal of American Judaism (Green 2015). Neo-Hasidism was the application of Hasidic theology to all aspects of contemporary life, including the environmental crisis, as we will elaborate in the following chapter. For now, let me continue to follow the intersecting lives and careers of Schachter and Green, who both moved to Philadelphia in the mid-1970s; Schachter accepted a position at the Department of Religion at Temple University, and Green accepted an academic position at the University of Pennsylvania.

In the 1970s programs and departments of Jewish studies were established in universities all over America, offering courses and degrees for undergraduate and graduate degrees. In part the growth of Jewish studies reflected the increasing self-confidence and pride of American Jewry after Israel's victory in the Six-Day War of 1967. American Jews began to feel at home in America, and Jewish philanthropy funded Jewish studies in American higher education. This development was also related to the emergence of ethnic studies, Black studies, women studies, Asian studies, and Hispanic studies during the 1970s, all of which demanded recognition and respect of their particularity and showed how they enrich the humanities and the social sciences. The rise of area studies was part of a large intellectual transformation brought about by postmodernism and its critique of modernist metanarrative. These intellectual changes also led to establishing departments of religious studies in secular universities and secularizing the study of religion. Although the method of religious studies was secular, the very inclusion of religion within the secular academy meant that the earlier rigid bifurcation of "religion" and "the secular" was no longer tenable. The academic study of Judaism benefited from all these developments as much as did American Jewry in general. Engagement in scholarship of the Jewish tradition and the Jewish past itself became a way to affirm one's Jewishness and infused the public life of American Jewry with new energy.

From the mid-1970s to the mid-1990s Schachter was active in Philadelphia, addressing American Jews' spiritual quest not so much in the context of the academic study of religion at Temple University (where the focus was more on interfaith dialogue, ecumenism, and the comparative study of religion) but rather by creating the Jewish Renewal Movement. This was not to be merely an act of restoring an old paradigm but of radically transforming Judaism for Jewish spiritual seekers (Magid 2013, 35–73). Schachter-Shalomi's project was to create "a Judaism that invites the infinite and operates at a higher level of spiritual consciousness. It senses the divine just beyond the surface of everyday existence and wants to connect to that" (Schachter-Shalomi and Segel 2013, xii).

This paradigm shift was to be experienced in the reconstituted B'nai Or, the liturgical community that welcomed Jews of all levels of learning, observance, and commitment. By the mid-1970s, however, Jewish feminism was thriving, and feminists would not endorse the gendered name of the fellowship. It was therefore changed from the masculine term B'nai Or (Sons of Light) to the gender-neutral term P'nai Or (Faces of Light). Feminist sensibilities and egalitarianism would play an important role in the Jewish Renewal Movement's structures and rituals. As the spiritual leader of this liturgical community, Schachter, who in the mid-1980s added "Shalomi" to his surname, became a Jewish guru who brought combined Old World Judaism with New Age spirituality. Many spiritual seekers flocked to be with and learn from him after experimenting with Eastern spirituality in India's Ashrams; others came seeking healing and solace in a world filled with personal tragedies and emotional challenges, and still others were attracted to Schachter-Shalomi's spiritual ecumenism and affirming personality.

Schachter-Shalomi infused Jewish learning with spirituality and spirituality with social activism. Through his tireless teaching, mentoring, lecturing, writing, networking, and outreaching, he established the Jewish Renewal Movement as an alternative to the other main Jewish denominations in America (Groesberg 2008). Reflecting Schachter-Shalomi's intellectual and spiritual proclivities, the Jewish Renewal Movement has blended Kabbalah, Hasidism, psychology, popular science, and the spiritual teachings of several world religions, especially Buddhism, Hinduism, and Islam. The Jewish Renewal Movement highlighted the wisdom of the heart and celebrated Judaism through singing, chanting, meditative exercises, breathing, and visualization, all within the framework of traditional Judaism. Exceptionally nonjudgmental and egalitarian, Schachter-Shalomi welcomed all Jews and encouraged all people to live in the presence of God.

Jewish Renewal anchors Jewish religious environmentalism in America, integrating ecological spirituality, social responsibility, and environmental activism. This unique blend was inspired by Heschel and Schachter-Shalomi, but it was also indebted to the collaboration between Schachter-Shalomi and another visionary, spiritual seeker, and community builder: Rabbi Arthur Waskow.

Arthur Waskow

Like Green, Arthur Waskow grew up as a secular Jew but over time moved toward more observant Jewish life, albeit in the Liberal and Progressive modes. Born in Baltimore, Waskow received his BA from Johns Hopkins University (1954) and a PhD in American History from the University of Wisconsin-Madison (1959). In the early 1960s, at the height of the Cold War and the threat of nuclear destruction, peace was but a precarious ideal. Waskow was committed to the causes of peace and nuclear disarmament. From 1961 to 1963 he served as Senior Fellow at the Peace Research Institute in Washington DC and co-founded the Institute for Policy Studies in 1963, where he served as Resident Fellow until 1977. Like many other Leftist American secular Jews, Waskow was involved in Democratic politics, the Civil Rights Movement, and the women's movement and participated in many protests and nonviolent actions against the Vietnam War, racial segregation, the oppression of Jews in the Soviet Union, Apartheid in South Africa, and other causes. In the 1960s he became involved with the Jewish Renewal Movement, signaling a desire to move away from secular politics to religiously grounded life. In 1971 he helped found the Fabrangen Havurah (the name was a corruption of the Yiddish word *farbrengen*, meaning, "gathering") in Washington DC, along with Jews for Urban Justice, an organization founded in 1966 by a cadre of Jewish political activists in the DC area. Jews for Urban Justice supported the grape boycott instituted by California farm worker unions led by Cesar Chavez and they appealed to the Talmudic ruling that it was unlawful to eat food grown by exploited labor (Groesberg, 50–55). The Fabrangen style of social activism was not to the liking of all Jews, but it was very successful in appealing to estranged youth who wanted to engage in meaningful Jewish life combined with social activism. This was a distinctive path for Jewish spiritual self-expression, alongside study or ritual observance.

In 1977 when his position at the Peace Institute was not renewed, Waskow moved from Washington DC to Philadelphia to collaborate with his spiritual mentor, Schachter-Shalomi with whom he had studied privately and from whom he would receive private ordination in 1995. Whereas Schachter-Shalomi paved the path for Jewish eco-theology, Waskow translated it into Jewish religious environmentalism, namely, a full-fledged program to anchor environmental action in the teachings and rituals of Judaism. The institutional setting of his environmental activism was the Reconstructionist Rabbinical

College (RRC), which also played a role in Arthur Green's academic and public career (he left the University of Pennsylvania in 1984 to become first its dean and then President). He held this position until 1993, when he returned to Brandeis University to assume the Philip W. Lown Professorship of Jewish Thought, the Chair that had been created for his mentor, Professor Altmann. While Green was living in Philadelphia, he was not associated with the Jewish Renewal Movement, perhaps to protect his scholarly reputation and commitment to rigorous academic scholarship. During Green's presidency of the RRC, the institution was torn by several controversies related to feminism, homosexuality, and openness to Goddess worship in American neo-paganism. The clashes were unavoidable since the institution harbored different conceptions of "spirituality" and "secularism," one articulated by Heschel and Schachter-Shalomi, the other by Mordecai M. Kaplan.

Mordecai M. Kaplan

The Reconstructionist Rabbinical College was established in the eventful year of 1968, and it too reflected the desire to revive Judaism by reconstructing it. This was the brainchild of another 20th century visionary, Mordecai M. Kaplan who developed a secularized Zionist interpretation of Judaism. Like Heschel and Schachter-Shalomi, Kaplan grew up as an Orthodox Jew in Europe and received traditional education in Vilna. The family migrated to the United States in 1889, when the father, Rabbi Israel Kaplan, was invited to serve as a rabbi in New York, and in the United States, the precocious teenager, Mordecai Kaplan, increasingly became disenchanted with Orthodox Jewish life. Kaplan fully understood the intellectual challenges that modernity posed to Judaism, especially in America, and he knew that to survive in America, Judaism would have to be reconstructed. For his undergraduate studies he enrolled in City College in New York, while pursuing traditional Jewish learning and rabbinic ordination at the Jewish Theological Seminary (1893–1902). He also studied at Columbia University and received an MA in sociology, and it was at Columbia that the sociology of religion of Emile Durkheim and American Pragmatism of Dewey influenced his reinterpretation of Judaism as a civilization. Kaplan's life captured the dilemmas of Jewish secularization in the 20th century to which he responded by the effort to reconstruct Judaism.

After his rabbinic ordination Kaplan first served as associate rabbi of Kehillat Jeshurun, an Orthodox synagogue in New York. In 1909 he also became a faculty member of the Jewish Theological Seminary, first heading its Teachers Institute, then becoming a professor of homiletics and the philosophy of religion until his retirement. Kaplan's intent to modernize Orthodoxy in America was expressed in the creation of the Young Israel Modern Orthodox movement with Rabbi Israel Friedlander, but as he fleshed out his view of Judaism as an evolving civilization. Kaplan was rejected by the movement which eventually treated him as a heretic (Scult 1993). In 1922

Kaplan founded the Society for the Advancement of Judaism, his synagogue in New York City, and through the 1930s he developed his vision for a Reconstructionist Judaism as an evolving religious civilization (Kaplan 1934). For Kaplan the renewal of Judaism was predicated not on a return and reinterpretation of Kabbalah and Hasidism but rather on the understanding that Judaism encompasses peoplehood, folkways, land, art, literature, music and philanthropy. Inspired by Durkheim's sociology of religion, Kaplan sought to revitalize Judaism by creating all-embracing "organic" Jewish communities that would ensure the self-perpetuation of Jewish identity and further both "secular" and "religious" components of Jewish heritage. In his approach the secular-religious divide had no meaning since Judaism is a this-worldly religion in which God is the "force that makes for salvation." Transcendence and self-transcendence are always within the "immanent frame."

The trajectory of Reconstructionist Judaism and the Jewish Renewal Movement intersected when the Reconstructionist Rabbinical College (RRC) wanted to establish a presence on the national scene by tackling serious socio-political issues facing the Jewish community and the nation. One of the burning issues of the time was the nuclear arms race which was about to be renewed under President Reagan. The Reconstructionist movement considered the possible nuclear annihilation of humanity a topic that deserved an appropriate Jewish response and it searched for a person who could direct the new center. Arthur Waskow, who was already well known for his work in the anti-nuclear movement, was selected to head the Shalom Center, which opened in 1983 as a think tank affiliated with RRC. The Shalom Center was to analyze policy related to the nuclear rearmament race and to devise strategies for preventing disastrous misadventures. The Shalom Center raised public awareness about nuclear issues and took a stand on the ongoing Israeli–Palestinian conflict, which frustrated all efforts to reach peace in the Middle East. Most importantly the Shalom Center highlighted the connection between world peace and the well-being of the Earth's ecosystem. Waskow and other environmentally concerned Jews began to mine the Jewish tradition for its ecological wisdom and argued that Judaism's laws, values, and rituals could be translated into sound environmental policies (Waskow 1996; 2000; Waskow, Elon, and Hyman 1999). Judaism is a down-to-earth religion that deeply cares about the well-being of the Earth and all its inhabitants, and the Jewish task is to heal the wounded world (Waskow and Berman 1996).

Eco-Kashrut

The concept that best captures this embodied spirituality is eco-Kashrut (or eco-kosher). The term was coined by Schachter-Shalomi in 1968, during the Cesar Chavez labor protests in California, but it was popularized and elaborated by Waskow through his environmental activism. As Schachter-Shalomi defined it, eco-Kashrut is "a new kind of kashrut" that "would combine the ancient ways of thought consumption and avoidance of cruelty and violence with the new awareness of the wider repercussions of some of our actions" (Schachter-Shalomi and Segel 2013, 157). A practice of eco-Kashrut adopts a very traditional Jewish way of applying consumption-related concerns to traditional Jewish life. But it also differs from traditional kashrut in three important ways.

> First eco-kashrut is concerned not only with the origin of the thing consumed—what animal the meat came from, say, or what dishes it was cooked in—but also with the results of our consumptions, such as the environmental and human toll of our actions.
>
> (ibid., 158)

Second, eco-Kashrut engages "all the interlocking costs and repercussions" but sees them as "a matter of degree rather than a yes-or-no answer" (ibid., 159). And third, an eco-Kashrut practice is "a matter of individual consciousness and decision, rather than a matter of legislation" (ibid., 159).

Waskow translated the concept of eco-Kashrut into a full-fledged program of environmental justice, connecting concerns about industrial agriculture, global warming, and fair treatment of workers with the Jewish dietary laws about food production, preparation, and eating. The ideal of eco-Kashrut suggests that Jews should only consume products that meet both Jewish dietary laws and Jewish ethical standards, and eco-kosher consumers should encourage food producers to care for the environment, animals, and workers (Waskow 1996). Although it pertains to food production and consumption, the ideal of eco-Kashrut is much broader, and although it pertains to physical reality, eco-Kashrut is a spiritual ideal. For Waskow and many Jewish environmentalists inspired by his work, projects that are designed to halt reckless squandering of the Earth's mineral wealth and heedless pollution of the Earth's biotic resources are invested with spiritual meaning because they pertain to human relations with God.

Eco-Kashrut is the method by which Jewish environmentalism seeks to achieve the religious value of *Tikkun Olam* (literally, "repair of the world"). The term *Tikkun Olam* can be traced to rabbinic texts (Mishna Gittin 4:2; Gittin 4:6; Babylonian Talmud Ketubot 52a; Gittin 36a; Gittin 3a) where it refers to addressing the limitation of the created world by observing the Torah. In medieval Kabbalah, especially its 16th century Lurianic version, *Tikkun Olam* became the goal of Jewish life, but it was given a metaphysical significance. When Jews observe the Torah with the proper kabbalistic intentions, they release "divine sparks" from their entrapment in corporeal reality. In Kabbalah, *Tikkun Olam* was not to be limited to earthly life. It had a cosmic and even theurgic meaning: the human performance of the commandments (*mitzvot*) repairs the brokenness of the divinity itself! I will explore this theological doctrine in the following chapter, but for now it is important to note that *Tikkun Olam* has become the buzzword of Jewish social activism, especially for Reform and Progressive Judaism (Krasner 2014). The term entered American Jews' consciousness in the late 1980s, and rabbis, educators, communal workers, and activists invoked it to justify their involvement in a variety of causes, including environmental repair of the damaged world (Sherwin 2014). For many minimally observant Jews, engagement in social action, science and medicine, and the arts is justified by appeal to the concept of *Tikkun Olam*.

Waskow and Schachter-Shalomi's collaboration resulted in Aleph: The Alliance for Jewish Renewal as well as in the merger of P'nai Or's newsletter with Waskow's journal *Menorah*, later renamed *New Menorah*, which Waskow continued to edit. In the 1990s the Jewish Renewal Movement began to proliferate outside of Philadelphia, with new congregations being established all over the United States. Aleph formed the Network of Jewish Renewal Communities (NJRC) and discussions for merger of P'nai Or, Aleph, and the Shalom Center began in 1993. The institutional merger did not come to pass because Waskow wished to retain the autonomy of the Shalom Center's initiatives. Nonetheless, Aleph became the new umbrella organization for the dissemination of Renewal ideology and practices, and NJRC, its central administrative body, "provided member congregations and Chavurot with resource materials related to the movement's ideology, worship practices, spiritual teachings and socio-political agenda" (Groesberg 2008, 69). Through the 1980s and early 1990s Jewish environmentalism began to make inroads with the organized Jewish community with the establishment of the Coalition of Environment and Jewish Life (COEJL) in 1993.

Since the 1990s Jewish environmentalism has gradually emerged as a concern of organized Jewish life in America with the "greening" of Jewish institutions (i.e., synagogues, schools, communal organizations, Jewish community centers, and youth movements). Today a variety of organizations, programs, and initiatives promote sustainable practices, reduce consumption and promote new eating habits, plant community gardens, link sustainable agriculture to urban Jewish life and education, include environmental issues in youngsters and adults' education, organize nature walks and outdoor activities, celebrate Jewish holidays with attention to environmental and agricultural themes; promote justice in food production with attention to sustainable agriculture and compassionate treatment of farm animals, and encourage Jews to live sustainably. These programs transcend congregational and denominational boundaries and are often carried out in interfaith settings in collaboration with non-Jewish organizations. Jews are also involved in various national protests and advocacy work related to climate change and opposing fracking technology.

The five individuals we have featured above were crucial to the emergence of eco-Judaism in America, but they were not the only people involved in promoting environmentalism. Women too contributed to the growth and spread of Jewish environmentalism in the United States as publishers, advocates, exegetes, theologians, educators, and activists. Their contribution will be highlighted in Chapter 4, which explores the link between environmentalism and feminism.

Jewish Environmental Organizations: Spirituality in Action

Today, in the middle of the third decade of the 21st century, Jewish environmentalism is a recognizable sensibility within Jewish life in America. I prefer to call is a "sensibility" rather than a "movement," because Jewish environmentalism is highly decentralized. There is no one governing body that unites all Jewish environmental activities (e.g., education, fundraising, community outreach, and public advocacy), but many overlapping and competing organizations and initiatives responding to various aspects of the eco-crisis. In 2026, Adamah: People, Planet, Purpose is the largest environmental organization, but Adamah does not manage or coordinate all environmental activities of one Jewish environmental movement, nor does it represent the environmental concerns of all American Jews. The very denominational organization of American Jewry contributes to the fragmentation of Jewish environmentalism. While Jews from across the denominational spectrum of modern Judaism engage in various environmental activities, environmental activism still expresses the distinct ideological orientation of the various Jewish denominations. Consequently, there is considerable duplication in Jewish environmental activism which diffuses the Jewish voice on environmental matters rather than strengthens it. Additionally, although environmental activists share certain values, differences concerning the observance of Jewish Law hamper the ability of organizations to work together and build institutional cohesion (Yoreh and King 2024).

Because Jewish environmental organizations are often grassroot initiatives of individual activists, they do not enjoy the same level of institutional stability and financial viability of the official Jewish denominations. Some organizations are supported by local Jewish federations, others receive funding from private foundations or philanthropists, and still others are supported by a specific Jewish denomination. The lack of stable funding means that organizations come and go and that environmental initiatives often last for only a limited number of years. Projects and initiatives may be supported by grants for a limited time, but the economic model that rewards creativity and risk-taking does not necessarily ensure stability or continuity. In short, while Jewish environmentalism is a recognizable voice within American Jewry, Jewish environmental organizations still face many challenges, chief among them funding.

Among the Jewish environmental organizations there is considerable diversity in emphasis but also a great degree of overlap and shared interest. Almost all of them are engaged in educating Jews about the environment, the pending climate crisis, and the relationship between Judaism and environmentalism. To impart environmental knowledge and raise environmental awareness, these organizations incorporate outdoor activities, be it hiking, gardening, or farming. They convey the environmental value of sustainability primarily through a focus on food production and consumption, since food is where humans, animals, and plants intersect. Several Jewish environmental organizations are engaged in or support farming and encourage urban gardens in Jewish community centers, synagogues, or summer camps. In so doing, Jewish environmentalists today return to and revive the earlier Jewish American agricultural colonies of the 1880s–1920s, especially in New Jersey (Eisenberg 1995), and that revival allows Jewish environmentalists who define themselves as "Eco-Jews" to claim that environmental activism in the United States is by no means novel but goes back to the early 20th century. In truth, Jewish agriculturalism in the United States at the turn of the 20th century had more to do with Socialist utopianism and the attempt to shift the economic patterns of Jewish existence (Herscher 2018) than with protection of the environment or the restructuring of the relationship between humanity and nature. We will revisit this point in Chapter 6.

Many environmental organizations differentiate between "leaders" and "rank-and-file" members. Leadership training fosters commitment to environmentalism as much as it generates more activism. Leaders/activists are expected to start their own projects and initiatives and continue to spread environmental commitment through changing consumption habits. Although Jewish environmental organizations are about creating a new lifestyle through education, there is quite a gap between theoretical (i.e., theological) literature produced by the visionaries and the depth of Jewish textual knowledge or familiarity with Jewish traditional rituals. References to Jewish values or sacred texts do appear in the organizations' activities, but the core of the activity is neither hermeneutical nor theological. Jewish environmental organizations practice "down-to-earth" Judaism literally speaking, and the references to Jewish sacred texts or Jewish values are often quite superficial. Of course, there are many Jewish environmental activists who are deeply rooted in the tradition and their written work integrates deep learning with their environmental activism (e.g., Benstein 2006), but it is fair to say that there is a gap between the Jewish learning of the eco-theologians and the Jewish learning of the activists who implement Jewish eco-theology through environmental education and activism.

The brief discussion of Jewish environmental organizations aims to convey development over time, major activities, and types of environmental engagement.

Shomrei Adamah

The first Jewish national environmental organization was Shomrei Adamah (Keepers of the Earth) founded by Ellen Bernstein in 1988. A biology teacher and a rafting instructor, she illustrates the move "back to the sources" that characterized the awakening Jewish spirituality in the late 1960s and 1970s. Under her leadership began a comprehensive engagement of the Jewish religious sources from an environmental perspective, and it was no coincidence that the activity took place in Mt. Airy, Philadelphia, where Schachter-Shalomi, Waskow, and the Reconstructionist seminary were all located. With Rabbi Daniel Fink (Bernstein and Fink 1992), Bernstein produced a "Guidebook for teaching Jewish perspectives on the human relationship with nature," whose intention was to raise Jewish awareness about environmental matters and to change Jewish attitudes, conduct, and sensibilities. Her later edited anthology (Bernstein 2000) illustrates how Jewish environmentalists in the late 20th century sought to bring nature to the center of Jewish self-understanding. The volume focuses on the agricultural foundation of biblical society, the naturalist imagery of prophetic oracles, the concern for nature in biblical legislation, the principle of respect toward nature in rabbinic law, organic symbolism and ritual practices in Kabbalah, the immanent theology of Hasidism, and the Zionist return to the ancestral Land of Israel, which places the land and the natural life it supports at the center of modern Jewish existence.

These themes have continued to engage Jewish environmentalism in the 1980s and 1990s as Shomrei Adamah spawned other chapters. The Shomrei Adamah of Greater Washington chapter, for example, produced a quarterly news journal, provided speakers for events, ran educational wilderness trips, and captured the imagination of many scholars, leaders, rabbis and young adults, for whom it developed a range of educational materials. These types of activities charted the path for many local Jewish environmental organizations whose concern for the well-being of the environment dictated the approach to the primary sources: texts were selected to support environmental sensibility and to prove that Judaism is compatible with environmentalism. Members of the organization are united in the common belief that Judaism obligates them to protect the Earth, and they devote their time to educate Jews and connect them to the ecological wisdom of the Jewish tradition.

Coalition on Environment and Jewish Life (COEJL)

In the early 1990s Jewish environmentalism became part of the larger environmental scene because of the emergence of religious environmentalism. In 1992 Al Gore and Carl Sagan extended an invitation to leaders of major Jewish organizations, eminent rabbis, denominational presidents, and Jewish US senators to gather in Washington, DC, to discuss how Jews could articulate a distinctive Jewish response. The result of that event was the establishment of the Coalition on the Environment and Jewish Life (COEJL) in 1993 whose mandate was to catalyze a distinctly Jewish pragmatic and policy response to the environmental crisis. COEJL was initially envisioned as a time-limited project to "jump start" an environmental program that would become permanently integrated into Jewish institutions. Uniquely in Jewish organizational life, COEJL started as an interdenominational initiative, when the Jewish Council for Public Affairs, the Religious Action Center for Reform Judaism and the Jewish Theological Seminary of America all joined forces in their attempt to enact a distinctively Jewish programmatic and policy response to the environmental crisis.

Representing twenty-nine Jewish civic organizations, COEJL has promoted environmental education, scholarship, advocacy, and action in the American Jewish community through publications, public education campaigns conferences, Jewish environmental leadership development, and a Legislative Advocacy Network. Given its interdenominational nature, COEJL served as the voice of the organized Jewish community on environmental issues in Washington DC and around the country. COEJL still exists, but its range of activities and public impact have much diminished since the 1990s. Sustaining the activity of Jewish environmental organizations seems to be rather challenging.

Hazon: Jewish Lab for Sustainability

Attention to environmentalism in the organized Jewish community during the 1990s inspired the emergence of several organizations in the early 2000s. Founded by Nigel Savage in 2000, Hazon: The Jewish Lab for Sustainability illustrated how entrepreneurialism, social networking, and leadership training emerged as the dominant modus operandi of Jewish environmentalism in the 2000s. Hazon's goal was "building a movement that strengthens Jewish life and contributes to a more environmentally sustainable world for all." The organization promised that people who participate in their program will "deepen their Jewish identities; experience the connection between inherited Jewish wisdom, food, climate and the natural world; and become linked with others who care about creating a more sustainable Jewish community and world" (Adamah 2025). The entrepreneurial orientation and lingo of social organization are evident in Hazon's proclamation that it "develops, networks, and provides resources for those growing into leadership—people changing existing institutions or building new organizations and community that live the values we cherish and the world needs." The Jewish renewal envisioned by Hazon is quite different from the ecological spirituality of the Jewish Renewal Movement of P'nai Or, but people associated with P'nai Or also took part in Hazon's activities. These organizations sought to renew commitment to Judaism through immersive experience in environmental activities.

To effect change in Jewish life, Hazon focused on leadership training. It created the program "Jewish, Outdoor, Food, Farming & Environmental Education" (JOFEE), which educates and supports leaders who are committed to change with a particular focus on food systems. The program emphasizes “vibrant experiential education” as well as “intellectual excellence and rigor” and “healthy and sustainable action.”

Hazon also sponsored and organized the Jewish Food Movement (Schorsch 2017; Silvern 2021), which in turn organized an annual Food Conference and an annual Jewish Food Festival involving several thousand people (Most 2016). Urban farming was one of Hazon’s major initiatives. In the Detroit region, Hazon organized the Seal of Sustainability Program, which strengthens local food systems and helps connect the suburbs to the city and the region’s Jewish and African American communities. In addition to farming, Hazon also focused on changing habits and raising awareness regarding transportation by sponsoring bike rides in the United States and in Israel as fund-raising efforts. These and other activities and initiatives that Hazon established and that currently take place in Adamah show how Jewish environmentalism is more interested in behavioral transformation than inward experience. Jewish farming today goes beyond urban farming. As Adrienne Krone has shown, since 2004 about 20 organizations of Jewish farmers “put environmental ethics into practice as they face the realities of our time” (Krone 2021). Jewish farming today presents itself as continuation to Jewish agricultural activities in the early 20th century in the United States, but the two projects are quite different. We will return to it in Chapter 6.

Isabella Freedman Jewish Retreat Center

By 2014, Hazon had grown immensely by absorbing preexisting environmental organizations and initiatives. The most important of these earlier projects was the Isabella Freedman Jewish Retreat Center, a certified-organic farm whose ten acres produce fresh, seasonal herbs, flowers, and vegetables for the retreat’s kitchen and CSA program. The farm’s sustainable growing methods include crop rotation, cover cropping, composting drip irrigation, creating habitat for pollinators, and growing diverse mixes of annual and perennial flowers. Many practices facilitate conservation of energy and water, limit the use of toxic chemicals, and reduce waste through compositing. The organization has a fascinating history that began long before environmentalism became a Jewish concern. In 1893 the Jewish Working Girls Vacation Society established a summer camp for Jewish working women, primarily immigrants in New York’s garment industry, offering them an affordable vacation. The camp paid for their vacation and reimbursed campers for lost wages. In 1936 the camp’s name was changed to Camp Isabella Freedman in honor of the philanthropist who was also a board member of the organization. In 1956 the camp moved to Falls Village, Connecticut, and instituted residential programs for Jewish senior adults. In 1994 Isabella Freedman developed the Teva Learning Center with Surprise Lake Camp, an innovative experiential learning program for Jewish elementary school students that combine environmental awareness, Jewish spirituality, and environmental activism.

Education is the focal point of the Teva Learning Center, currently renamed Teva Learning Alliance, whose programs are designed to immerse participants in the natural world through structured activities that sensitize them to nature's rhythms and inspire them to develop a more meaningful relationship with nature and with their own Jewish practices through precepts such as *Tikkun Olam* (Repair of the World) and *Bal Tashchit* (Do Not Destroy) that will be further discussed in Chapter 7. Through traditional Jewish activities of Torah study, prayers, and meditation, participants become aware of nature as divine creation. Each year, the programs of Teva Learning Center involve about 4,000 people, but there is no way to determine the depth of the engagement or its long-term impact. Former staff members have spawned several other groups, the most important of which are the Eden Village (an eco-summer camp), the Jewish Farm School, an environmental educational organization, and Kayam, an educational farm on the grounds of the Pearlstone Retreat Center and Camp Milldale in Maryland.

Another environmental program generated by the Isabella Freedman Center is Adamah: The Jewish Environmental Fellowship. This is a leadership training program for young Jewish adults, teaching them the vital connection between Judaism and environmental stewardship. This six-month residential program integrates agriculture and sustainably living skills, Jewish learning and living, and leadership commitment to the value of *Tikkun Olam*. Its emphasis on communalism is undoubtedly influenced by the kibbutz movement in Israel, the intentional agricultural communes established first in 1910 in Deganya. Many of the leaders as well as rank-and-file members of Jewish environmental organizations have visited Israel or lived for a while in a kibbutz. In some cases, the attempts to recreate the spirit of the kibbutzim in America's Jewish summer camps and environmental leadership programs reflect the awareness that American Jews are not going to settle in Israel, but that they can take some of the best "crops" of the Zionist movement and transplant them to the United States.

The environmental educational projects of the Isabella Retreat Center took a more distinctive spiritual turn in 2006, when the organization merged with the Elat Chayyim Jewish Retreat Center of Accord, New York. This organization was a project of Aleph: Alliance for Jewish Renewal, which we mentioned previously. Elat Chayyim (literally, "The Goddess of Life") was founded by Rabbi Jeff Roth and Joanna Katz to offer weekly spiritual retreats year-round in New York's Catskill Mountains. It sought to serve as a resource and retreat center for the development, experience, and promotion of a comprehensive Jewish spiritual practice, including study, vegetarianism, yoga, meditation, and prayer. These retreats cultivate awareness of the Divine Presence in all aspects of life. Elat Chayyim welcomes people from all backgrounds, including individuals with limited or no Jewish education, seekers who have walked other spiritual and religious paths, and those who are traditionally observant. "Finding God in Nature" is presented as a form of Jewish spirituality that is rooted in traditional Jewish liturgy, where "images of nature are an expression and embodiment of the Divine." Today, Elat Chayyim functions within the framework of Adamah: People, Planet, Purpose.

Adamah: People, Planet, Purpose

In 2023, Hazon merged with Pearlstone Retreat Center under the new name of Adamah: People, Planet, Purpose. Adamah currently boasts

> 32,000 people who participated in immersive retreats, Jewish environmental education, and community engagement; 400 nonprofit organizations that joined Adamah Jewish Climate Leadership coalition, recognizing the urgency of climate change, and 7,627 students engaged through Jewish Youth Climate Movement and Adamah on campus, empowering the next generation.
>
> (Adamah 2025a)

These are not insignificant numbers, suggesting the emergence of environmental awareness among American Jews who express their Jewishness through environmental action or who bring their Jewish values to environmental action. In pro-environment activities Jews find *belonging, identity, meaning, and purpose*, which are precisely what religion provides its adherents.

Since Adamah: People, Planet, Purpose has absorbed Hazon: Jewish Lab of Sustainability (which itself absorbed earlier environmental organizations and initiatives), the range of activities offered by Adamah is very broad indeed. The goal is to create "vibrant Jewish life in deep connection with the Earth." The goal is attained through "immersive experiences [that provide] a vital remedy for mind, body, and soul," with the intent to "foster connection, rejuvenation, and inspiration for diverse Jewish participants from all over the world" (Adamah 2025). Adamah's retreats are led by rabbis from across the Jewish denominational spectrum, including rabbis from Yeshivat Chovevei Torah, representing the liberal/progressive strand within Modern Orthodoxy. One Adamah program—Torah Yoga—shows that contemporary Jewish nature spirituality developed out of the New Age movement of the 1960s, which had previously inspired the Isabella Freedman Center and Elat Chayim. Torah Yoga is "an annual summer ingathering of practitioners, leaders and seekers" whose purpose is "to connect to and strengthen our shared ancestral roots through Jewish text study, prayer and yoga practice, enriched by genealogy, art, dance, nature and food" (ibid.). The religious focus of Adamah and the amalgamation of Jewish and Buddhist practices illustrate the hybridity characteristic of post-secularism.

Today, there are many Jews who are interested in fusing their Judaic and environmental commitments. Calling themselves EcoJews, they promote environmental spirituality through environmental education and activism, and their focus varies, including urban agriculture (e.g., Urban Adamah), ethical treatment of animals (e.g., Farm Forward), or finding God in nature (e.g., Torah Trek, Wilderness Torah, and Organic Torah). Aytzim: Ecological Judaism is the organization that serves as the hub of information and networking for Jewish environmental activists. The organization exemplifies how technology revolution helped spread the message of environmentalism, and it is unique in its Zionist commitment. Founded as the Green Zionist Alliance by Alon Tal, Rabbi Michael Cohen, and Eilon Schwartz in 2001, who will be discussed in Chapter 5, Aytzim is an American Zionist organization that addresses Israel's environmental challenges. The organization disseminates information about environmental matters, offers educational programing for teens and adult groups, and supports Israel's environment and environmental movement through advocacy campaigns and eco-trips to Israel. Although Aytzim defines itself as "secular," one of its projects was Shomrei Bereshit: Rabbis and Cantors for the Earth, organized by the late Rabbi Lawrence Troster, for whom environmentalism was a religious activity. This initiative was intended to organize Jewish participation in OurVoice.net, the international multifaith campaign for a strong climate agreement, encouraging Jews around the world to participate in a global vigil in support of a climate treaty. The project offered environmental education and training opportunities for Jewish leaders and organizes an environmental voice at the World Zionist Congress. Aytzim is the parent organization of Jewcology, a web-based resource about Jewish environmentalism that makes available to readers relevant primary sources, scholarly studies, and information about activities. The materials available on that website make it clear that in Jewish environmentalism, earthly activities are invested with religious or spiritual meaning and spiritual and religious values are translated into social and political activism.

Climate Advocacy

Since 1998, after the United States refused to sign the Kyoto Agreement, climate change became the "face" of the ecological crisis, and new Jewish organizations emerged to focus the attention of American Jews toward the challenges of climate change. Rabbi Arthur Waskow, who was mentioned above, was at the forefront of that effort. Evoking Tractate Avot 1:14, "If not now, when?" Waskow persuasively claimed that climate change is the Jewish issue of our time and appealed to Jews and non-Jews "to reflect upon the impact of climate change on this Holy temple that is our Planet Earth, to turn and take action, large and small, toward a better future" (Waskow 2014; Waskow and Laur 2014). His call to tackle climate change did not go unheeded, and new organizations devoted to climate advocacy were founded with women in leadership roles. In 2016 Dr. Mirele B. Goldsmith, a psychologist and environmental educator, co-founded the Jewish Climate Action Network in New York (and later another chapter in Boston), whose goal is to "mobilize communities to take leadership in bold climate campaigns through education and reducing carbon footprint" (Jewish Climate Action

Network n.d.). In 2018 Goldsmith also founded the Jewish Earth Alliance as a grassroots coalition that seeks "to mobilize Jews to stand up and raise a loud, collective voice for policies that reflect Jewish values" (Jewish Earth Alliance 2025). In 2019 Rabbi Jennie Rosenn, a Reform Rabbi and social activist who served as the Director of the Life and Values Program at the Nathan Cummings Foundation, founded Dayenu: A Jewish Call to Climate Action with the intent "to mobilize Jewish support for climate solutions and to raise up a spiritual, religious, and moral voice in the national and global movements confronting the climate crisis" (Dayenu 2025). The organization's mission is "to secure a just, livable, and sustainable world for all people for generations to come by building a multigenerational Jewish movement that confronts the climate crisis with spiritual audacity and bold political actions." To stimulate Jewish climate activism, Hazon has invoked the religious value of *teshuvah* (i.e. repentance) as the right frame of mind that "offers us an opportunity to face our negative actions and create a new path forward" (Adamah 2025). As climate change has emerged to become an urgent Jewish social issue, it has overshadowed earlier environmental commitments to conservation, preservation, finding God in nature, or appreciating the Earth-based roots of Judaism.

The variety of initiatives, projects, and programs make clear that Jewish environmental activism is very diverse indeed. But how are we to characterize Jewish environmentalism: is it "religious" or "secular?" What makes Jewish environmentalism "Jewish?" Jewish environmental organization is both religious and secular, and these dimensions are accommodated by the broad and fluid term "spirituality." The precise meaning of "environmental spirituality" varies from one organization to another and from one practitioner to another. Some Jewish environmentalists are very rooted in the sources of Judaism and offer a genuine synthesis of Judaic and environmental commitments; others who are less informed about the Jewish textual tradition take part in these environmental activities not because these activities are uniquely Jewish but because they can be experienced together with other Jews. In these regards, environmental commitment to the Earth's well-being provides a spiritual meaning to their Jewish ethnicity. *Tikkun Olam* is indeed a deeply religious Jewish value, but it applies to the entire world and not only to Jews. We will revisit *Tikkun Olam* as the slogan of environmental spirituality and environmental activism in Chapter 7 to explain how it illustrates the dynamics of the post-secular age.

References

Adamah. 2025. https://adamah.org/.

Adamah. 2025a. "*2024 Year in Review.*" https://adamah.org/.

Ariel, Yaacov. 2003. "Hasidism in the Age of Aquarius: The House of Love and Prayer in San Francisco." *Religion and American Culture: A Journal of Interpretation* 13(2): 139–165.

Ariel, Yaacov. 2011. "Jews and New Religious Movements: An Introductory Essay." *Nova Religio: The Journal of Alternative and Emergent Religions* 15 (1): 5–21.

Batnitzky, Leora Faye. 2011. *How Judaism became a Religion: An Introduction to Modern Jewish Thought*. Princeton: Princeton University Press.

Benstein, Jeremy. 2006. *The Way into Judaism and the Environment*. Woodstock, VT: Jewish Lights Publishing.

Bernstein, Ellen and Dan Fink. 1992. *Let the Earth Teach You Torah: A Guide to Teaching Jewish Ecological Wisdom*. Wyncote, PA: Shomrei Adamah.

Bernstein Ellen (ed.). 2000. *Ecology and the Jewish Spirit: Where Nature and the Sacred Meet*. Woodstock, VT: Jewish Lights Publishing.

Biehl, Janet. 1997. *The Murray Bookchin Reader*. London: Cassell.

Biehl, Janet. 2015. *Ecology or Catastrophe: The Life of Murray Bookchin*. New York: Oxford University Press.

CCAR (Central Conference of American Rabbis). 1977. "*CCAR Resolution on Energy*." www.ccarnet.org/ccar-resolutions/energy-1977.

CCAR (Central Conference of American Rabbis). 1998. "*CCAR Resolution on Confronting the Challenge of Climate Change*." https://www.ccarnet.org/ccar-resolutions/confronting-the-challenge-of-climate-change-1998/.

Cohen, Jeremy. 1989. *"Be Fertile and Increase, Fill the Earth and Master It": The Ancient and Medieval Career of a Biblical Text*. Ithaca, NY: Cornell University Press.

Dayenu: A Jewish Call to Climate Action. 2025. https://dayenu.org/.

Dresner, Samuel and Byron L.Sherwin. 1978. "To Take Care of God's World: Judaism and Ecology." In *Judaism: The Way of Sanctification*, 131–144. New York: United Synagogue of America.

Eisenberg, Ellen. 1995. *Jewish Agricultural Colonies in New Jersey, 1882–1920*. Syracuse: Syracuse University Press.

Ellingson, Stephen. 2018. *To Care for Creation: The Emergence of the Religious Environmental Movement*. Chicago: University of Chicago Press.

Finer, Shmuel. 2010. *The Origins of Jewish Secularization in Eighteenth-Century Europe*. Philadelphia: University of Pennsylvania Press.

Foltz, Richard C. (ed.) 2003. *Worldviews, Religion, and the Environment: A Global Anthology*. Belmont, CA: Wadsworth.

Fox, Matthew. 2014. *Mesiter Eckhart: A Mystic-Warrior for Our Times*. Novato, CA: New World Library.

Fox, Steve. 1981. *The American Conservation Movement: John Muir and His Legacy*. Madison, WI: University of Wisconsin Press.

Geller, Jay. 2018. *Bestiarum Judaicum: Unnatural Histories of the Jews*. New York: Fordham University Press.

Gerstenfeld, Manfred. 1998. *Judaism, Environmentalism and Environment*. Jerusalem: Rubin Mass and the Jerusalem Institute for Israel Studies.

Gerstenfeld, Manfred. 1998a. "Neo-Paganism in the Public Square and Its Relevance to Judaism." *Jerusalem Letter/Viewpoints*, no. 392.

Gerstenfeld, Manfred. 2001. "Jewish Environmental Studies: A New Field." *Jewish Environmental Perspective*, no. 1. Jerusalem Center for Public Affairs.

Gordis, Robert. 1971. "Judaism and the Spoliation of Nature." *Congress Biweekly*, April 2, 9–12.

Gottlieb, Roger S. 2010. "Religion and the Environment." In *Routledge Companion to the Study of Religion*, 2nd ed., ed. John Hinnells, 492–508. New York and London: Routledge.

Green, Arthur. 2010. *Radical Judaism: Rethinking God and Tradition*. New Haven: Yale University Press.
Green, Arthur. 2015. *The Heart of the Matter: Studies in Jewish Mysticism and Theology*. Philadelphia: Jewish Publication Society.
Groesberg, Sholom. 2008. *Jewish Renewal: The Movement's History, Ideology and Future*. New York: iUniverse, Inc.
Hargrove, Eugene C. (ed.). 1986. *Religion and Environmental Crisis*. Athens, GA and London: University of Georgia Press.
Held, Shai. 2013. *Abraham Joshua Heschel: The Call of Transcendence*. Bloomington and Indianapolis: Indiana University Press.
Helfand, Jonathan. 1971. "Ecology and the Jewish Tradition." *Judaism: A Quarterly Journal* 20: 330–335.
Herscher, Uri D. 2018. *Jewish Agricultural Utopias in America, 1880–1910*. Detroit: Wayne State University Press.
Heschel, Abraham Joshua. 1951. *Man Is Not Alone: A Philosophy of Religion*. New York: Farrar Straus, and Giroux.
Heschel, Abraham Joshua. 1965. *Between God and Man: An Interpretation of Judaism*. Selected, edited, and introduced by Fritz A. Rothschild. New York: The Free Press.
Heschel, Abraham Joshua. 1996. *Moral Grandeur and Spiritual Audacity: Essays*, ed. Susannah Heschel. New York: Farrar, Straus, Giroux.
Jenkins, Willis, and Christopher Key Chapple. 2011. "Religion and Environment." *Annual Review of Environment and Resources* 36: 441–463.
Jewish Climate Action Network. n.d. https://www.jewishclimate.org/.
Jewish Earth Alliance. 2025. https://www.jewishearthalliance.org/.
Jewish Eco-Seminars. 2025. https://www.jewishecoseminars.com/.
Kamenetz, Rodger. 1994. *The Jew in the Lotus: A Poet's Rediscovery of Jewish Identity in Buddhist India*. New York: Harper.
Kaplan, Edward K. and Samuel H. Dresner. 1998. *Abraham Joshua Heschel: Prophetic Witness*. New Haven: Yale University Press.
Kearns, Laurel. 2011. "The Role of Religions in Activism." In *The Oxford Handbook of Climate Change and Society*, ed. John S. Dryzek, Richard B. Norgaard, and David Schlosberg, 413–428. Oxford: Oxford University Press.
Krasner, Jonathan 2014. "The Place of Tikkun Olam in American Jewish Life." *Jewish Political Studies Review* 25(3/4): 59–98.
Krone, Adrienne. 2021. "Farming on the Front Lines: Jewish Environmentalism and Kinship in the Chthulucene." *Worldviews: Global Religions, Culture, and Ecology* 26(1–2): 148–161.
Labendz, Jacob Ari and Shmuli Yanklowitz (eds.). 2019. *Jewish Veganism and Vegetarianism: Studies and New Directions*. Albany, NY: State University of New York Press.
Lamm, Norman. 1971. "Ecology and Jewish Law and Theology." In *Faith and Doubt: Studies in Traditional Jewish Thought*, 162–186. New York: Ktav Publishing House.
Lamm, Norman. 1974. "Nature." In *The Good Society, Jewish Ethics in Action*. Selected and edited by Norman Lamm, 213–222. New York: Viking Press.
Leopold, Aldo. 1949 [1968]. *The Sand County Almanac*. New York: Oxford University Press.
Magid, Shaul. 2013. *American Post-Judaism: Identity and Renewal in Postethnic Society*. Bloomington, IN: Indiana University Press.

Marmur, Michael. 2016. *Abraham Joshua Heschel and the Sources of Wonder.* Toronto, Buffalo, and London: University of Toronto Press.

Mayse, Evan Ariel. 2015. "Arthur Green: An Intellectual Profile." In *Arthur Green: Hasidism for Tomorrow*, ed. Hava Tirosh-Samuelson and Aaron W. Hughes. Leiden: Brill.

Merkle, John C. 1978. "The Sublime, the Human and the Divine in the Depth-Theology of Abraham Joshua Heschel." *The Journal of Religion* 58 (4): 365–379.

Most, Andrea. 2016. "The Contemporary Jewish Food Movement in North America: A Report from the Field(s)." In *Jews and their Foodways*, ed. Anat Helman, 204–217. Oxford: Oxford University Press.

Nathans, Benjamin. 2002. *Beyond the Pale: The Jewish Encounter with Late Imperial Russia*. Berkeley, Los Angeles, and London: University of California Press.

Neril, Yonatan and Evonne Marzouk (eds.). 2013. *Uplifting People and Planet: Eighteen Essential Lessons on Judaism and the Environment*. San Francisco: Canfei Nesharim.

Prell, Riv-Ellen. 1989. *Prayer & Community: The Havurah in American Judaism.* Chicago: University of Chicago Press.

RA (Rabbinical Assembly). 2001. "*Resolution on Energy and Environment.*" https://www.rabbinicalassembly.org/resolution-energy-and-environment.

RA (Rabbinical Assembly). 2008. "*Resolution on Environment: Green Covenant.*" https://www.rabbinicalassembly.org/resolution-environment-green-covenant?tp=110.

RAC (Religion Action Center of Reform Judaism). 2023. "*Position of the Reform Movement on the Environment.*" https://www.rac.org/position-reform-movement-environnment.

Rakover, Nahum. 1982. "Ecology in Judaism." *Encyclopedia Judaica, 1973–1982*, 227 Rakover, Nahum. 29. Jerusalem: Keter.

RCA (Rabbinic Council of America). 1991. "*The Environment.*" https://rabbis.org/the-environment-1991.

RJ (Reconstructing Judaism). 2001/2002. "*Environmental Activism and Jewish Spirituality.*" https://www.reconstructingjudaism.org/files/resources/document/environmental_activism_and Jewish_spirtuality.pdf.

RJ (Reconstructing Judaism). 2017. "*Movement Resolution on the Environment.*" Last modified January 25, 2017. https://www.reconstructingjudaism.org/article/movement-resolution-environment.

RRA (Reconstructionist Rabbinical Association). 2008. "*Resolution on the Global Climate Crisis.*" https://therra.org/resolutions/rra_resolution_climate_2008.pdf.

Schachter-Shalomi, Zalman and Joel Segel. 2013. *Jewish with Feeling: A Guide to Meaningful Jewish Practice.* Woodstock, VT: Jewish Lights Publishing.

Schachter-Shalomi, Zalman with Edward Hoffman. 2012. *My Life in Jewish Renewal: A Memoir.* Lanham, MD: Rowman and Littlefield Publishers.

Schachter-Shalomi, Zalman and Nathaniel M. Miles-Yépez. 2003. *Wrapped in a Holy Flame: Teachings and Tales of the Hasidic Master.* San Francisco: Jossey-Bass.

Schorsch, Jonathan. 2017. *The Food Movement, Culture, and Religion: A Tale of Pigs, Christians, Jews and Politics.* Wiesbaden, Germany: Springer.

Schwartz, Richard H. 2001 [1982]. *Judaism and Vegetarianism.* New revised edition. Woodstock, NY: Lantern Publishing and Media.

Schwartz, Richard H. 2002. [1984]. *Judaism and Global Survival.* New revised edition. Woodstock, NY: Lantern Books. [New York: Atara Publishing Co.]

Schwarzschild, Steven S. 1984 [2001]. "The Unnatural Jew." *Environmental Ethics* 7: 347–362. Reprinted in *Judaism and Environmental Ethics: A Reader*, ed. Martin A. Yaffe. Lanham, MD: Lexington Books.

Scult, Mel. 1993. *Judaism Faces the Twentieth Century: A Biography of Mordecai M. Kaplan*. Detroit: Wayne State University.
Seidenberg, David. 2005. "Jewish Environmentalism in North America." In *Encyclopedia of Religion and Nature*, ed. Bron R. Taylor, 909–913. London: Continuum.
Sherwin, Byron. November 2014. "Tikkun Olam: A Case of Semantic Displacement." *World Jewry*. Available at: https://Jcpa.Org/-Main-Issues/World-Jewry; https://jcpa.org/researcher/Byron-sherwin/.
Sideris, Lisa and Kathleen Dean Moore (eds.). 2008. *Rachel Carson: Legacy and Change*. Albany: SUNY Press.
Silvern, Steven E. 2021. "The Jewish Food Movement: A Sustainable and Just Vision for Place, Identity, and Environment." In *Religion, Sustainability and Place: Moral Geographies of the Anthropocene*, ed. Steven E. Silvern and Edward H. Davis, 327–354. New York: Palgrave Macmillan.
Sleeper, James A. and Alan L. Mintz (eds.). 1971. *The New Jews*. New York: Vintage Books.
Stern, Eliyahu. 2018. *Jewish Materialism: The Intellectual Revolution of the 1870s*. New Haven and London: Yale University Press.
Stoll, Mark. 2015. *Inherit the Holy Mountain: Religion and the Rise of American Environmentalism*. New York: Oxford University Press.
Taylor, Bron R. 2005. "Environmental Ethics." In *Encyclopedia of Religion and Nature*, ed. Bron R. Taylor. London: Continuum.
Tirosh-Samuelson, Hava (ed.). 2002. *Judaism and Ecology: Created World and Revealed Word*. Cambridge, MA: Harvard University Press.
Tirosh-Samuelson, Hava. 2023. "Jewish Environmentalism in the United States: Achievements, Characteristics, and Challenges." *Religion and Development* 2: 381–417.
UAHC (Union of American Hebrew Congregations). 1965. "*Resolution on Conservation and Development of Nature Resources*." https://www.urj.org/what-we-believe/resolutions/conservation-and-developlemt-natural-resources.
UAHC (Union of American Hebrew Congregations). 1969. "*Resolution on Environmental Pollution*." https://www.urj.org/what-we-believe/resolutions/environmental-pollution.
UAHC (Union of American Hebrew Congregations). 1979. "*Resolution on Energy*." https://www.urj.org/what-we-believe/resolutions/energy.
UAHC (Union of American Hebrew Congregations). "What Is Eco-Kosher?" 1996. In *This Sacred Earth: Religion, Nature, Environment*, ed. Roger Gottlieb, 297–300. New York: Routledge.
UAHC (Union of American Hebrew Congregations). (Autumn 2014). "Notes from a Small Planet: Move Our Money, Protect Our Planet." *Jewish Currents*. Available at: https://jewishcurrents.org/notes-small-planet-move-money-project-plant.
Waskow, Arthur and Phyllis Berman. 1996. *Tales of Tikkun: New Jewish Stories to Heal the Wounded World*. Jason Aronson.
Waskow, Arthur, Ari Elon, and Naomi Mara Hyman (eds.) 1999. *Trees, Earth, and Torah: A Tu B'Shvat Anthology*. Philadelphia: Jewish Publication Society.
Waskow, Arthur (ed.). 2000. *Torah of the Earth: Exploring 4,000 years of Ecology in Jewish Thought*. Woodstock, VT: Jewish Lights Publication.
Waskow, Arthur and Joe Laur. 2014. "*Tackling Climate Change: If Not Now, When?*" https://kolaleph.org/2014//04/22/tackling-climate-change-if-not-now-when/.
Winkler, Gershon. 1998 [2004]. *The Way of the Boundary Crosser: An Introduction to Jewish Flexidoxy*. Lanham, MD: Rowman and Littlefield.

Winkler, Gershon. 2003. *Magic of the Ordinary: Recovering the Shamanic in Judaism.* Berkeley, CA: North Atlantic Books.

Wyschogrod, Michael. 1991. "Judaism and the Sanctification of Nature." *The Melton Journal* 24: 5–7. Reprinted in *Judaism and Environmental Ethics: A Reader*, ed. Martin D. Yaffe, 289–296. Lanham, MD: Lexington Books.

White, Lynn Jr. 1967. "The Historical Roots of Our Ecologic Crisis." *Science* 155: 1204–1207.

Worster, Donald. 2008 [2011]. *A Passion for Nature: The Life of John Muir.* Oxford: Oxford University Press.

Yanklowitz, Shmuli. 2012. *Jewish Ethics and Social Justice: A Guide for the 21st Century.* New York: Derusha Publishing, LLC.

Yanklowitz, Shmuli (ed.). 2015. *The Jewish Vegan.* New York: CreateSpace Independent Publishing Platform.

Yanklowitz, Shmuli (ed.). 2018. *A Torah Giant: The Intellectual Legacy of Rabbi Dr. Irving (Yitz) Greenberg.* New York: Ktav Publishing House.

Yoreh, Tanhum. 2010. "Ultra-Orthodox Recycling Narratives: Implications for Planning and Policy." *Journal of Enterprising Communities* 4 (4): 323–345. doi:17506201011086129.

Yoreh, Tanhum. 2018. "Consumption, Wastefulness and Simplicity in Ultra-Orthodox Communities." *Studies in Judaism, Humanities, and the Social Sciences* 2 (2): 137–152. doi:10.26613/sjhss.2.2.49.

Yoreh, Tanhum. 2019. *Waste Not: Jewish Environmental Ethics.* Albany, NY: SUNY Press.

Yoreh, Tanhum. 2019a. "Rethinking Jewish Approaches to Wastefulness." *The Review of Rabbinic Judaism: Ancient, Medieval and Modern* 22 (1): 31–45. doi:10.1163/15700704-12341350.

Yoreh, Tanhum and Grace King. 2024. "Convergence and Divergence of Rhetoric and Frames in Faith-based and Secular Environmentalism." *Environmental Sociology* 11 (2): 239–258. doi:10.1080/23251042.2024.2419329.

3 God and Nature in Jewish Eco-Theologies

The Re-enchantment of Nature and the Dialogue of Religion and Science

This chapter explores how contemporary Jewish theologians have responded to the ecological crisis by re-enchanting nature, rejecting the notion that nature is inert matter devoid of intrinsic moral value. Since Jewish environmental spirituality is not monolithic, the chapter seeks to describe the major approaches to the re-enchantment of nature rather than argue for the veracity or usefulness of a particular approach. Most Jewish theologians uphold traditional Jewish theism according to which the personal God who created the world is the same who revealed the Torah to Israel. Between the Torah and the natural world is a correlation, so that nature must be cared for according to the Torah's instructions. The ethics of care and responsibility endows nature with moral significance. For others, reenchanting nature means the pantheistic view, according to which the entire cosmos is a living organism rooted in divine consciousness, the ground of all Being. This pantheism spiritualizes the material and materializes the spiritual. A third, panentheistic approach re-enchants nature by insisting that the world is "in God" but that God, the One, is more than the totality of all processes in the physical world. The world of nature with its infinite multiplicity is the material "garb" of the One who is both transcendent and immanent. Humans worship God through the corporeality of nature, which is understood symbolically. Finally, there are those who take up Process Thought, endorsing the view that God is an evocative, persuasive force that calls us to improve and mend through love and social action. Regardless of the nuanced differences among these theologies, they all view nature, of which humans are an integral part, as a meaningful, valuable, and purposeful reality that reflects God's presence. No Jewish theologian regards nature as inert, dead "stuff" that serves as a mere resource for humanity's benefit.

The language we use to express the relationship between God, and the world is necessarily metaphorical: we create pictures of reality, namely, models that use metaphors to express human values, norms, and ideals. Sallie McFague, a Christian ecofeminist theologian, explains: "a model is a

DOI: 10.4324/9781351123945-4

metaphor that has gained sufficient stability and scope so as to present a pattern for relatively comprehensive and coherent explanation" (McFague, 1987, 34). Models "order the images of a tradition so that they may become intelligible patterns for life" (McFague, 1982 [1975], 103). Like other feminists, McFague rejected the traditional hierarchical models of Christian theology and offered new, relational, egalitarian models of God—God as mother, God as friend, God as lover. These models are more appropriate to our ecological challenges because they frame relations between God and the world not as subject–object but as subject–subject relationship (McFague, 1997, 100–103). McFague's feminist, relational, egalitarian eco-theology has greatly influenced Jewish eco-feminists, who will be discussed in the next chapter, but she is rarely referred to by (male) Jewish theologians. Their theological rethinking was inspired not by contemporary feminism (although they acknowledge its existence and relevance) but by Kabbalah, the Jewish mystical tradition, which imagines the divine Presence in the world in feminine terms: the Shekhinah. Indeed, Kabbalah plays a central role in contemporary Jewish eco-theology.

For over two millennia Jewish theologians, philosophers, and mystics have reflected on God's relationship with the natural world. These reflections do not constitute Jewish eco-theology. I define eco-theology rather narrowly as *theology that explicitly refers to the current ecological crisis.* According to this narrow definition, reflections on nature in rabbinic, philosophic, kabbalistic, and Hasidic texts, articulated long before the ecological crisis, do not constitute Jewish eco-theology, although these texts serve as resources for contemporary Jewish eco-theology. In previous publications I have demonstrated how Jewish conceptions of nature evolved over time in response to changing historical circumstances and intellectual transformations of Western culture. Here I discuss Jewish theologians whose reflections on God and the world explicitly respond to the ecological crisis of our time.

Characteristics of Jewish Eco-Theology

Four general observations will serve as our introduction to this inquiry. First, although Jewish eco-theology exists today, the ecological crisis is not high on the Jewish theological agenda. Only a small number of Jewish theologians have made the environmental crisis the center of their work, although no Jewish theologian doubts the existence or severity of the ecological crisis. Other topics, such as the Holocaust, the survival of the Jewish People, sexuality and gender, the Jewishness of the State of Israel, or the relationship between Israel and its Arab neighbors take precedence. Therefore, it is not surprising that in the Library of Contemporary Jewish Philosophers, a set of 21 volumes of which I was the editor-in-chief (Tirosh-Samuelson 2012–2018), only three theologians—Jonathan Sacks, Lenn E. Goodman, and Arthur Green—have made the environmental crisis central to their theological projects. In part that reflects the fact that contemporary Jewish theologians have focused on the Jewish People rather than the natural world. A good example is Michael Wyschogrod, *The Body of Faith: God in the People of Israel* (1983 [1989]). As we shall see below, for Wyschogrod the very focus on the natural world's integrity, well-being, or sacredness is the mark of paganism, which Judaism has fought against since antiquity. Other Jewish theologians may not go as far as Wyschogrod, but it is fair to say that placing nature at the center of Jewish theology remains contested.

Second, unlike their Christian cohorts, Jewish eco-theologians do not write systematic eco-theology. The various strands of Christian eco-theology—biblical hermeneutics, theologies of liberation, ecofeminist theologies, eco-justice theologies, and process theologies (e.g., Dalton and Simmons 2010; Deane-Drummond 2008; 2017)—all share the assumption that theology bridges religious and secular cultures. By contrast, Jewish eco-theology is decidedly hermeneutical, offering close reading of authoritative texts, be they biblical, rabbinic, philosophic, kabbalistic, or Hasidic. The audience of Jewish eco-theology is predominantly if not exclusively Jewish and the goal is to show that the Jewish textual tradition is relevant to the ecological crisis. Instead of saying how the world should be managed, Jewish eco-theologians instruct Jews how to live the Jewish way of life with its rituals, sacred narratives, festivals, and life-cycle celebrations in an environmentally responsible way. Because to be Jewish is to stand in some relationship to the Torah, Jewish eco-theology offers an "environmental Torah," so to speak, namely instructions how to live an environmentally sensitive Jewish life in the age of environmental degradation. Jewish environmental theology is hermeneutical rather than speculative, pragmatic rather than systematic.

Third, serious engagement with the ecological crisis requires one to be aware of the science of ecology and attentive to scientists' documentation of the environmental crisis. Jewish eco-theologians have shown limited attention to the science of ecology, and scientific information does not drive Jewish eco-theology, even though the relevance of science is presumed (Tirosh-Samuelson 2012). This too is a clear difference between Jewish and Christian eco-theologians. For example, Christian ecofeminists such as Sallie McFague (McFague 1993, 38–47) and Heather Eaton (2007; 2008; 2011) summarize contemporary evolutionary science to critique and reject the atomistic, mechanistic, and materialistic worldview of modern science. Christian theologians have engaged contemporary science, especially the sciences of physics and biology, in detail to resolve the so-called "conflict of religion and science," whereas Jewish theologians have been only marginally engaged with this conflict. The prominence of science in Christian eco-theology is most evident in the discussion of climate change and climate engineering (Clingerman 2014; Clingerman and O'Brien 2014; 2016), topics which Jewish eco-theologians have barely begun to address, although Jewish environmental activists are very involved in climate advocacy (Tirosh-Samuelson 2018). The recent, most elaborate Jewish theological engagement with climate change, Laurie Zoloth's *Ethics for the Coming Storm: Climate Change and Jewish*

Thought (2023), offers homiletical ethics, namely, a Jewish ethic based on rabbinic midrashim to address the challenges of living in the age of climate change.

The limited attention to science reflects the fact that Jews by and large have accepted and even welcomed the separation of religion and science, since that separation enabled them to integrate into Western (Christian) society as equal citizens. During the late 19th and 20th centuries, Jews who entered the European universities from which they had been previously excluded flocked to the natural sciences. Jewish scientists were attracted to modern (secular) science as a replacement of their traditional Judaism, although they brought to their scientific pursuit skills honed by traditional Jewish education. Modern Jewish scientists either endorsed the separation of religion and science or held that in principle there can be no conflict between them (Efron 2007; 2014). Interest in the "dialogue" between religion and science, namely, deliberate attempts to reconcile commitment to Judaism and openness to secular knowledge, characterizes Modern Orthodoxy, which promotes "Torah Umadda" (literally, "Torah and science," more broadly, "Torah and secular knowledge").

Other strands of modern Judaism are less interested in the dialogue of religion and science because they do not consider the issue to be so problematic: either there is no conflict between Judaism and science at all, or Judaism and science operate on different planes. It is no coincidence that the person who coined the phrase "non-overlapping magisteria" (NOMA) to describe how science relates to religion, the renowned biologist Stephen Jay Gould (1997), was a Reform Jew! Jewish eco-theologians presuppose that science is relevant to their reflections and refer to scientific theories such as the theory of evolution or the theory of emergence, but the details of scientific information remain largely outside their theological discourses. Thus, the new cosmology, new physics, and new biology are accepted as new "maps of reality," but the details of these maps remain unexplored. Jewish eco-theologians engage science only superficially.

Finally, there is no one Jewish eco-theology that expresses Jewish environmental spirituality and no one paramount model of God. How a given Jewish theologian interprets the relationship between God and the world depends on their approach to the authority of the Jewish textual tradition, or certain strands within it. For example, theologians associated with the Jewish Renewal Movement promote pantheistic or panentheistic models of God based on Kabbalah and Hasidism, strands of Judaism that are less authoritative for Modern Orthodox, Conservative, or Reform theologians, who follow the footsteps of medieval Jewish philosophers, especially Moses Maimonides, and their modern heirs. This does not entail that to be a Jewish environmentalist one must be associated with the Jewish Renewal Movement or that one must endorse the pantheist or panentheistic models of God. Indeed, the theistic model of God, which the Jewish Renewal Movement critiques, supports environmentalism by calling for respect for nature, an ethics of responsibility, and protection and conservation of nature. Similarly, the notion that God is a dynamic, ever-changing, relational deity is prominent among Conservative or Reconstructionist theologians influenced by Process theology, but it is mistaken to conclude that only they talk about divine becoming and relationality, since these ideas are also shared by Jewish theologians not committed to Process Thought. In other words, the models of God–world relations presented below should not be taken rigidly as mutually exclusive systems identifiable with specific strands of modern Judaism but as heuristic devices that highlight certain themes in contemporary Jewish eco-theology.

The Theistic Model: Torah and Nature in Modern Orthodoxy

Nature in Judaic Theism

Judaism is commonly characterized as a theistic religion because the Bible and the rabbinic tradition posit a personal God. In traditional Judaism, God is the Creator of the world who also revealed His Will in a form of Law, the Torah, to the Chosen People, Israel. Between God and Israel there is everlasting bond, a Covenant (*brit*), which expresses the love of God to his Chosen People and their commitment to live in accordance with God's commandments (*mitzvot*) as articulated in the Torah and interpreted by the authoritative rabbinic tradition. The collateral of the Covenant between God and the People of Israel is the Land of Israel, which was promised to the People of Israel on condition that they treat it properly in accordance to divine commands. The Covenant makes Israel, collectively and individually, *responsible to* God, who rewards or punishes human deeds, and *responsible for* the world, which ultimately belongs to God. So long as the People of Israel observes the revealed God's will, the Land of Israel is fertile and fecund and the people of Israel flourish; but when Israel sins by not observing God's will, the Land of Israel loses its fertility, and the people of Israel suffer. When the sins of Israel become too egregious, God exiles the people from the Holy Land. In the sacred narrative of Judaism, the condition of the Land of Israel reflects the quality of the relationship between God and Israel.

The theistic model of God is inherently relational: God interacts with the created world, especially with human beings, and among them especially with the People of Israel, the Chosen People. The transcendent Creator manages His creation with care and concern for its well-being, and Israel, who worships God, thanks Him in daily prayers and blessings. Since the Earth belongs to the Lord (Ps. 24:1), caring for God's creation is a religious obligation. However, human relationship with God's created world is rooted in tension. On the one hand, humans have the capacity to control the natural world; on the other, humans have the responsibility to take care of the natural world. Genesis 1:28 and Genesis 2:15 respectively articulated this tension: the first verse authorizes human control over the Earth; the second imposes the obligation to cultivate and guard it. The two verses are complementary rather than contradictory. Precisely because humans were created in the "image of God" (*tzelem Elohim*), we have the creative ability that makes it possible to benefit from nature as well as to transform and even destroy it. Therefore, human creative capacities must be curtailed. For that reason, the Bible limits human action regarding nature: humans must observe divine commands that specify how to treat the natural environment; human actions are to be undertaken for the sake of improving the created world. And humans must regularly cease their creative activities by resting on the Sabbath and periodically in the Sabbatical Year, emulating God, who rested on the seventh day of Creation. Jewish environmental legislation instructs humanity how to act responsibly toward the Earth, increasing order in the created world and protecting it from destruction.

The theistic model is quite problematic from the perspective of contemporary environmentalism. First, the transcendence and radical otherness of God is not only an ontological description but a value judgment on the relative merit of the physical world of nature. Since the created world owes its existence to the supernatural, transcendent God, the natural order derives its value from the Creator, rather than possesses intrinsic value. Second, the theistic model is unabashedly anthropocentric, viewing humans as fundamentally different from and superior to all other creatures, because only we were created in the "image of God" (*tzelem Elohim*). Moreover, the theistic model justifies human "mastery" or "dominion" over nature (Gen. 1:26–28), which environmentalists consider the very cause of the environmental crisis. Finally, the theistic model differentiates between temporal life in "this world" (*ha-olam ha-zeh*) and eternal life in the "world to come" (*ha-olam ha-ba*) and looks at the latter as the goal of the former. Earthly life in the physical created world is preparation for the spiritual life in the world-to-come, where death abolishes all physical characteristics. In the theistic model the natural world either lacks intrinsic value or at best is that which must be transcended to experience the spiritual telos of life. On the surface, at least, it appears that divine, spiritual reality conflicts with earthly physical reality.

Jewish religious environmentalists have rejected these claims by arguing that in Judaism the spiritual and the physical are closely intertwined: Jewish rituals have emerged out of agricultural celebrations and are rooted in the very rhythms of nature, the cycles of the seasons, and the embodied human being (*adam*), created from the Earth (*adamah*) to which humans return at death (Gendler, 1971, 1989; Bernstein 2000; Waskow 2013). Thus, Jewish religious environmentalism stands at the point "where nature and the sacred meet," to cite the subtitle of Ellen Bernstein's foundational book. In her later book, *Splendor of Creation: A Biblical Ecology* (2005), Bernstein offered an ecological midrash linking ecology, spirituality, and a Jewish environmental activist's personal experience to articulate a biblical Jewish environmental spirituality. We will say more about Ellen Bernstein in the following chapter. But if Judaism is Earth-bound, does it mean that Judaism considers nature holy? Does

Judaism worship nature? If so, what is the difference between Judaism and paganism? Here lies a major conundrum for Jewish eco-theology.

Judaism versus Paganism

From the start, Jewish environmentalists had to face the problem of paganism. Eilon Schwartz, a leading Jewish environmentalist associated with Conservative Judaism, juxtaposes Jewish theism and paganism. "Paganism, both in its biblical and Hellenistic manifestations, understood God as being contained within Nature, Jewish monotheism distanced the Holy from paganism and its concept of nature" (E. Schwartz 2001 [1995], 298).

Modern Orthodox theologians have been critical of contemporary environmentalism precisely because it represents the revival of paganism. For them the notion that nature is holy is idolatrous because it eliminates the radical ontological difference between God and creation presupposed by the argument that God is transcendent (Lichtenstein 1971; Lamm 1971; Helfand 1971). Michael Wyschogrod, for example, categorically rejected the intrinsic holiness of nature, a position he calls "upper ecology," characteristic of ancient polytheistic nature religions as well as contemporary neo-paganism. Echoing Weber, Wyschogrod asserts that if nature is holy, then modern science would have been impossible "since it is the desacralization of nature that makes science possible" (Wyschogrod 2001 [1991], 294). If one uncritically re-enchants nature by reviving pagan nature religions, one undermines the very scientific outlook of modernity. The crux of the matter, Wyschogrod insightfully argues, lies in the doctrine of creation: "The moment the doctrine of creation emerges, nature is no longer self-sufficient, no longer eternal, no longer perpetual, no longer all engulfing" (ibid.).

The doctrine of creation stands at the center of Jewish theology, classical and modern, because it is the core of the theistic model. In rabbinic Judaism creation is understood to mean that nature is intrinsically good, but it is neither holy nor perfect. In imitation of God, human beings can and should intervene in the natural world in order to perfect it, and in this regard, humans are seen as "partners" of God. For that reason, Modern Orthodox theologians have been very open to and supportive of modern technology, including even the cloning of humans. However, Modern Orthodox theologians hold that human interaction with nature invests nature with religious meaning because it is governed by divine commands. The natural world cannot be seen as separate from God and cannot be worshipped for its own sake.

Torah and Nature

In the previous chapter we noted that Rabbi Norman Lamm, the past President of Yeshiva University, the flagship educational institution of Modern Orthodoxy, acknowledged that since modern science and technology can destroy the world, the "moral imperative incumbent upon all mankind [is] to curb its arrogant and mindless devastation of nature" (Lamm, 1971, 162). Yet, Lamm differentiated between the Judaic respect for creation and the pagan worship of nature, stating categorically that "Judaism, in contradistinction to paganism, refuses to ascribe the quality of holiness to nature and natural objects as such. Nature is profane ... The God of the Bible is beyond, not within, nature" (ibid., 173). Since Lamm is fully aware that Kabbalah and Hasidism posited the holiness of nature, he moves on to critique "immanentism" and aligns himself with the 19th century Opponents (Mitnagedim) of Hasidism in the 19th century. God can be both transcendent and immanent because of divine revelation; the revealed law as interpreted by the authoritative tradition bridges the gap between God and the created world. Lamm concludes his essay by saying that "to appeal to contemporary man to revert, in this twentieth century, to a pagan-like nature worship in order to retrain technology from further encroachment and devastation of the resources of nature, is a piece of atavistic nonsense" (ibid., 182). Instead, he declares that Judaism "possesses the values on which an ecological morality may be grounded" (ibid.) and these values are to be found in Jewish Law (Halakhah). We will return to biblical "ecological morality" as the Judaic response to the eco-crisis in Chapter 7.

How does Torah, the revealed Law, relate to nature? In continuity with Samson Raphael Hirsch (d. 1888), the founder of Neo-Orthodoxy in Germany, Lamm promotes the approach of *Torah Umadda* and insists that there is no conflict between Judaism and science and no tension whatever between Torah and nature. He declares:

> We affirm that both Torah and nature are the results of divine revelation; and even as God is One, so is there no split between His self-revelation in Torah (His word) and His self-disclosure in nature (His world).
>
> (Lamm 1990, 123)

Indeed, traditional Jewish liturgy affirms this dual revelation of God in nature (cosmos) and in the Torah (logos) in the morning blessings that refer to God as Creator of nature and the Teacher of Torah. Both cosmos and logos are "legitimate sources of religious inspiration," Lamm argues (ibid., 124) and there is no contest between them. Since God reveals Himself in the Torah as well as in nature, God is transcendent and immanent, remote and unknowable as well as intimately close. Like Wyschogrod, Lamm asserts the centrality of the doctrine of creation and explains how it differs from the doctrine of emanation:

> A created world essentially means a desacralized cosmos, the handiwork of the Creator that does not, however, share in His holiness; whereas an emanated world implies that nature retains albeit in highly diminished form, ontological traces of the sanctity of the Source.
>
> (Lamm 1971, 172)

Here lies the difference between the theistic model of God, endorsed by Modern Orthodoxy and Conservative Judaism, and the pantheistic/panentheistic models, promoted by contemporary Neo-Hasidism and the Jewish Renewal Movement.

Joseph B. Soloveitchik: Halakhic Spirituality

The most elaborate theistic eco-theology was articulated by Joseph B. Soloveitchik (d. 1993), the intellectual leader of Modern Orthodoxy for most of the 20th century. A scion of an old dynasty of rabbis that led Eastern European Jews, Soloveitchik combined the rigor of Talmudic culture in the Lithuanian Yeshiva of Brisk with the liberal arts and the humanities he studied at the Free Polish University of Warsaw and the Friedrich Wilhelm University in Berlin, where he graduated with a degree in Philosophy in 1932. Soloveitchik "not only engaged with various currents of modern thought—neo-Kantianism, existentialism, phenomenology, philosophy of science—but also he has attempted to make them part and parcel of his self-understanding as an Orthodox Jew" (Berger, 1993, 326). After Kant, the philosopher most quoted by Soloveitchik was Max Scheller, whose *Von Ewigen im Menschen* (On the Eternal in Man) greatly influenced Soloveitchik's attempt to reconcile the perceived tension between the biological and the ethical dimensions of the human condition (Rynhold and Harris 2018, 9). Invited to become the Chief Rabbi of Boston, Soloveitchik settled in the United States in 1932 and for over half a century trained, ordained, and inspired scores of Orthodox rabbis through his senior Talmud seminar in Isaac Elchanan Theological Seminary, which was affiliated with Yeshiva University in New York and the course on Jewish philosophy at the university's Bernard Revel Graduate School.

Soloveitchik was a dialectical thinker deeply attuned to the conflicts and tensions characteristic of human experience, especially modern humans who are torn between secularity and religiosity. Soloveitchik embraced the conflict and held that normative Judaism (i.e., Judaism governed by Halakhah) offers the best way to resolve the inevitable conflicts of the human condition, exacerbated but not occasioned by modernity. For Soloveitchik, Halakhah is not just "a group of norms guiding our activities but also as a conceptual theoretical structure, addressed toward our cognition" (Ravitzky 1986, 160). As a revelation of divine reality, it pertains to the infinite divine intellect understanding of that which is, through the process of halakhic reasoning, is the source of endless creativity. It is in the process of legal interpretation that we become "a partner with the Almighty in the act of creation" (Soloveitchik 1983, 81). By virtue of the halakhic life, the life lived through endless choices and creative innovations of legal rulings, the religious personality perpetually creates itself.

Although Halakhah is an abstract legal system, Soloveitchik insists that "the Halakhah is not at all concerned with a transcendent world;" instead,

Halakhah concerns itself primarily with life in this world, where "man is given the opportunity to create, act, [and] accomplish" (Soloveitchik 1983, 32). The halakhic life lived through performance of commandments "is confined to this world, to physical concrete reality, to clamorous, tumultuous life, pulsating with exuberance and strength" (ibid., 33–34). That Jewish Law concerns the very physicality of life in this world expresses the life-affirming quality of Judaism, in contrast to Christianity, which negates life by negating the body and advocating asceticism. Unlike Christianity, "Judaism is almost entirely involved in real bodily existence, in physiological functions" (Soloveitchik 2008, 110) but the prescriptions of Halakhah facilitate the sanctification of the physical and the embodied by placing limits and boundaries on the physical. The entire purpose of Halakhah is "the sanctification of biological existence" (ibid., 119).

Within the Halakhic system the deepest conflict that characterizes the human condition, the conflict between "Is" and "Ought," or between "nature" and "ethics," is best resolved. Soloveitchik opens his essay "Majesty and Humility" with the following sentence: "Man is a dialectical being; an inner schism runs through his personality at every level" (Soloveitchik 1978, 37). Like Max Weber, Soloveitchik analyzes the human condition by identifying ideal types that shed light on existing human tendencies or orientations. The human personality mirrors two conflicting orientations—majesty and humility—and both are attributes of God. They are sources of human dignity but in opposing ways: whereas God's majesty leads humans to "rule, to be king, to be victorious" (ibid., 33), resulting in scientific knowledge and technological inventions, humility is the divine attribute of "withdrawal and retreat" that facilitates love relations and the creation of faith communities. In the modern era human beings mistakenly reduce human dignity solely to the realm of secular activity, forgetting the other divine attribute, humility, since God is not only a victorious ruler but also the Creator who was able to impose limits on Himself to make room for non-divine reality.

The dialectical condition of the modern person is elaborated in Soloveitchik's famous essay "The Lonely Man of Faith" (1965). There Soloveitchik presents his philosophical anthropology as a typology of two foundational ideal types: "Adam the First" (Adam I) and "Adam the Second" (Adam II). These two ideal types, or dispositions, correspond to the two creation narratives in Genesis. In the first creation narrative, God (Elohim) created the human after "his own image" at the end of the six days of creation as a unity of male and female; according to the second narrative, God (YHWH) made the human of the dust of the Earth, breathed in him a "breath of life" (*nishmat hayyim*) and created a female counterpart. Soloveitchik reads the two biblical narratives as complimentary aspects of one creation narrative, which he interprets as a religious existentialist. The two narratives typify two different ways that we can understand ourselves and our relationship to the world and to God. Soloveitchik's typology is reminiscent of Augustine's differentiation between the City of Man and the City of God, but as we already noted Soloveitchik was quite critical of

Christianity because it negated or denigrated the natural, embodied, human existence.

Written before the ecological crisis was fully recognized, it is no surprise that Soloveitchik's famous essay does not speak about the ecological crisis explicitly. Yet, the essay reflects on the human predilection to master the natural world by means of technology, which is indeed the cause of the ecological crisis. For Soloveitchik, technology expresses the human proclivity of Adam I, who "sees the world as an object to be mastered," and technology makes it possible for humans to overcome their natural limitations and vulnerabilities. Desiring to overcome helplessness, human beings interact with each other on a utilitarian, functional basis, mastering the external world, overcoming poverty, hunger, and other natural limitations. The technological Adam I reflects what Buber characterized as an I-It relation to the world, namely, an instrumental, conditional, and utilitarian relation. Adam I senses his humanity and dignity through the unleashing of energy and power, but his posture has little to do with awareness of God; it reflects instead the awareness of the human as being part of the cosmos. The first creation narrative refers to God as Elohim, because that name expresses God's power and governance of nature.

By contrast, the orientation of Adam II leads humans to experience the world not as a reality to be controlled "but a reality to be experienced with a sense of wonderment, puzzlement, and surprise" (Hartman 2001, 107), an attitude elaborated by Abraham Joshua Heschel, discussed in the previous chapter. From this perspective, human dignity is derived not from mastery and control of nature but from "the quest for purpose, meaning, and relationship" which the human finds in the encounter with another person—a Thou—through friendship, passion, and love. The second creation narrative, therefore, refers to God not with the impersonal name Elohim but in the proper name of God, YHWH, signifying a personal relationship. The dignity of Adam II is thus derived from a personal relationship with God and with other humans, beginning with the relationship with the woman.

The thoroughly modern Soloveitchik did not lament the rise of secularism per se but the loss of our ability to truly imitate *humilitas Dei*. The modern person suffers from hubris that comes with the victory of modern science and technology, which has caused the loss of the "metaphysical polarity implanted in man as a member of both the majestic and covenantal community" (Soloveitchik 1978). Without the ability to imitate God's humility, we are unable to make room for meaningful interpersonal relations, as much as we are not able to "to withdraw from rationalistic position" in order to freely assume the burden of divine laws. According to Soloveitchik, "it is God who wants the man of faith of oscillate between the faith community and the community of majesty, between being confronted by God in the cosmos and the intimate immediate apprehension of God through the covenant" (Hartman 2001, 120). Modern humans have undermined those creative dialectics, resulting in spiritual impoverishment. Put differently, the disenchantment of nature characteristic of the modern secular world is the result of privileging Adam I at the expense of Adam II.

In the posthumously published *The Emergence of Ethical Man* (2005), Soloveitchik's theistic eco-theology is explicated most fully. Based on his public lectures, this work illustrates how a Modern Orthodox theologian weaves biblical exegesis, philosophy, and science to reflect on the relationship between God, humanity, and nature. Soloveitchik opposed naturalistic interpretations of Judaism, proposed for example by Mordecai Kaplan. Against Kaplan's views (to be explored below), Soloveitchik presented God's creative act as a law-like activity, because in creation God sets up boundaries and marked differences. In so doing, God "separates existence from 'nothingness,' the ordered cosmos from the void" (Soloveitchik 1983, 103). According to Soloveitchik, the biblical narrative of creation is not a story about a transcendent deity "who creates a world with which He will never come in contact," but rather as "the powerful being who dominates all, and who is not at an infinite distance from His creatures" (Soloveitchik 2005, 9). The biblical creation story is "the biography of nature" (ibid., 9) that teaches "the unity of the created universe" and its internal, hierarchical order of "vegetative life, animal and finally man" (ibid., 10). Endorsing the "scientific view that man emerged very late in the process of organic evolution" (ibid., 4) Soloveitchik declares that man "is an integral part of nature."

But if humans are merely natural beings, how can they respond to the moral imperative, which is not natural? Where does the moral imperative come from? After all, animals are not guided by moral principles, and nature does not take care of the weak and the sick. For Soloveitchik, ethics comes from the living God in whose infinite consciousness there is no tension between "ethics" and "nature." The living God who created nature is the God who grants life, so "life" is "an attribute of God" (Soloveitchik 2005, 41). Indeed, "if there is life in a creature, it is because God grants a spark of His own living being to the world" (ibid.).

Since life belongs to God, between God and living things there is a juridical relationship; briefly put, "All living beings enjoy only the rights of tenancy; they do not exercise dominion over their holdings …" (Soloveitchik 2005, 42). Here lies the difference between the God of Israel, who is the "owner of all life," and the "pagan gods" that are "non-living, dead, hewn out of stone or carved out of wood" (ibid., 40). It is because the God of Israel is a living God that He has the capacity to "perform within the cosmic and social drama, but also the ability to respond, to react to certain events taking place in this world and to act accordingly" (ibid., 40). As Soloveitchik interprets the theistic model, in God there is no tension between "is" and "ought," between the law of cosmic nature and the moral law.

Halakhah establishes an ethics of stewardship toward the natural world. "Stewardship" and "dominion" are complementary rather than mutually exclusive, because "dominion" does not mean "domination." Soloveitchik teaches that

> Nature surrenders voluntarily to man's control and rule, she entrusts man with her most guarded secrets. It is more cooperation than dominion, more partnership than subordination. [...] By the slightest error, man forfeits his rights to dominate and becomes an outcast.
>
> (Soloveitchik 2005, 60)

For Soloveitchik, the covenant between man and nature means that "earth, nature, and man flow into each other." But the covenant is conditional and depends on human behavior. The environmental laws of Judaism, such as the prohibition on wanton destruction (*Bal Tashchit*), especially as explicated by Samson Raphael Hirsch (Hirsch 1981, 62; Rosenbloom, 1976, 315–36) enable humans to live in harmony with nature. Living according to nature is an old principle articulated already by the Stoics in antiquity, whose views greatly influenced rabbinic Judaism (Tirosh-Samuelson 2003a, 51–53). In writing that "Sin is detachment from nature and non-compliance with her dicta," Soloveitchik (2005, 57) implies that the current environmental degradation is a sin against nature, against God, and against other humans, especially those of future generations. The only way to avoid environmental sin is to observe the environmental laws of the Torah by acting as stewards of creation. Caring for the environment is the morally right, ethical choice that only human beings can make because they are endowed with the divine image.

Judaic theism as articulated by Lamm and Soloveitchik supports environmental spirituality as well as environmental activism without assuming that nature is intrinsically holy and without denying the religious value of created embodiment and physicality. In the theistic model nature is good and religiously meaningful because it is created by God who also instructed humanity how to care for the created order by following the commandments. Although the theistic model celebrates the moral distinctiveness of the human being who alone has self-awareness, it considers the human to be embedded in the natural world, highlights the human responsibility for the well-being of created nature, and specifies environmental duties and obligations toward the earth, plants, and animals. The task of humans is to live in harmony with created nature rather than exploit nature. The theistic model frames environmental spirituality not in contrast to "religion" but as an expression of the deepest religious values and norms of the Jewish tradition, which negates paganism, on the one hand, and Christianity, on the other hand.

Soloveitchik is a major example of the theistic model endorsed by Modern Orthodox theologians, but non-Orthodox Jewish eco-theologians also reflect on the relationship of God in theistic terms. For example, the differentiation between 'dominion' and 'domination' has also been stressed by Rabbi Ellen Bernstein, the non-Orthodox environmental activist who founded Shomrei Adamah, who is considered the "birth mother of Jewish environmentalism." Rabbi Evert Gendler and Rabbi Lawrence Troster, eco-theologians associated with Conservative Judaism, held similar views. Rabbi Gendler pioneered

Jewish religious environmentalism in the late 1960s by advocating organic farming and vegetarianism. He expanded the concept of "covenant" to "God's relationship with other living creatures" (Gendler 2000, 60) and argued that "Buber's 'I-Thou' approach to the natural world must complement and constrain our more ordinary more instrumental 'It-It' approach" (ibid., 61). Rabbi Troster, who was active in interfaith environmentalism through GreenFaith, showed how the doctrine of creation coheres with contemporary cosmological theories, the Big Bang, and the Anthropic Principle (Troster 1987/1988; 1997) and explicated the implications of the doctrine of creation for Jewish religious life (Troster 2000). And Lenn E. Goodman (b. 1944), a historian of Jewish philosophy and a constructive philosophical theologian, has explained how the Judaic understanding of the doctrine of creation is compatible with the theory of evolution if we realize that "values are not explained mechanically" (Goodman 2010, 157). Goodman integrates natural teleology with an objectivist theory of value, reminding us that "theists find meaning in purposes. They see the same nature as scientists but focus on different facets of the jewel. They rebel at any fusing of objectivity with a dead neutrality, or stolid denial of purposes" (ibid., 173). For Goodman, moral value inheres in hierarchically ordered natural order so that there is commensurability between what is and what ought to be (Goodman 1991; Mittleman 2015).

Pantheism and Panentheism: Spinoza, Kabbalah, and Hasidism

The theistic model continues to dominate Jewish theological discourse, but the Jewish theologians associated with Neo-Hasidism—Zalman Schachter-Shalomi, Arthur Green, and David Mevorach Seidenberg, who will be discussed below—have been critical of it, endorsing instead a pantheistic or panentheistic model of God. If theism highlights the ontological gap between God and the physical world, pantheism, by contrast, is "the belief that all things are divine"; it follows that "The spiritual task is to get beneath the surface illusion and discover the deeper unity, to realize that even you as an apparently independent self are at one with the All, the divine whole of reality" (Peters 2013, 52). Ted Peters, a Lutheran theologian, insightfully notes that pantheism is on the rise today because "New Age Spirituality has incorporated pantheism," emphasizing "the sacredness of all things" and that "translates into ecological ethics," in which "we should acknowledge that the natural world is intrinsically valuable and protect the ecosphere from further deterioration" (Peters 2013, 53). Peters goes on to inquire whether "a Christian theologian can be a pantheist" and he rejects this possibility because "The world is not a manifestation of the divine, for Christians. The world is a creation, something God created from nothing" (ibid). The question Peters poses about Christianity applies as well to Judaism. But the story of pantheism within Judaism requires a lengthier discussion because the pantheistic model of God has been employed by "religious" Jews as well as by "secular" Jews. In fact, pantheism is inseparable from the story of Jewish modernity, its discontents, and the ongoing efforts to renew Judaism from within. We need to explore this point in some detail if we are to understand how Neo-Hasidic eco-theologians think about God and the world.

Spinoza and Spinozism

The Jewish philosopher most associated with pantheism was Baruch (Benedict) Spinoza (d. 1677), whose famous formula "*Deus, sive Natura*" equates God and nature. For Spinoza "God is not only the immanent or dwelling cause of all things. The separation of things from God is a metaphysical absurdity" (Seeskin 2002, 125). Put differently, God is the totality of all existents and God "exists in infinite ways as one substance" (Smith 1997, 47). Spinoza was excommunicated from the Amsterdam synagogue on January 27, 1656, and his writings were banned. However, the ban did not prevent his monistic pantheism from becoming profoundly influential and as highly controversial among Jews and Christians alike. Less than thirty years after Spinoza's death, John Toland, the English philosopher, coined the term "pantheism" in 1705 to describe the philosophy of Spinoza. Spinoza's reception among Jews was complicated. After his excommunication Spinoza refused to convert to Christianity and lived on the margins of both faith communities, communicating with the scientists and like-minded thinkers. The Jewish promoters of Enlightenment (the Maskilim) presented Spinoza primarily as a pantheist rather than an atheist and they reappropriated the excommunicated Spinoza back into the Jewish fold, although he had broken decisively with the Jewish community. Spinoza became regarded as the first modern, secular Jew, a hero for some and a despised heretic for others (D. Schwartz 2015).

In the 19th century, the Maskilim who admired Spinoza and translated the *Ethics* into Hebrew interpreted him in variety of ways, reflecting an internal debate between moderate and radical variants of Haskalah about "potentialities immanent to Judaism" (D. Schwartz 2015, 39). In the 1870s, absorbing ideas from Marxism and Darwinism, Jewish secularism emerged as a consistent interpretation of the Jewish historical experience and Spinoza played a heroic role for Marxist and Zionist thinkers (Stern 2018, 107). Spinoza would continue to symbolize the contentious encounter between Judaism and modernity and the ongoing debate about the meaning and validity of Jewish secularism (Biale 2011). All modern Jewish secularists can be considered "children of Spinoza," so to speak.

Neoplatonism and Kabbalah

Although Spinoza's pantheism signaled discontinuity with rabbinic Judaism, his ideas had deep roots in medieval Jewish thought, both rationalist Aristotelian philosophy and Kabbalah (Ravven and Goodman 2002). Medieval

Jewish philosophers restated traditional Jewish theism in philosophic, impersonal categories. The personalist conception of God was expressed not in philosophy but in religious poetry that eventually entered canonic Jewish liturgy. As Jewish philosophers, chief among them Moses Maimonides, became familiar with Aristotle's philosophy, they wrestled with the tensions between the personal and impersonal models of God, between the doctrine of creation and the doctrine of emanation, between divine free will and determinism, questions that engaged Jewish philosophy from the 13th to the 16th centuries. Although the Maimonidean tradition dominated official Jewish theology during the late Middle Ages it was severely criticized by Jewish mystics who presented themselves as recipients of an ancient, esoteric tradition, the Kabbalah.

Kabbalah highlighted God's presence in the world, namely, divine immanence, and it did so through the elaborate symbolism of the Shekhinah. Expressed in gendered symbolism, the Shekhinah is a feminine power that receives the divine efflux (*shefa*) of the nine masculine Sefirot above her. Together, the ten Sefirot represent the revealed aspect of God, apart from the transcendent and unknowable. This erotic energy pulsates through the Sefirotic world and from there emanates to all levels of reality, ensuring their vitality. As much as the Shekhinah governs all of the processes, the Shekhinah is also impacted by these processes. Human beings, and especially the People of Israel, impact the Shekhinah's well-being and affect her relationship with Sefirah Tif'eret (Beauty) that functions as her "husband." Whereas human sins cause the separation of the masculine and feminine aspects of God, the human performance of the commandments with the proper intentions can bring about their reunification. Indeed, it is the union of the (male) kabbalist's soul with the (feminine) Shekhinah that enables the kabbalists to redeem the extra-deical world and reunite the Shekhinah and her husband Tif'eret (Tirosh-Samuelson 2011). The well-being of the world and the well-being of the deity are determined by human deeds and intentions.

The gendered language of Kabbalah, and especially the Shekhinah symbolism, has been most attractive to Jewish eco-feminists who challenge the theistic model and the radical differentiation between Judaism and paganism. We will explore Jewish ecofeminism in the following chapter. For now, let us note that Kabbalah could be interpreted either pantheistically or panentheistically. The pantheistic interpretation states that there is only one substance, God, so that "God" and "nature" are different names of the same reality. According to the panentheistic interpretation, "God relates to the world like our mind relates to our body … The world is God's body, and God is the mind of the world" (Peters 2013, 57). Panentheism affirms that although God and the world are ontologically distinct and God transcends the world, the world is "in" God ontologically (Cooper 2013, 18). In panentheism, "all things exist within God's being. The entire world of nature and history exist within God's being, but they do not exhaust God's being" (Peters 2013, 57). Accordingly, panentheists

> reject the idea of creation ex-nihilo, and instead prefer the idea of continuing creation to emphasize the shared temporal relations between the world and God … The world and God are mutually inter-dependent. Like pantheism, panentheists believe that everything in the world is connected to everything else, and everything is connected to God. God's being and the being of the world are inseparable.
>
> (ibid., 58)

Despite their subtle differences, both models of God highlight the unity that underlies the world's multiplicity.

In Kabbalah, nature is truly enchanted because Kabbalah views the natural world as the mirror of God (Wolfson 2002). Yet precisely because the kabbalists regarded nature as a linguistic construct, namely, a text authored by God, Kabbalah regarded the physical and the material rather negatively as that which must bc transcended to cling to God (Tirosh-Samuelson 2002). Moreover, in Kabbalah (as in medieval Jewish philosophy in general) matter is metaphysically associated with evil and the material body is the source of passions that lead people to sin, removing them from God. The human task is therefore to transcend embodied materiality by spiritualizing the body through observance of divine rituals performed with mystical intentions (*kavanot*). Thus, medieval Kabbalah exacerbated the tension between the spiritual and physical dimensions of life (especially in the case of human beings) and encouraged Jews to adapt an ascetic lifestyle.

Kabbalah further evolved in the 16th century in interpretation of a mystical fraternity in Safed, Israel, then under Ottoman rule. Under the leadership of Isaac Luria (d. 1572) these mystics sought to explain the origins of the universe and the purpose of human life, especially Jewish life. In Lurianic Kabbalah, the first act of creation was divine "contraction" (*tzimtzum*) in which God withdrew into Himself to make room for the non-divine world. The next phase of the creative process was the expansion of a divine ray of light into the Void, the empty space (*tehiru*), created by God's self-withdrawal. The ray of light constructed the infrastructure of the Sefirotic world arranged as the primordial human (*Adam Kadmon*), but these divine containers, or vessels (*kelim*) were unable to contain the infinite divine substance (*atzmut*) and they shattered. As a result of this "breaking of the vessels" (*shevirat ha-kelim*), the divine sparks (*nitzotzot*) were scattered throughout the cosmos, entrapped in the "shards" or "barks" (*kelippot*) of the material world. Lurianic kabbalists debated whether these events were to be interpreted literally or metaphorically, and these debates would determine whether Lurianic Kabbalah was to be understood pantheistically or panentheistically. But on both interpretations the Lurianic myth of creation enchants nature: all levels of reality, including the physical, material world, are enlivened by the presence of the divine "sparks." The task of the human being, or more precisely, the task of the Jew, is to uplift the sparks through the performance of the commandments with the proper kabbalistic intentions. The ultimate goal of this human activity is to repair the world, which has been sundered by the primordial divine catastrophe, exacerbated by human sinful behavior. The kabbalist who knows the code with which the world was created can "repair" the world and the "repair of the world" (*Tikkun Olam*) pertains not only to the material and social world we experience but to the divine realm of the Sefirot. Thus, through knowledge of Kabbalah and kabbalistic performance of the commandments, humans can participate in redemptive activities, mending the broken cosmos and the broken deity.

Lurianic Kabbalah profoundly enchanted the material world. As Lurianic spirituality spread through Europe during the 17th and 18th centuries, a peculiar fusion of Kabbalah, mysticism, and magic came to dominate Jewish religious life. In the early modern period, the boundary between speculative Kabbalah (*kabbalah iyyunit*) and practical Kabbalah (*kabbalah ma`asit*) collapsed and practitioners of Kabbalah insisted on the new continuity between the mental world of the practitioner and the external physical world. Practitioners of Kabbalah were believed to be able to heal the sick and transform the material world because they possessed knowledge of the linguistic code in which nature was written: the divine Name in its infinite permutations. Based on their technical knowledge of kabbalistic intentions and their practices, these practitioners, known as Masters of the Name (*ba'aley shem*) functioned as healers and miracle workers. A new type of Jewish religious leadership emerged that was not based on intellectual prowess and textual reasoning but on the ability to heal the body and the mind. The Masters of the Name were Jewish shamans who addressed the material, social, and emotional needs of Jews, especially for those living away from urban centers that were dominated by the rabbinic intelligentsia and communal institutions.

Hasidism and the Enchantment of Nature

Out of this shamanic piety emerged Hasidism, a Jewish revival movement that challenged traditional authority and gave rise to new communal organizations. Founded by Israel Baal Shem Tov (d. 1760), Hasidism translated the highly elaborate cosmic myth of Lurianic Kabbalah into a way of life to be practiced by every Jew regardless of formal education. A healer who practiced Kabbalah, the Baal Shem Tov (or the Besht, as he was commonly known) was active in the Ukrainian countryside rather than in urban centers and he became the prototype of the shamanic leader who operates "between ecstasy and magic," to use the apt subtitle of Moshe Idel's analysis of Hasidism (Idel 1995). During the 19th century the Hasidic community witnessed the "routinization of charisma," precisely as Weber astutely analyzed, and Hasidism's shamanic qualities were muted to be replaced with dynastic structures and more formal learning. Hasidism became more textually focused as Hasidic masters communicated their theology through homilies and biblical commentaries first orally and then in a written form (Wiskind-

Elper 2018). Though Hasidism emerged as a challenge to traditional rabbinic learning, Hasidism strengthened Jewish traditionalism, precisely when European Jews were undergoing secularization and acculturation.

Like Kabbalah, the Hasidic attitude toward the natural world was paradoxical. On the one hand, Hasidism enchanted the world by declaring, along with *Sefer ha-Zohar*, the locus classicus of medieval Kabbalah, that "no place is devoid of God." On the other hand, since the spiritualization of the material is done by means of mental activity whose goal is mystical union with God (*devekut*), the practitioner was removed from the physical world of nature. Furthermore, attaining mystical union with God required the "annihilation of the self" (*bittul ha-ani*) through meditative practices that imagined the "annihilation of [material] reality" (*bittul ha-yesh*). The more mystically oriented forms of Hasidism, especially Chabad, advocated acosmism rather than celebrated the physicality and materiality of nature (Elior 1986). Hasidic immanentist theology did not cherish the physical world for its own sake, which is the core of environmentalism, but spiritualized the physical world through contemplation or through textuality.

The Re-enchantment of Nature in Neo-Hasidism

Since the 1960s Neo-Hasidic theologians promoted Hasidism and its underlying kabbalistic worldview as the proper Judaic response to the ecological crisis. Rejecting the theistic model, they advocated either pantheism or panentheism as new reality maps, new models of God that are more compatible with contemporary science. If we are to address the ecological crisis, so they claim, we must bring about a "paradigm shift" from theism to pantheism or panentheism that coheres with the evolutionary story of the universe and the evolution of human consciousness. Neo-Hasidism can thus be viewed as a distinct Jewish variant of New-Age Spirituality, exhibiting the features discussed in the previous chapter. Advocating the spiritual revival of Judaism, Neo-Hasidism is critical of institutionalized forms of Judaism, focuses on the inward life of the believer, and encourages the creation of spiritual practices often influenced by other spiritual traditions. The three eco-theologians discussed in this section have promoted Neo-Hasidism as a Jewish response to the environmental crisis.

Zalman Schachter-Shalomi: Toward a Gaian Consciousness

Zalman Schachter-Shalomi, the founder and leader of Jewish Renewal introduced in the previous chapter, launched Jewish New-Age Spirituality. In continuity with his Hasidic teachers, he was a storyteller, a spiritual healer, and ritual innovator whose effusive personality touched and transformed all those who met him. For Schachter-Shalomi, there can be no genuine conflict between what is true in "science" and what is true in "religion." Using different terminology, "science" or "religion" generates different "reality maps" (that is, worldviews) whose truth must cohere even though it can be expressed through different languages. The scientists and theologians who best understood this point were the Jesuit priest and paleontologist Teilhard de Chardin, the ecological cultural historian Thomas Berry, and the environmentalist and futurist James Lovelock, who authored the Gaia Theory. They inspired Schachter-Shalomi to view the universe as an evolving "living organism" whose parts are interconnected and interdependent, although Schachter-Shalomi does not engage them in any detail.

A uniquely syncretistic thinker, Schachter-Shalomi fuses Hasidism, Kabbalah, Jewish philosophy, religious history, astrology, psychology, and popular science into an idiosyncratic, Neo-Hasidic variant of Judaism that seeks to address all people and not only Jews. Shaul Magid has presented Schachter-Shalomi as a post-ethnic, radical theologian who offered "an alternative to conversion, undermining the notion that Judaism is the exclusive property of Jews" (Magid 2013, 50). The Jewish Renewal Movement was meant to appeal to "non-Jewish Jews" (namely, to secular Jews whose spiritual quest led them to other traditions) as well as to gentiles who marry Jews or who are simply attracted to Judaism. In a series of essays in *Tikkun Magazine* (Magid 2006a; 2006b; 2006c) Magid demonstrated that the Jewish Renewal Movement can be categorized as a New Religious Movement that transformed Jewish spirituality in America. In his later work Magid offered a brilliant analysis of Schachter-Shalomi's theology and of the Jewish Renewal Movement, presenting them as expressions of American pragmatism and pluralism (Magid 2013, 60–61).

By situating Schachter-Shalomi in the context of American pragmatism, Magid may have imposed greater philosophic coherence on Schachter-Shalomi's worldview than is warranted, but the gist of Magid's analysis is correct. Schachter-Shalomi rejected traditional theism, which sees God as radically other than the world, in favor of pantheism, according to which "God is everything and everything is God" (Schachter-Shalomi and Gan Kagan 2006, 157). Accepting the metaphoric interpretation of the doctrine of *tzimtzum* (Magid 2013, 65; Schachter-Shalomi and Miles-Yépez 2013, 21) Schachter-Shalomi moves the focus from ontology to psychology: "contraction" and "self-withdrawal" refer not to an objective, cosmological event but to conceptual shift within human consciousness. Accepting the story of evolution as a framework for the history of the universe, the history of the human species, and the history of human consciousness, Schachter-Shalomi depicts human history as epochal shifts in human consciousness from polytheism (4000 BCE–2000 BCE), to monotheism (2000 BCE–70 CE), to Theism/Deism (70 CE–2000 CE), and finally to our contemporary age in which pantheism prevails (2000 CE–4000 CE). Each epoch is symbolized by a sign of the Zodiac (Taurus, Aries, Pisces, and Aquarius, respectively) since Schachter-Shalomi, like other adherents to New Religious Movements, takes astrology seriously.

In the Age of Aquarius, a "paradigm shift" from theism to pantheism has taken place because human beings reached the realization that "body and soul are not separate from one another, and we must not reduce one to the other" (Schachter-Shalomi and Gan Kagan 2006, 158). Instead of seeing the world as a duality of matter of form, or mind and body, we should come to see the world as one, undivided, "living organism" (ibid., 178) that has its own functional integrity. Using the language of Lurianic Kabbalah, Schachter-Shalomi refers to this new "reality map" as a new "constellation," or "configuration" of God (*partzuf*). All the parts of the organismic universe are interconnected and interdependent, and each one must properly perform its function for the organismic whole to thrive. In the Jewish parlance, so claims Schachter-Shalomi, the pantheistic model is expressed in the expression "the bundle of life" (*tzeror ha-hayyim*), a Jewish analogue of the phrase "web of life." Therefore, Schachter-Shalomi prefers to talk about the "Great Chain of Life" rather than the "Great Chain of Being" (Schachter-Shalomi 2003).

Schachter-Shalomi's pantheism has much in common with the scientific theory known as "biocentrism," which holds that "there is no separate physical universe outside life and consciousness" because "nature and mind are correlative" (Lanza and Berman 2009, 159) and draws upon quantum physics, evolutionary biology and evolutionary psychology, and system theory. Schachter-Shalomi was familiar with works of popular science advocating biocentrism (especially the writings of Ken Wilber and Fritjof Capra), but instead of engaging contemporary science, Schachter-Shalomi refers to premodern Lurianic Kabbalah, especially the doctrine of the four worlds (*olamot*)—"emanation" (*atzilut*), "creation" (*beriah*), "formation" (*yetzirah*), and "making" (*asiyah*)—which he interprets as levels of consciousness that enable us to experience the world and God. Despite his superficial engagement with contemporary science, Schachter-Shalomi shares the ecological outlook of biocentrism as well as its inherent pluralism: all scientific disciplines, all religious and spiritual traditions, and all ways of life and cultural habits express the plurality and diversity of life.

Schachter-Shalomi's pantheism has obvious ecological ramifications. The high level of awareness he calls Gaian consciousness "compels us to ask what does the animal kingdom require of us, we, who are one of its members? What should we do for the sake of the Earth?" (Schachter-Shalomi and Gan Kagan 2006, 178). Since humans are "sons and daughters of the Earth," they "must contribute to the renewal and healing of life (corporeal as well as spiritual)" (ibid.). For Jews, Gaian consciousness must be translated into Eco-Kashrut, and conversely, Eco-Kashrut is "a concept coined to offer a framework for spiritual-social discussion about the relevance of Halakha to new realms of consciousness" (ibid., 212). Eco-Kashrut bridges the gap between thought and action, between text and social context, between "Is" and "Ought," between "nature" and "Torah." Schachter-Shalomi's Jewish Renewal thus goes beyond traditional Hasidism and beyond the textualization of nature characteristic of Kabbalah and Hasidism. Eco-Kashrut is a this-worldly religious environmental practice that is justified by pantheistic eco-theology. Some of the disciples of Schachter-Shalomi (e.g., Rabbis

Gershom Winkler or Eliezer Benzion) reconnected with the shamanistic dimension of Hasidism and fused it with other shamanic traditions.

Arthur Green: Nature as the "Garbing" of the One

Rabbi Arthur Green, the close friend and disciple of Rabbi Zalman Schachter-Shalomi, is both a historian of Jewish thought who focuses on Kabbalah and Hasidism and a constructive theologian. If Schachter-Shalomi has enhanced contemporary Jewish spirituality by focusing on pastoral counseling, institution building, and ritual practices, Green's contribution has been anchoring Jewish spirituality in sound historical, academic scholarship. The author of two monographs on Hasidic masters (as well as numerous academic essays on Kabbalah and Hasidism, e.g., Green 2015), two volumes on Jewish spirituality and, most relevant to this discussion, three works of systematic theology (Green; 2002; 2003; 2010), Green has promoted Neo-Hasidism (Green 2019; 2019a) as the "Kabbalah for the Environmental Age" (Green 2002a). Like Schachter-Shalomi, Green re-enchants nature by highlighting the intersection of Being and consciousness, but whereas Schachter-Shalomi does so and characterizes his position as pantheism, Green defines his own view as panentheism. Schachter-Shalomi himself, however, denied that there is much difference between pantheism and panentheism, and indeed all the themes of Schachter-Shalomi's pantheistic theology are present in Green's works: the critique of theism (or more precisely deism) (namely, the impersonal view of God as the designer of the world), the identification of "creation" with "emanation," the acceptance of evolution as the framework for the story of the universe and the story of humanity, including human consciousness, and the emphasis on the unity that underlies multiplicity.

Like Schachter-Shalomi, Green exemplifies the hybridity characteristic of the post-secular age. His this-worldly spirituality fuses "religious" and "secular" elements. In *EHEYE: Kabbalah for Tomorrow* (2002) Green presents himself as a "seeker" (Green 2002, x), the marker of the New Spirituality of the 1960s and a person who "in a certain sense" is a "mystic" (ibid., xii) inspired by the Neo-Hasidic, Polish-Jewish author, Hillel Zeitlin (d. 1942) and by Abraham Joshua Heschel (Green 2015a). Looking retrospectively at his life, Green describes himself as "an old revolutionary" (ibid., xvi), referring to his activities in the 1960s which included experimentation with psychedelics, a practice "he is still in favor of." Taking his own personal religious experience as a point of departure for theorizing, Green tells us that this experience convinced him that "religious life exists in order to keep both the individual and the community in touch with our own deepest moments of experience and insights" (Green 2002, xv). In *Radical Judaism* (2010), Green's book of systematic theology, he defines his "theological position as that of mystical panentheist, one who believes that God is present throughout all of existence, that Being of YHWH underlies and unifies all that is"

(Green 2010, 18). A few pages earlier, however, Green states: "I think of myself as a religious humanist" (Green 2010, 7), explaining that "religious humanism means that we will fulfill that awesome role only by realizing that we are part of a reality infinitely more ancient, more profound, and more unified than any of us can express or know" (Green 2010, 7).

"Religious humanism" commonly describes the views of Martin Buber and Abraham Joshua Heschel, but it can be interpreted in a variety of ways. If "religious humanism" means living in the presence of a relational God who interacts with human beings in their worldly affairs, what makes Green's theology "radical?" Green explains that the adjective "radical" in the title of the book is an allusion to the "Radical Theology movement of the late 1960s" (Green 2010, 13), also known as the Death-of-God theology, which Green's teacher, Heschel, considered a "blasphemy." In fact, it was Heschel who instructed the young Green to study Hasidism, where he could encounter a truly "radical theology" that focuses not on the death of God but on the living God who is the One underlying all aspects of reality. Green's Neo-Hasidism is "radical" because it challenges the doctrine of creation, the belief in a transcendent God, and the separation between God, World, and Self characteristic of traditional Judaic theism. For Green, "transcendence thus dwells *within* immanence. There is no ultimate duality here, no 'God and world,' no 'God, world and Self,' only one Being and its many faces" (Green 2010, 18). In Green's mystical panentheism the world is "in" God, but the world does not exhaust God.

Green has reflected on the ecological crisis of our generation, arguing that in Kabbalah one could find the worldview that could best address it. He states,

> I long for Judaism that teaches us how to live in harmony with the natural world, one whose most basic teaching will demand of us that we position ourselves at the cutting edge of sensitivity toward relieving the suffering and pain of all God's creatures. God's name is inscribed in all that is.
>
> (Green 2002, 16)

These are the main elements of Green's mystical panentheism. In Kabbalah, Green finds "inspiration and wisdom for what is essentially a Jewish mysticism for a post-kabbalistic age" (Green 2002, 17). He is not interested in a naïve, uncritical return to the premodern sources, but rather in the reconstruction of the premodern texts to address our postmodern, post-secular situation. Green is a post-secular thinker because he "questions whether science is the right way to pursue some of our great and eternal questions about life's purpose" (ibid., 17). Instead of worshipping secular, modern science, Green recommends that we complement science with insights from other cultural traditions, such as Kabbalah, which "were cast aside with the advent of modernity" (ibid., 17). The insights of Kabbalah about the "secret unity of all Being, hidden within multiplicity and diversity of life as we experience it" are compatible with the teaching of science, especially the theory of evolution, which is "an account of the emergence of multiplicity" (ibid., 113).

Green's mystical panentheism is "in no way opposed to science but rather seeks to encompass it" (Green 2002, 32). In fact, he calls contemporary scientists and physicists

> the new kabbalists of our age who claim even to know the black hole of which being itself came to be, speculating on the first few seconds of existence as some of the mystics once did on the highest triad of the ten Sefirot.
> (ibid., 110)

He maintains that between Kabbalah and contemporary science there is total congruence, since evolution, the dominant scientific paradigm, explains "the unity of all species, the shared life energy that flows within them and links them to common ancestry and common sources of nourishment." In the process of evolution

> each creature embodies the life energy and hence the presence of the One, but only humans are called 'God's image' in our tradition. We are the first to have the mental capacity to recapitulate the process, to be self-conscious about our roots within the One.
> (ibid., 114)

For this reason, "science and religion should be seen to complement one another, rather than stand in conflict" (ibid., 113).

Green is not an environmental activist, but his theology is shaped by awareness of the ecological crisis and environmental degradation. In *Radical Judaism,* Green once again calls us to offer a "new and powerful harmonistic vision, one that will allow even the weakest and most threatened of creatures a legitimate place in this world and will call upon us not to wipe it out by careless whim" (Green 2010, 24). In this retelling of "the entire course of evolution as the infinitely varied self-garbing of an endless energy flow," Green teaches that

> we humans represent a significant step forward in the evolutionary path toward the self-articulation and self-fulfillment of the One … but I do not view us humans—surely not as we are now—as the end or purpose of evolution … If existence survives on this planet, Mind will one day be manifest to a degree far beyond our present ability to comprehend or predict.
> (ibid., 27)

Alluding to new technological advancements, especially artificial intelligence and genomics, Green seems to relish the march toward higher forms of self-knowledge, without mentioning the disturbingly problematic aspects of contemporary technology or how contemporary religious tradition has engaged these technological developments.

Green's environmental commitment is based on his monistic cosmology "of God as the One, the inner force of all being, present within each thing that exists, as it is present within each moment of time" (Green 2010, 127). He retains the traditional tropes of the Judaic tradition because that is the culture and language to which he is deeply committed as a born Jew. Using the tropes of the Judaic tradition Green articulates a clear environmental vision: "We are 'the image of God' not because of our superiority to other beings but because we can articulate the notion that all beings bear God's presence and we can create ways to respond to that presence." (ibid., 128).

In terms of human obligation to and responsibility for nature, there is little difference between the mystical panentheism of Green and the Halakhic theism of Soloveitchik or other Jewish theists who are deeply concerned about the ecological crisis. How Neo-Hasidism could function as "Kabbalah for the environmental age," that is, as an environmental theory and praxis rooted in the sources of Judaism, will be explicated by David Mevorach Seidenberg, a disciple of Arthur Green, who is a promoter of Neo-Hasidism and an environmental activist.

David Mevorach Seidenberg: Sacralization of Nature and Eco-Justice

Rabbi David Mevorach Seidenberg (b. 1963) is the most systematic and prolific Jewish eco-theologian today. Deeply anchored in Jewish studies, the discourse of religion and ecology, and Jewish environmentalism, Seidenberg has offered the most elaborate and comprehensive ecological interpretation of Judaism that builds on the works of Schachter-Shalomi and Green but goes well beyond them. Ordained as a Conservative rabbi at the Jewish Theological Seminary (1994), Seidenberg also holds a PhD from the Jewish Theological Seminary (2003) and an MA in Education from Harvard (1987). Over the years he has held various temporary academic posts but he mostly functions as an independent scholar who offers his expertise as a religious environmentalist to various academic and nonacademic institutions, organizations, and programs. Seidenberg disseminates his ecological interpretation of Judaism primarily through his website NeoHasid.org where he posts his academic and semi-academic essays. He also writes quite prolifically for various Jewish environmental organizations, such as Hazon, for the web-based list "The Jew and the Carrot" administered by Aytzim: Ecological Judaism, mentioned in Chapter 2, and for media outlets such as the *Huffington Post*, the *Times of Israel, Tikkun Magazine*, and Tikkun Daily Blog. The information revolution of the digital age has allowed Seidenberg to disseminate Neo-Hasidism, bypassing traditional communal and academic structures and conventions.

Unlike his Neo-Hasidic teachers, Seidenberg is also an environmental activist and very connected to other activists and organizations discussed in the previous chapter. Seidenberg is a founding member of Green Hevra, a consortium of Jewish environmental organizations; a fundraiser and rabbinic advisor for Bustan, an environmental initiative in Israel that focuses on sustainable development and the rights of Bedouins; and a founder and organizers of EcoJews of the Bay, among other projects. As founder and director of the NeoHasid.org in 2005 Seidenberg is "devoted to: 1) bringing the resources of Chasidic song and liturgy beyond the Orthodox world, 2) teaching and expanding the connection between Torah and ecology, and 3) creating and disseminating new liturgy that fuses traditional practice with ecological concern and with gender issues." This statement illustrates how Seidenberg integrates eco-theology with Neo-Hasidism, Jewish Renewal, and Jewish feminism.

Neo-Hasidism is a distinct Jewish variant of contemporary New Spirituality. On his website Seidenberg defines Neo-Hasidism as

> an amalgam of trends and of people: liberal Jews craving something of the old-world that's richer than what they grew up with, ex-Chasidim looking for a place to share *nigunim*, women and men experimenting with deep davening in egalitarian contexts, anyone who is a Chasid without a rebbe, on pilgrimage to Uman, people who daven Shlomo *nusach* at Burning Man or Rainbow, artists exploring Chasidic forms in secular settings, and, of course, Chasidim who are making connections with the wider world.
>
> (Seidenberg 2006a)

Neo-Hasidism then appeals to different audiences and fuses "secular" and "religious" agendas, sensibilities, and ways of being in the world. To this eclectic audience Seidenberg offers his ecological exposition of Jewish sources, ecological interpretation of Jewish rituals, and systematic eco-theology.

Seidenberg is deeply anchored in the academic discourse of religion and ecology and is most familiar with Christian eco-theology and ecofeminism. His *Kabbalah and Ecology: God's Image in the More-Than-Human World* (2015) exemplifies the hermeneutical style of Jewish eco-theology. Defining himself as a "constructive theologian" for whom "theology is an act of rereading" (Seidenberg 2015, 35), Seidenberg wrote the book "to open up ways of interpreting the tradition that may affect both the way we understand the past and the way we live in the present" (ibid., 36). Based on close reading of rabbinic, philosophic, kabbalistic, and Hasidic sources and the vast modern scholarship on these sources, Seidenberg argues that "creation in the image of God" does not apply merely to human beings but to all other created beings and indeed to the cosmos in its totality. Since the "image of God" (*tzelem Elohim*) is present in all levels of reality, it is wrong to differentiate between God and the world, between the "spiritual" and the "material" dimensions of reality, between human beings and nonhuman beings, or between "nature" and "culture." Borrowing from David Abram's influential work, *The Spell of the Sensuous: Perception and Language in a More-than-Human World* (1996), Seidenberg prefers to refer to "nature," or reality writ large, as "the more-than-the-human world," whose multiple dimensions, both "material" and "spiritual," are intrinsically interconnected.

Seidenberg's eco-theology has two complementary dimensions: the historical-philological dimension and the constructive theological dimension. The former leads him to re-examine the canonic texts to see them with fresh eyes; the latter leads him to tease out the theological and ethical lessons of the canonic texts for contemporary religious environmentalism. In Part I of *Kabbalah and Ecology*, Seidenberg examines rabbinic uses of the term "image of God" to show that the rabbis did not adhere to anthropocentrism. The human being was created "for the sake of all creation" rather than the other way around (Seidenberg 2015, 57). The rabbinic worldview was indeed hierarchical, but the rabbinic "Great Chain of Being" was very different from the medieval one so that it is a mistake to read the rabbinic texts through the prism of medieval philosophers. Seidenberg's eco-theology seeks "to liberate rabbinic theology from medieval theology" (ibid., 65). Medieval Jewish theologians, beginning with Maimonides, identified the "image of God" with intellect or will, contrary to the view of the rabbis for whom the "image of God" pertained to the totality of the embodied, gendered human. Seidenberg argues that both medieval and modern Jewish philosophers misinterpreted rabbinic views because they failed to realize that the "image of God" pertained to the entire human psycho-physical unity rather than to one aspect, such as the intellect or the will. Of all the Jewish philosophers, however, Seidenberg singles out Maimonides as an example to follow because he avoids "the most egregious aspects of medieval and modern anthropocentrism and provides a critical foundation for a Jewish and monotheistic eco-theology" (ibid., 148). According to Seidenberg, Maimonides shows that "there is no inherent contradiction between adopting a thoroughly rationalist interpretation of Judaism and rejecting anthropocentrism" (ibid.). Seidenberg's constructive eco-theology attempts to combine Maimonides's rationalism with the holistic anthropology of Kabbalah, in which "the physical image equals the spiritual image" (ibid., 187).

Seidenberg holds that in Kabbalah "one finds a fertile ground in which to root contemporary Jewish eco-theology" (Seidenberg 2005, 945). Kabbalah offers two fundamental principles as a strong foundation of Jewish eco-theology: the kabbalistic formula "there is no place empty of God" (Tikkuney Zohar 70:122b; 57:91b) means that "the presence of God can be found in every single creature and being" and the formula "the world is blessed because of us" (Zohar 3:144b) means "that the acts of the righteous bring blessing to the whole of creation, to the earth and all its creatures, as well as to God" (Seidenberg 2005, 945). Seidenberg highlights the interpretation of Kabbalah by Moses Cordovero (d. 1570), for whom "all creatures deserve and require respect" and "mercy and respect and bringing beneficence upon every aspect of creation is what it means to become like the Creator" (ibid., 949). For that reason, Seidenberg maintains that Kabbalah could serve

> to develop an ecological ethics including views regarding the ethical treatment and moral standing of other animals and other species, the contemplation of the natural world as a revelation of divine presence, and the extension of the idea of God's image from humanity to creation itself.
> (ibid.)

Kabbalistic texts are "resonant for any theology of Nature that attempts to account for contemporary science," but Seidenberg (like Green and Schachter-Shalomi) does not provide the scientific evidence for this claim beyond referring to David Bohm.

In Part II of *Kabbalah and Ecology* Seidenberg shows that Kabbalah vastly expanded the meaning of "the image of God" by applying it to all levels of hierarchical reality, namely to the "more-than-the-human-world." Because the Sefirot manifest the image of God, since they are paradigms of all existents, all levels of reality similarly reflect the "image of God." Proving this claim through myriad textual examples, Seidenberg concludes that "The 'more-than-human' should include souls, divinity, and worlds, both physical and spiritual, just as it should include all beings, upper and lower" (Seidenberg 2015, 207).

Seidenberg's interpretation of Kabbalah makes him closer to the biocentrism of Schachter-Shalomi than to the mystical panentheism of Green. Precisely because Green defines himself as a "religious humanist," his review of Seidenberg's book was quite critical of Seidenberg's biocentrism because it erases human uniqueness and the boundaries between humans and all other beings (Green 2017). In his response to Green's review, Seidenberg agreed that "it would be wrong to deny that anthropocentrism is a major theme of Jewish thought, but it was the medieval radicals like Saadia Gaon who distorted the tradition to say that this was the truth of our religion" (Seidenberg 2017, 63). Seidenberg reiterates his claim that "Jewish thinkers saw the universe as the greatest image of God and human beings as an image of that greatness … Kabbalah almost universally saw fruit trees, rainbows and certain other natural phenomena as images of God" (ibid., 64). Even if one does not accept the environmental implications of this eco-theology, Seidenberg insists that his reading of kabbalistic sources is textually warranted.

For Seidenberg the concern for equality and justice has a necessary environmental dimension. In his essay on ecology and human rights he states, "The notion of rights, beautiful in isolation, appears to rest on the essential inequality of all other species and non-human individuals, of ecosystems, even of Earth itself, making everything else subservient to human rights" (Seidenberg 2008, 49). In Judaism human uniqueness is expressed in the concept of "image of God," which Seidenberg interprets to mean that "our humanity emerges from our relationship with all life—not just with other human beings—and from our connection to the earth." Thus, if human rights include the right for education or the right for free speech, then "must it not also be a human right to live connected to the world that teaches and nurtures us to become human?" (ibid). Many environmentalists have reflected on the ecological challenge in terms of rights, especially the rights of future generations, but Seidenberg takes a more present-oriented approach by focusing on the rights of one community in Israel, the Bedouins in the Negev. With the move of the Israeli Defense Forces (IDF) to the Negev, Bedouin tribes who have been living there since the 1950s have been removed from certain areas that were now declared closed military zones. Seidenberg's concern for the rights of the Bedouin community in Israel fits with the work of other ecologically minded "spiritual progressives" associated with the community of *Tikkun Magazine*, to which we shall return in Chapter 6.

Seidenberg is a prime example of a Jewish religious environmentalist who integrates Judaica scholarship, Neo-Hasidic eco-theology, and environmental activism into a holistic, Torah-centered way of life. In numerous blogs and short essays Seidenberg has reflected on and explicated the ecological significance of key Jewish rituals and principles, including the Sabbath, the Sabbatical Year (*Shemitah*), the Jubilee, the festival of Sukkot, Tu Bi'Shvat, the fast of Tishah be-Av, the principle of *Bal Tashchit* (Do Not Destroy), and eco-Kashrut. By linking these standard themes of Jewish religious environmentalism to Kabbalah and Hasidism, Seidenberg gives his Neo-Hasidism a decidedly ecological orientation and conversely gives the sacralization of nature a distinctive kabbalistic/Hasidic flavor.

This is most evident in his exposition of Seder Tu Bi'Shvat, a kabbalistic ritual innovated in the 16th century by Isaac Luria and his disciples in Safed. Elaborating the rabbinic celebration of the New Year for trees, the ritual involves eating ten specific fruits of the Land of Israel and drinking four cups of wine (as in the Passover Seder) in a specific order while reciting the appropriate blessings was to bring about the desired spiritual perfection. The kabbalistic ritual was published in the 18th century book *Hemdat Yamim* in the section known as *P'ri Etz Hadar*, and through it was absorbed into Hasidic texts. Today Jewish environmentalists, mainly those associated with the Jewish Renewal Movement and Neo Hasidism, have revived the ritual, recommending it to all those who see "nature (along with the human body) [as] in a sense the most available arena of divine revelation … Nature may serve as a mirror in which all of the mysteries of the concealed Godhead are reflected" (Krassen 2000, 137). In his extended meditation on *P'ri Etz Hadar*, Seidenberg notes that the text "teaches that both trees' fruits and human offspring are expressions of the fertility of the earth, and manifestations of divine blessing." In this text the

> earth is the model of the upper worlds … and the earth is patterned in God's image … In Kabbalah, fruit trees, birds, rainbows, and other natural phenomena are also seen as connectors between upper and lower, manifesting the image of the upper worlds here in the physical earthly realm. Thus, the image of the Tree of Life, which is an image of God, is found in the trees which give us fruit.
>
> (Seidenberg 2006; Cf. Seidenberg 2015, 229)

In short, Seidenberg shows what "Kabbalah for the environmental age" looks like as a Neo-Hasidic variant, but whether this would capture the imagination of Jews who are not associated with Neo Hasidism or with the Jewish Renewal Movement remains to be seen.

God and the World in Jewish Process Thought

Neo-Hasidism, a Jewish variant of contemporary New Spirituality, is a dominant voice within contemporary Jewish environmental spirituality, but this is not to say that Jewish environmental spirituality is necessarily Neo-Hasidic. We began the chapter by explaining how the theistic model supports respects for nature and environmental activism, and we conclude the chapter by focusing on the relational model of God in Jewish Process Thought. Liberal (i.e., non-Orthodox) theologians associated with Reconstructionist, Conservative, or Reform strands of contemporary Judaism have reflected on God and the world not in terms of Neo-Hasidism and Kabbalah but in terms of Christian or post-Christian Process Thought. In Process Thought both God and nature are viewed as processes of becoming that exemplify the same metaphysical laws. Thus, God is an integral aspect of nature and nature is an aspect of God. Reframing Judaism in terms of Process Thought has enabled Jewish process theologians to integrate the "secular" and "religious" aspects of Judaism while responding to the ecological crisis of our time in distinctly Judaic terms.

Process Thought gained prominence between 1930 and 1955 when a group of Protestant theologians in the University of Chicago's Divinity School, who were linked to the Social Gospel Movement, developed various aspects of Alfred North Whitehead's metaphysical cosmology and philosophy of organism articulated in the 1920s. In the second half of the 20th century Process theologians including John B. Cobb, David Ray Griffin, Clark M. Williamson, Catherine Keller, Jay McDaniel, Schubert Ogden, and Marjorie Hewit Suchocki offered complex answers to cultural social, political, religious, and intellectual questions and problems, among them the environmental crisis. The institutional center of Process Thought moved from the University of Chicago to Claremont School of Theology in California, where John B. Cobb and David Griffin taught and the Whitehead Research Project was founded by Ronald Faber, who has greatly contributed to the explication of Process Thought (Faber 2017). In the late 1960s John B. Cobb became acutely aware that traditional Christian theology of personal salvation, justification, sanctification, election, faith, works, and sacraments focuses on what happens after death rather than on the natural world where the human drama takes place. Critiquing the dualism of traditional Christian theology (i.e., the dualism of history and nature, soul and body, humans and nonhumans), Cobb urged fellow Christians to reconnect to the land by "recovering our Jewish heritage … and the Jewish sense that faithfulness to God is bound up with the way we treat the land" (Cobb 1990). In championing the ecological interpretation of Process Thought and the Process style of eco-theology (Cobb 1971; 1980), Cobb advocated "secularizing Christianity." He has inspired many Christian and Jewish eco-theologians, especially ecofeminist theologians such as Carol P. Christ and Judith Plaskow, who will be discussed in the following chapter.

The key insight of Process Thought is that "reality is organic and social, creative and communal—change and interrelatedness are part of its nature—and God is intimately involved in all the events of reality" (Lubarsky 1996, 8). According to Process Thought, God is both present in and transcendent to the world. Process Thought is not pantheistic since it does not equate God and the world, nor does it consider nature to be God. Process Thought is panentheistic because it views God as both in the world and yet distinct from the world; God is viewed as the "soul" or "mind" of the world.

Process Thought begins with experience since experience is the source of knowledge, and even more, it is the source of existence. Because of the irreducible multiplicity of such experiences, Whitehead concluded that reality cannot consist in anything else but the complexity, relativity, relationality, and mutuality of experiences. Change is constant and it pertains not only to the world but also to God who is understood not as immutable Being but as the ever-changing process of becoming. In Process Thought, God is not a supernatural entity that breaks the laws of nature, because God is part of a comprehensive system of reality in which everything is interconnected, and God must live by the same fundamental set of principles describing all that exists. Thus, everything that exists is not inert "stuff" or isolated "things" but interconnected, related processes and events that are experienced and experiences arise out of that which is lived.

Since Process Thought views reality as inherently relational, it reinterprets the notion of divine power. Rejecting the concept of an omnipotent God who acts through coercive power, Process Thought speaks about the "lure" of God's persuasive power. Relational power is persuasive in the sense that in every situation God is the ground of novelty. The God who is always becoming is part of the relational processes in which God experiences every drop of becoming. As such, God shares all experiences of every creature, every pain, joy, hope, despair, failure, and triumph. Thus, the God of Process Thought is not an impartial, disinterested observer of the world but a deeply involved, totally interested, and partial participant of each and every relationship. As perfect relational power, God is love in the fullest sense. God is the only one who has the strength, the ability, to be open to every single experience in the world. God is the only one who can take everything in it, integrate it with God's own infinitely ancient wisdom and create God's self out of that relationship in each moment. God is the only one who can then feedback to every creature in the world a lure and call toward those possibilities that are best for it. All the possibilities are there, good and bad, but they come to us, with God's call toward the better.

Mordecai M. Kaplan: Naturalistic Theism or Transnaturalism

Process Thought has greatly influenced Jewish theology in the 20th century beginning with Mordecai Kaplan, the founder of Reconstructionist Judaism. His famous slogan, "God is the power that makes for salvation," echoes Whitehead's notion of divine lure. Kaplan's naturalistic theism, also referred to as transnaturalism, captures the dilemmas of modern secularism for Jews and especially the secularization of Jewish life in America during the 20th century. Kaplan's naturalist reconstruction of Judaism as a civilization and his emphasis on creativity laid the groundwork for Jewish theologians to be open to Process Thought and to reinterpret Judaism using its categories. Kaplan's project of reconstructing Judaism has something in common with Schachter-Shalomi's Jewish Renewal Movement, because both projects were concerned with the survival of Judaism considering modern secularism, even though Kaplan and Schachter-Shalomi differed in terms of their intellectual orientation and religious temperaments.

Kaplan was not an eco-theologian in the narrow sense defined in this chapter. Nonetheless, Kaplan's naturalistic reconstruction of Judaism reframed the interplay between the "religious" and the "secular" dimensions of Judaism, laying the groundwork for the absorption of Process Thought among Jews.

Interpreting Judaism as a civilization, the overarching theme of Kaplan's rationalist reconstruction of Judaism was the rejection of supernaturalism in favor of religious naturalism, or transnaturalism. Although Kaplan observed Jewish traditional rituals, his radical reinterpretation brought upon him the wrath of Ultra-Orthodox Jews in New York who in 1945 issued a decree of excommunication against him, as the Jews of Amsterdam decreed against Spinoza nearly three hundred years earlier. Kaplan himself, as his authoritative biographer, Mel Scult, states, "despised Spinoza because of his disdain for the Jewish people, [but] he nonetheless believed that Spinoza 'achieved the highest conception of God,' a concept that centered on the naturalist rather than on the supernatural super 'self' of traditional religion" (Scult 2014, 21). Kaplan was especially in agreement with Spinoza's understanding of salvation, since he too "identified salvation with the fulfillment of that which constituted human nature, which for him [Kaplan] consisted in a life of the body and of the emotions in the service and under the control of reason" (ibid., 21).

Unlike Whitehead and his followers, Kaplan was not a precise philosopher or a systematic theologian, yet he wrestled with all the philosophical and theological questions posed by modern secularization, especially for Jews in America. As a Jew brought up in an Orthodox household, Kaplan could not give up the traditional, theistic conception of God, but as a modern Jew who endorsed the natural and social sciences, Kaplan could not subscribe to the notion of a supernatural, providential deity. Although Kaplan became "the most famous naturalist the Jewish people has ever produced," he was no atheist or agnostic. Belief in God informed every aspect of Kaplan's life, but as Kaplan reinterpreted or reconstructed the key concepts, beliefs and practices of Judaism he eliminated any reference to supernaturalism, magic or theurgy characteristic of Kabbalah and Hasidism. In so doing, Kaplan could be considered an example of a theologian who "disenchanted" Judaism, but the story is more complex precisely because Kaplan was very taken by Henry Bergson's philosophy of vitalism, by Whitehead's Process Thought, and by American Pragmatism, especially the predicate theology of John Dewey. Under their influence Kaplan thought about God as a force or power that makes for salvation, luring people to act according to the ideal. Ideals are more than merely physical, but they are not supernatural; ideals point beyond themselves to the larger cosmos as coherent reality, but ideals do not abrogate the laws of nature. For Kaplan, the natural world includes more than the empirical, the factual, and the actual; the natural also includes that which is possible and ideal.

Kaplan has much in common with Process Thought that embraces "both the finite world and divinity, defends freedom, purpose and reason as inherent in the structure of reality, argues against both pure objectivity and pure subjectivity, and upholds the goals of truth, beauty and goodness" (Lubarsky 1996a, 8). Kaplan's Reconstructionist Judaism focuses on immanence without dismissing transcendence, it recognizes the communal aspects of religion but does not reduce God to the ethical or the communal; it asserted belief in an ordered universe where all the parts are interrelated, but it also recognizes God as the source of novelty and creativity. As Kaplan put it succinctly, "the world is more than nature; it is nature with a soul. That soul is God."

Kaplan's naturalist reconstruction of Judaism had to address an obvious question: if Judaism is a civilization, where is its geographic center? The story of the Jewish People is linked to one geographic location, the Land of Israel, but the story of the Jews is also one of dispersion and diasporic existence outside the Land of Israel. As Kaplan was reconstructing Judaism in the 1920s and 1930s, Zionism emerged as an important vehicle for the modernization and secularization of Jews and Judaism. Like other American Zionists, Kaplan supported the building of the Jewish community in Palestine and regarded Zionism "as a way to revitalize the Jewish people and stemming the tide of assimilation" (Scult 1994, 312). The Jewish community in Palestine (Yishuv) (and thereafter the State of Israel) was to be a haven for oppressed Jews, but Kaplan along with most American Jews did not regard the Jews in America as oppressed. Although the Jewish civilization could be experienced most fully and creatively in the Land of Israel, Kaplan's own efforts focused on the reconstruction of ethno-religious Jewish societies in America where Jews could belong to two civilizations simultaneously.

Mordecai Kaplan left an indelible mark on American Jewry both theologically and organizationally, since he innovated the Jewish community center as a "Shul with a school and a pool," a model that guided the organization of the Jewish community in America. Kaplan's pithy slogan, "the past has a vote but not a veto," would also shape the attitude toward the Jewish religious tradition common not only among Reconstructionist Jews but among practically all non-Orthodox Jews in America: one takes the tradition lovingly but not literally; one does not forsake the tradition but also one is not bound to the same conclusions of the Jewish ancestors. That approach has enabled Reconstructionist rabbis to reinterpret the tradition creatively in response to contemporary challenges, including the environmental challenges.

Fred Scherlinder Dobb: Dialogical Spirituality

Rabbi Fred Scherlinder Dobb, the spiritual leader of Adat Shalom Reconstructionist Congregation in Bethesda, Maryland, is an environmental activist that chaired COEJL (the Coalition on the Environment and Jewish Life). For him, Jewish religious environmentalism is part of the ongoing work of reconstruction of Judaism and exploration of the implications of Kaplan's intentional borrowing of the bio/eco metaphor of Judaism as a "Tree of Life." As Dobb explains, Kaplan's "brilliant appellations for God: 'the Life, the Love and the Intelligence of the Universe'" are metaphors that remain "grounded solidly within nature" (Dobb 2012, 200). Rabbi Dobb considers the differentiation between "problematic pantheism" and "kosher panentheism" to be less than compelling, a position similar to that of Schachter-Shalomi. According to him,

> to say that God dwells within the natural world or even that ongoing revelations comes only from nature, is like any metaphor for the Ineffable, partially true ... It may also be a new imperative, one whose time has come, if the universe feels more sacred and more worthy of our love and protection when we assign to it all that previous generations had understood as God.
>
> (ibid., 200–201)

He therefore endorses the metaphorical theology of Sallie McFague that speaks about the world as "God's body" and holds that

> nature itself serves as a meeting-point for religion and science in our day, as a key resource in our efforts to find a theology for an ecological age, a theology at once philo-scientific, meaningfully Jewish and green-oriented.
>
> (ibid., 202)

Committed to the re-sacralization of nature, Rabbi Dobb has helped his community to become a green spiritual center by explaining how the "Torah pushes the Sustainability envelope," how Judaism anchors climate change activity, how Jewish texts and sources offer "wisdom on global climate change," and how "Judaic teachings support ecological consciousness by seeing the world as sacred space." These themes overlap with the Neo-Hasidic re-sacralization of nature proposed by Neo-Hasidic theologians but unlike them the Reconstructionist rabbi is not wedded to Kabbalah and Hasidism.

Bradley Shavit Artson: The God of Becoming and Relationship

Today non-Orthodox rabbis are

> searching for ways to think and talk about God that make sense (that is, that are both internally consistent and consistent with our current best understanding of how reality operates), that resonate with Jewish texts and traditions, and that promote wise and ethical behavior
>
> (Spitzer 2012, 87)

and they find Process Thought to be the most intellectually satisfying framework. As articulated by Whitehead, Process Thought "sees God not only *in* process, but also the Source of novelty, that which makes possible something other than the eternal repetition of that which has already existed" (ibid., 88). Process Thought enables liberal Jewish theologians to reconstruct Judaism in a philosophically sophisticated and scientifically coherent way. Rabbi Bradley Shavit Artson, who holds the Abner and Roslyn Goldstine Dean's Chair of the Ziegler School Rabbinic Studies and serving as Vice President of American Jewish University in Los Angeles, has shown how Judaism can be rendered in terms of Process Thought. Arston is best equipped to do so because he earned a PhD in Process Thought at the Claremont School of Theology, where he engaged the eco-theology of John B. Cobb and became acquainted with the detailed analysis of Process Thought by David Ray Griffin and Ronald Faber, who are on the faculty of Claremont School of Theology, and the ecofeminism of Catherine Keller, the eco-theologian on the faculty of Drew University.

An early supporter of Jewish environmentalism, Rabbi Artson called for fusing Judaism with ecological awareness and investing Jewish and Zionist particularism with humanistic universalism. In his essay "Our Covenant with Stones: A Jewish Ecology of the Earth" Artson argued that the "Jewish assessment of our tradition and our God … should emerge organically from the categories of the Jewish civilization and its writings, rather than from any wholesale adoption of contemporary environmentalism's standards" (Artson 1991–1992 [2001], 161). He went on to say that

> although our tradition may say little directly about air pollution or about the polar ice caps, it does dwell at great length on how Jews are to live with the soil, on the sanctity of the Earth and its produce, the

> holiness of one particular place (*Eretz Yisrael*), and on particular times (*Shabbat, Yovel*, and the *Shemittah*).
>
> (ibid., 161)

For Artson,

> if we begin with intrinsically Jewish categories, we might construct a Jewish ecology, enriching the traditional structure of Judaism with a consciousness of environmental issues rather than simply tailoring Jewish religion to fit within the procrustean bed of a dismembered ecological Judaism.
>
> (ibid., 162)

Exploring the "value concepts" of the Jewish religion as expressed in biblical and rabbinic sources and teasing out their environmental import, Rabbi Artson concludes the programmatic essay saying, "As Jews our loyalty begins with Eretz Yisrael. But we are not only Jews. As human beings, we also owe fealty and love to the entire planet, as well as to that corner of it which is home" (ibid., 168). Zionism, environmentalism, and the sacralization of nature reinforce each other: covenant (*brit*) applies to human relationship with the Earth as well as to the relationship of Jews with the Land of Israel.

Two decades later, Artson fleshed out his message by translating Judaism into the concepts of Process Thought (Artson 2013; Cf., Artson 2015). Intended for Jewish and non-Jewish readers alike, Part I of his first book introduces the reader to the major concepts of Process Thought in comparison to other Western theologies, and Part II shows that "significant strands within Judaism have always been inherently Process" (Artson 2013, xviii).

Applying the insights of Process Thought to key concepts of Jewish ethics—*tzedek* (justice), *shalom* (wholeness/integrity), *berakhah* (blessing/well-being), and *chesed* (loving-kindness) (Artson 2013, 69–73). As he explicates the values of the Jewish tradition it becomes clear that Jewish spirituality is necessarily embodied. If God is "to be found in each and every flower, in every breath we take, in the people who are sitting next to us" it means that there is no such thing as inert matter. Bodies are always soulful, and souls are always embodied. Therefore, we cannot experience life or love, the force that integrates emotions and deeds, in abstraction. Evoking Spinoza, Artson insists that "love must be grounded in the entirety of who we are—our memories, character, experiences, body, temperament and aspiration. Our love must also be grounded in our integrity—in our authentic selves we are in private and in public" (ibid., 71–72). For Jews, this insight has distinctive environmental and Zionist implications. As human beings, Jews must relate "to the holiness of planet Earth" (ibid., 62), but as members of the Jewish People, Jews have a special relationship to one geographic region, the Land of Israel, notwithstanding their dispersion. Artson explains

> to confuse the bounty of our living globally with a rejection of our yearning for Zion and distinctive wonder of a revived Jewish state in our age would be an error of historical proportions and a betrayal of millennia of Jewish aspiration.
>
> (ibid., 108)

In another essay included in an early anthology of Jewish environmental writings, Artson argued that the rabbis broadened "the notion of sanctity to other lands … thereby making holiness 'portable.' Jews could bring their laws, or at least most of them, wherever they went. And therefore, Jews would make any land in which they lived holy" (Artson 2000, 45). By adapting "an attitude of expansive holiness … Israel, the holy people, could encounter holiness anywhere and everywhere. The entire earth offered an encounter with God and godliness" (ibid, 46–47). The early rabbis, as Arston concludes, broadened

> the notion of holiness to include the whole world. Through focusing the blessings for food on 'fruit of the ground,' by adding liturgical petitions on behalf of the weather outside the Land of Israel, by providing for multiple models of land sanctity, the Mishnah and the Talmud established a claim for the holiness of the whole earth, without relinquishing a special Jewish estimation for Israel itself … That schema sanctifies all the earth, summoning us, as Jews and as humans, to enter into a relationship of love and piety with our beleaguered planet. It is incumbent upon us to make of this earth a place where *Shechina*, God's presence, can inhabit, so that the whole world becomes a holey dwelling place, a *mishkan*.
>
> (Artson 2000, 48–49)

Artson's embodied spirituality is environmentally oriented and place centered, recognizing the uniqueness of the Jewish people as a diasporic people with special ties to one territory. In practice, there is little difference between Artson's Process eco-theology and the eco-theology of other Jewish environmentalists who are not immersed in Process Thought.

Sandra B. Lubarsky: Process Thought and Centrality of Beauty

Our discussion of Jewish Process Thought will not be complete without a reference to Sandra B. Lubarsky, a scholar of Process Thought, Jewish philosophy, sustainability studies, and religion and ecology. Lubarsky is an academic and an environmental activist, but she is not a constructive Jewish eco-theologian who writes for practicing Jews nor is she an ordained rabbi who works within the Jewish communal structure. Yet, her work is relevant here because she showed how Process Thought can help Jewish environmentalists and Jewish eco-theologians develop a philosophical framework for integrating the main insights of environmental philosophy: relationality, interconnectedness, and sustainability. Lubarsky too was trained at the Claremont Graduate University with the leading Process philosopher, David Ray Griffin, with whom she co-edited her first book, *Jewish Theology and Process Thought*. The book is important not only because it demonstrates the impact of Process Thought on Jewish theologians but also because it invites Jewish theologians to engage with Process Thought. In her main contribution to the anthology, "Judaism and Process Thought: Between Naturalism and Supernaturalism," Lubarsky argues that "Whitehead's thought has not received serious consideration primarily because it has been equated with the naturalistic theology of Mordecai Kaplan, which was rejected by those who now dominate Jewish theological discussion" (Lubarsky 1996a, 47). Clarifying the differences between Kaplan's naturalism and Whitehead's Process Thought, she shows that

> panentheism is a form of naturalism that affirms the existence of a transcendent being but without belittling or violating the natural order. God acts within the natural world, not because God is nature or nature is God, but because God is not "an exception to all metaphysical principles invoked to save their collapse. He is their chief exemplifications."
>
> (Lubarsky 1996a, 56)

Going beyond the analysis of Jewish philosophers as they relate to Process Thought, Lubarsky's original contribution to Jewish eco-theology is her focus on the concept of beauty in the context of environmental activism. Environmental degradation destroys the inherent beauty of the created world and only when we learn to appreciate beauty can we become environmentally engaged. In her beautiful essays for a book that memorializes the life and work of Douglas Rainsford Thompkins (d. 2015), a businessman, outdoorsman, philanthropist, agriculturalist, and filmmaker, Lubarsky movingly teaches that "beauty has its own independent worth and usefulness. In a world where beauty is not an active ingredient, life sustaining patterns are eroded and processes of degeneration soon inexorable; indeed, they are allowed to be so" (Lubarsky 2017, 61). Since beauty is a "value intrinsic to life [it] requires that we replace mechanism with descriptions of reality that capture the vibrancy, wholeness, and abiding value of the world" (ibid., 66). She concludes her essay by saying that

> it is beauty that holds us to our bodies and the body of the Earth. Beauty is like oxygen, pumping our limbs with vitality, enabling us to seize the moment of a springtime burst of life in the desert. Its importance is cosmological, biological, ecological, psychological, social, and ethical. In calling us to articulate a different metaphysics and a different understanding of how to live in relational proportion to the universe, beauty testifies to its redemptive capacity to save the world.
>
> (ibid., 71)

Lubarsky's nature writings are not commonly cited by Jewish eco-theologians, but she inspired Rabbi Ellen Bernstein, the founder of Shomrei Adamah, who was mentioned in Chapter 2 and to whom we shall return in the following chapter. Lubarsky, who is currently the President of Flagstaff College, but who has spent her entire academic career at Northern Arizona University in Flagstaff, illustrates the difference between being a Jewish scholar who writes on religion and ecology and being a constructive Jewish eco-theologian who is active within the organized Jewish community as an ordained rabbi.

References

Abram, David. 1996. *The Spell of the Sensuous: Perception and Language in a More-Than-Human World*. New York: Vintage Book.

Artson, Bradley Shavi. 1991–1992 [2001]. "Our Covenant with Stones: A Jewish Ecology of Earth." *Conservative Judaism* 44 (1991): 25–35. Reprinted in *Judaism and Environmental Ethics: A Reader*, ed. Martin D. Yaffe, 161–171. Lanham, MD: Lexington Books.

Artson, Bradley Shavi. "Is There Only One Holy Land?" 2000. In *Ecology and Jewish Spirit: Where Nature and the Sacred Meet*, ed. Ellen Bernstein, 41–49. Woodstock, VT: Jewish Lights Publishing.

Artson, Bradley Shavi. 2013. *The God of Becoming and Relationship: The Dynamic Nature of Process Theology*. Woodstock, VT: Jewish Lights Publishing.

Artson, Bradley Shavi. 2015. *Renewing the Process of Creation: A Jewish Integration of Science and Spirit*. Woodstock, VT: Jewish Lights Publishing.

Berger, David. 1993. "Joseph Soloveitchik." In *Interpreters of Judaism in the Late Twentieth Century*, ed. Steven T. Katz, 325–342. Washington, DC: B'nai Brith Books.

Bernstein, Ellen (ed.). 2000. *Ecology and The Jewish Spirit: Where Nature and the Sacred Meet*. Woodstock, VT: Jewish Lights Publishing.

Biale, David. 2011. *Not in the Heavens: The Tradition of Jewish Secular Thought*. Princeton: Princeton University Press.

Clingerman, Forrest. 2014. "Geoengineering, Theology, and the Meaning of Being Human." *Zygon: Journal of Religion and Science* 49 (1): 6–21.

Clingerman, Forrest and Kevin J. O'Brien. 2014. "Playing God: Why Religion Belongs in the Climate Engineering Debate." *Bulletin of the Atomic Scientists*, 70 (3): 27–37. Available at: doi:10.1177/0096340214531181.

Clingerman, Forrest and Kevin J. O'Brien (eds.). 2016. *Theological and Ethical Perspectives on Climate Engineering: Calming the Storm*. Lanham, MD: Lexington Books.

Cobb, John B. 1971 [1995]. *Is It Too Late: A Theology of Ecology*. Minneapolis: Fortress Press.

Cobb, John B. 1980. "Process Theology and Environmental Issues." *The Journal of Religion* 60 (4): 440–458.

Cobb, John B. 1990. "*Theology and Ecology*." Available at: https://www.religion-online.org/article/theology-and-ecology/.

Cooper, John W. 2013. *Pantheism: The Other God of the Philosophers*. Grand Rapid, MI: Baker Academics.

Dalton, Anne Marie and Henry C. Simmons. 2010. *Ecotheology and the Practice of Hope*. Albany: SUNY Press.

Deane-Drummond, Celia. 2008. *Ecotheology*. London: Darton, Longman & Todd.

Deane-Drummond, Celia. 2017. *A Primer in Ecotheology: Theology for a Fragile Earth*. Eugene, OR: Cascade Books.

Dobb, Fred Scherlinder. 2012. "Science and Judaism in Light of Today's Ecological Imperative." *CCAR Journal: The Reform Jewish Quarterly* 59 (1), special issue edited by Philip Cohen and Hava Tirosh-Samuelson, 194–209.

Eaton, Heather. 2007. "The Revolution of Evolution." *Worldviews: Environment, Culture, Religion* 11 (1): 6–31.

Eaton, Heather. 2008. "Responding to Climate Change: Reflections on Scientific Realities, Spiritual Imperatives." *The Ecumenist: A Journal of Theology, Culture and Society* 45 (2): 1–4.

Eaton, Heather. 2011. "An Ecological Imaginary: Evolution and Religion in an Ecological Era." In *Ecological Awareness: Exploring Religion, Ethics, and Aesthetics*, ed. Sigurd Bergmann and Heather Eaton, 7–22. Berlin: LIT Verlag.

Efron John. 2014. *A Chosen Calling: Jews and Science in the Twentieth Century*. Baltimore: John Hopkins University Press.

Efron John. 2007. *Judaism and Science: A Historical Introduction*. Westport, CT: Greenwood Publishing Group.

Elior, Rachel. 1986. "Habad: The Contemplative Ascent to God," in *Jewish Spirituality II*, ed. Arthur Green, 157–205. Minneapolis: Fortress Press.

Faber, Roland. 2017. *The Becoming of God: Process Theology, Philosophy and Multireligious Engagement*. Eugene, OR: Cascade Books.

Gendler, Evert L. 1971. "On the Judaism of Nature." In *The New Jews*, ed. James A. Sleeper and Alan T. Mintz, 233–243. New York: Random House.

Gendler, Evert L. 1989. "The Earth's Covenant." *The Reconstructionist*, 54: 28–31.

Goodman, Lenn E. 2010. *Creation and Evolution*. London and New York: Routledge.

Goodman, Lenn E. 1991. *On Justice: An Essay in Jewish Philosophy*. New Haven: Yale University Press.

Gould, Stephen Jay. 1997. "Non-Overlapping Magisteria." *Natural History* 106: 16–22.

Green, Arthur. 2002. *EHYEH: A Kabbalah for Tomorrow*. Woodstock, VT: Jewish Lights Publications.

Green, Arthur. 2002a. "A Kabbalah for the Environmental Age." In *Judaism and Ecology: Created World and Revealed Word*, ed. Hava Tirosh-Samuelson, 3–16. Cambridge, MA: Harvard University Press.

Green, Arthur. 2003. *Seek My Face: A Jewish Mystical Theology*. Woodstock, VT: Jewish Lights Publishing.

Green, Arthur. 2010. *Radical Judaism: Rethinking God and Tradition*. New Haven: Yale University Press.

Green, Arthur. 2015. *The Heart of the Matter: Studies in Jewish Mysticism and Theology*. Philadelphia: Jewish Publication Society.

Green, Arthur. 2015a. "God's Need for Man: A Unitive Approach to the Writings of Abraham, Joshua Heschel." *Modern Judaism* 35 (3): 247–261.

Green, Arthur. 2017. "'Is Everything in God's Image'? Review of Kabbalah and Ecology: God's Image in the More-Than-Human World, by David M. Seidenberg (New York: Cambridge University Press, 2015)." *Tikkun* 32(1): 58–61.

Green, Arthur and Ariel Evan Mayse (eds.). 2019. *A New Hasidism: Roots*. Philadelphia: Jewish Publication Society; Lincoln: University of Nebraska Press.

Green, Arthur and Ariel Evan Mayse (eds.). 2019a. *A New Hasidism: Branches*. Philadelphia: The Jewish Publication Society; Lincoln: University of Nebraska Press.
Hartman, David. 2001. *Love and Terror in the God Encounter: The Theological Legacy of Rabbi Joseph B. Soloveitchik*. Woodstock, VT: Jewish Lights Publication.
Helfand, Jonathan. 1971. "Ecology and the Jewish Tradition." *Judaism: A Quarterly Journal* 20: 330–335.
Hirsch, Samson Raphael. 1981. *Horeb: A Philosophy of Jewish Laws and Observations*, trans. Isidor Grunfeld. London: Soncino Press.
Idel, Moshe. 1995. *Hasidism: Between Ecstasy and Magic*. Albany, NY: State University of New York Press.
Krassen, Miles. 2000. "Peri Eitz Hadar: A Kabbalist Tu B'Shevat Seder." In *Trees, Earth, and Torah: A Tu B'Shvat Anthology*, ed. Ari Elon, Naomi Mara Hyman, and Arthur Waskow, 135–153. Philadelphia: Jewish Publication Society.
Lamm, Norman. 1971. "Ecology and Jewish Law." In *Faith and Doubt: Studies in Traditional Jewish Thought*, 162–186. New York: Ktav Publishing House.
Lamm, Norman. 1990. *Torah Umadda: the Encounter of Religious Learning and Worldly Knowledge in the Jewish Tradition*. New Milford CT: Magid Books.
Lanza, Robert and Bob Berman. 2009. *Biocentrism: How Life and Consciousness are the Keys to Understandings the True Nature of the University*. Dallas, TX: BenBella Books.
Lichtenstein, Aaron. 1971. "Man and Nature: Social Aspects." In *Judaism in our Modern Society*. Jerusalem: Israeli Ministry of Education.
Lubarsky, Sandra B. 1996. "Introduction." In *Jewish Theology and Process Thought*, ed. Sandra B. Lubarsky and David Ray Griffin, 1–21. Albany NY: SUNY Press.
Lubarsky, Sandra B. 1996a. "Judaism and Process Thought: Between Naturalism and Supernaturalism." In *Jewish Theology and Process Thought*, ed. Sandra B. Lubarsky and David Ray Griffin, 47–58. Albany, NY: SUNY Press.
Lubarsky, Sandra B. 2017. "The Redeeming Power of Beauty." In *On Beauty: Douglas R. Tompkins—Aesthetics and Activism*, ed. Tom Butler and Sandra Lubarsky, 59–71. Berkeley, CA: The David Brower Center.
Magid, Shaul. 2006a. "Jewish Renewal—A New American Religion?" *Tikkun Magazine* 21 (1) (January/February).
Magid, Shaul. 2006b. "The Holocaust and Jewish Renewal: A Theological Response." *Tikkun Magazine* 21 (2) (March/April).
Magid, Shaul. 2006c. "Jewish Renewal, American Spiritualism, and Postmonotheistic Theology." *Tikkun Magazine* 21 (3) (May/June).
Magid, Shaul. 2013. *American Post-Judaism: Identity and Renewal in a Postethnic Society*. Bloomington and Indianapolis: Indiana University Press.
McFague, Sallie. 1982 [1975]. *Metaphorical Theology: Models of God in Religious Language*. Philadelphia: Fortress Press.
McFague, Sallie. 1987. *Models of God: Theology for an Ecological Nuclear Age*. Philadelphia: Fortress Press.
McFague, Sallie. 1993. *The Body of God: An Ecological Theology*. Minneapolis: Fortress Press.
McFague, Sallie. 1997. *Super, Natural Christians: How We Should Love Nature*. Minneapolis: Fortress Press.
Peters, Ted. 2013. "Models of God." In *Models of God and Alternative Ultimate Realities*, ed. Jeanine Diller and Ada Kasher, 43–62. Dordrecht: Springer.

Ravitzky, Aviezer. 1986. "Rabbi J. B. Soloveitchik on Human Knowledge: Between Maimonidean and Neo-Kantian Philosophy." *Modern Judaism* 6 (2): 157–168.

Ravven, Heidi M. and Lenn E. Goodman (eds.). 2002. *Jewish Themes in Spinoza's Philosophy.* Albany: SUNY Press.

Rosenbloom, Noah H. 1976. *Tradition in an Age of Reform: The Religious Philosophy of Samson Raphel Hirsch.* Philadelphia: Jewish Publication Society of America.

Rynhold, Daniel and Michael J. Harris. 2018. *Nietzsche, Soloveitchik, and Contemporary Jewish Philosophy.* Cambridge: Cambridge University Press.

Schachter-Shalomi, Zalman. 2003. *Wrapped in a Holy Flame*, ed. Netanel M. Miles-Yépez. San Francisco: Jossey-Bass.

Schachter-Shalomi, Zalman and Ruth Gan Kagan. 2006. *Jewish Renewal: Integrating Heart and World* (in Hebrew). Tel Aviv: Yediot Aharonot Books and Chemed Books.

Schachter-Shalomi, Zalman with Natanel Miles-Yépez. 2013. *God Hidden: Whereabouts Unknown: An Essay on the "Contraction" of God in Different Jewish Paradigms.* Boulder, CO: Albion–Andalus Inc.

Schwartz, Daniel B. 2015. "'Our Rabbi Baruch': Spinoza and Radical Jewish Enlightenment." In *Secularism in Question: Jews and Judaism in Modern Times*, ed. Ari Joskowitz and Ethan B. Katz, 25–47. Philadelphia: University of Pennsylvania Press.

Schwartz, Eilon. 2001 [1995]. "Judaism and Nature: Theological and Moral Issues to Consider while Renegotiating a Jewish Relationship to the Natural World." *Judaism* 44(4): 437–447, reprinted in *Judaism and Environmental Ethics: A Reader*, ed. Martin D. Yaffe, 267–282. Lanham, MD: Lexington Books.

Scult, Mel. 1994. *Judaism Faces the Twentieth Century: A Biography of Mordecai M. Kaplan.* Detroit: Wayne State University Press.

Scult, Mel. 2014. *The Radical American Judaism of Mordecai M. Kaplan.* Bloomington and Indianapolis: University of Indiana Press.

Seeskin, Kenneth. 2002. "Maimonides, Spinoza and the Problem of Creation." In *Jewish Themes in Spinoza's Philosophy*, ed. Heidi M. Ravven and Lenn M. Goodman, 115–130. Albany, NY: SUNY Press.

Seidenberg, David Mevorach. 2005. "Kabbalah and Eco-theology." *Encyclopedia of Religion and Nature*, ed. Bron R. Taylor, 945–950. London: Continuum.

Seidenberg, David Mevorach. 2006. "*Longer Medication on Pri Etz Hadar.*" NeoHasid.org.

Seidenberg, David Mevorach. 2006a. "Spiritual Journeys." NeoHasid.org.

Seidenberg, David Mevorach. 2008. "Human Rights and Ecology." *Tikkun* 23 (4): 48–50.

Seidenberg, David Mevorach. 2015. *Kabbalah and Ecology: God's Image in the More-than-Human World.* New York: Cambridge University Press.

Seidenberg, David Mevorach. 2017. "Being Here Now: This Creation Is the Divine Image: Response to Arthur Green." *Tikkun* 32 (1): 62–64.

Smith, Steven B. 1997. *Spinoza, Liberalism, and the Jewish Question.* New Haven: Yale University Press.

Soloveitchik, Joseph B. 1965. "The Lonely Man of Faith." *Tradition: Journal of Orthodox Thought* 7 (2): 5–67.

Soloveitchik, Joseph B. 1978. "Majesty and Humility." *Tradition: Journal of Orthodox Though* 17 (2): 25–37.

Soloveitchik, Joseph B. 1983. *Halachic Man*, trans. Lawrence Kaplan. Philadelphia: Jewish Publication Society.

Soloveitchik, Joseph B. 2005. *The Emergence of Ethical Man*, ed. Michael S. Berger. Jersey City, NJ: Ktav Publishing House.
Soloveitchik, Joseph B. 2008. *And From There You Shall Seek*, trans. Naomi Goldblum. New York: Ktav/Toras Horav Foundation.
Spitzer, Toba. Winter 2012. "Why We Need Process Theology." *CCAR Journal: The Reform Jewish Quarterly* 59 (1): 84–95.
Stern, Eliyahu. 2018. *Jewish Materialism: The Intellectual Revolution of the 1870s*. New Haven: Yale University Press.
Tirosh-Samuelson, Hava. 2002. "The Textualization of Nature in Jewish Mysticism." In *Judaism and Ecology: Created World and Revealed Word*, ed. Hava Tirosh-Samuelson, 391–404. Cambridge, MA: Harvard University Press.
Tirosh-Samuelson, Hava. 2018. "Redemptive Activism: Jewish Responses to Climate Change." In *Playing God? Multifaith Responses to the Prospect of Climate Engineering*. A Green Faith Report. Edited by Forrest Clingerman and Gary Gardner.
Tirosh-Samuelson, Hava and Aaron W. Hughes (eds.). 2012–2018. *The Library of Contemporary Jewish Philosophers*. Leiden and Boston: Brill.
Tirosh-Samuelson, Hava. 2012. "Judaism and the Science of Ecology." In *The Routledge Companion of Religion and Science*, ed. James Haag, Gregory R. Peterson, and Michael L. Spezio, 345–355. New York: Routledge.
Troster, Lawrence. 1987/1988. "Love of God and the Anthropic Principle." *Conservative Judaism* 40 (2): 43–51.
Troster, Lawrence. 1997. "From Big Bang to Omega Point: Jewish Response to Recent Theories of Modern Cosmology." *Conservative Judaism* 49 (4): 17–31.
Troster, Lawrence. 2000. "'In Your Goodness, You Renew Creation': The Creation Cycles of the Jewish Liturgy." In *Judaism and Ecology: Where Nature and the Sacred Meet*, ed. Ellen Bernstein, 107–111. Woodstock, VT: Jewish Lights.
Waskow, Arthur. 2013. "Jewish Environmental Ethics: Intertwining *Adam* with *Adamah*." *The Oxford Handbook of Jewish Ethics and Morality*, ed. Elliot N. Dorff and Jonathan K. Crane, 401–418. Oxford: Oxford University Press.
Wiskind-Elper, Ora. 2018. *Hasidic Commentary on the Torah*. London: The Littman Library of Jewish Civilization and Liverpool University Press.
Wolfson, Elliot R. 2002. "Mirror of Nature Reflected in the Symbolism of Medieval Kabbalah." In *Judaism and Ecology: Created World and Revealed Word*, ed. Hava Tirosh-Samuelson, 305–331. Cambridge, MA: Harvard University Press.
Wyschogrod, Michael. 1983 [1989]. *The Body of Faith: God in the People of Israel*. San Francisco: Harper and Row.
Wyschogrod, Michael. 1991 [2001]. "Judaism and the Sanctification of Nature." *The Melton Journal* 24: 5–7. Reprinted in *Jewish Environmental Ethics: A Reader*, ed. Martin D. Yaffe, 289–296. Lanham, MD: Lexington Books.
Zoloth, Laurie. 2023. *Ethics for the Coming Storm: Climate Change and Jewish Thought*. Oxford: Oxford University Press.

4 Earth-Based Judaism

Feminist Variants

Earth-Based Spirituality

Jewish environmental spirituality illustrates how concern for the future of the Earth and concern for the future of Jews and Judaism are closely intertwined. Jewish environmentalists maintain that Judaism's deep ecological insights are most relevant to the ecological crisis, and if humanity adopts them, we could save the world from ecological destruction. They also hold that for Judaism to survive and flourish in the future it must be reinterpreted as a nature religion, namely, one that teaches that God is revealed in the created world, that the worship of God is rooted in the rhythms and cycles of nature, and that Jews should interact with God not only through the study of their sacred texts but through embodied, physical ritual activities outdoors that look at nature itself as sacred.

Bron R. Taylor, a leading scholar of religion and ecology, coined the term "dark green religion" to name a religious ecological consciousness "flowing from a deep sense of belonging to and connectedness in nature, while perceiving the earth and its living systems to be sacred and interconnected" (Taylor 2010, 13). Taylor identifies four types of dark green religion: Spiritual Animism, Naturalistic Animism, Gaian Spirituality, and Gaian Naturalism. Taylor's typology enables him to analyze diverse cultural expression of religious environmentalism from John Muir, Aldo Leopold, and Rachel Carson to Deep Ecology, Earth First!, ecofeminism, and Neo-Paganism. This chapter expands the scope of "dark green religion" by looking at earth-based Judaism, but instead of simply imposing Taylor's categories and analysis on Judaism, this chapter insists that Judaism must be understood on its own terms historically, sociologically, and religiously.

Earth-based Judaism reflects the landscape of American culture since the late 1960s witnessing the rise of New-Age spirituality, Neo-Paganism, feminism, the women spirituality movement, environmentalism, and ecofeminism. The chapter begins by depicting the cultural landscape of the United States during the 1960s and asking how Jews and Judaism fit into it. Jewish women significantly contributed to the cultural trends, without reference to their Jewish identity or their Jewish upbringing. These women could be

DOI: 10.4324/9781351123945-5

labelled as "secular" Jews, but the label is ambiguous because the term 'secular' does not necessarily mean 'nonreligious' or 'a-religious.' A "secular Jew" could be a person who grew up in a traditional Jewish home and identify as a Jew but without observing the rituals of Judaism; or that person could be alienated from the Jewish religious tradition, while translating Jewish religious values into a non-Jewish idioms (e.g., communism or socialism); or that person could be a Jew who deliberately integrates Judaic values with the sensibilities of non-Judaic ideologies (e.g., feminism). Endorsing all sorts of ideologies and belief systems, the modern "secular" Jew is a nonobservant Jew who defines himself ethnically rather than religiously; he/she is a "non-Jewish Jew," as Isaac Deutscher famously put it (Deutscher 1968).

However, in the post-secular world, a non-Judaic ideology can also inspire, energize, or even reframe the religious identity of "secular" Jews. This chapter shows how Jewish acculturation in America brought the revival of Judaism when Jews employed the insights of Neo-Paganism, feminism, the women spirituality movement, and ecofeminism to rethink and reconstruct Judaism. The chapter presents leading individuals and organizations that endorse earth-based spirituality by either reclaiming what has been repressed or marginalized in the long history of Judaism, or by reinterpreting Judaism to show that it expresses ideas and ideals held dear by non-Jewish cultural movements. Earth-based Judaism seeks to reconstruct Judaism to revitalize it and make it meaningful to Jews today, thereby ensuring the survival of Judaism for the future. Expressing a Jewish variant of New-Age spirituality, earth-based Judaism reflects a critique of conventional Jewish practice and an embrace of a less cerebral, less textual, less political way of being Jewish, replacing it with a more embodied, more erotic, more artistic, and more sensual way of being Jewish. Some forms of earth-based Judaism are distinctly feminist, celebrating the Goddess as the symbol of Nature's vitality and creativity and offering new ritual, new liturgy, and new symbolism. Other forms of earth-based Judaism are not centered on the Goddess but are more interested in practicing Judaism outdoors, appreciating the agricultural roots of Jewish festivals, and developing ecological consciousness that recognizes the sacredness of the earth.

Earth-based Judaism expresses the spiritual needs and preferences of some Jews today, and as we indicated in Chapter 2, it is impossible to determine the number of people involved in it. As I do in other chapters, I tell the story of earth-based spirituality by focusing on certain individuals, because spirituality is always personal and idiosyncratic, emerging from the inner life of individuals who seek to find meaning in life, overcome personal traumas and challenges, resolve conflicts, or experience greater level of inner peace. Such self-based spirituality characterizes the New Age, so that earth-based Judaism exemplifies larger cultural trends. My selection of individuals to be featured is not arbitrary but reflects my judgment that they represent a certain cultural identity among contemporary Jews. The chapter focuses on feminist variants of earth-based Judaism, although there are also non-feminist variants.

Earth-based Judaism enables us to understand the interplay of science and religion in the post-secular age by exploring the theme of re-enchantment: in the case of Judaism this means the re-enchantment of the physical world, on the one hand, and the re-enchantment of Judaism, on the other hand.

As we saw in Chapter 1, the trope of "re-enchantment" expresses the notion that modern science caused the disenchantment of nature, which in turn resulted in the environmental crisis we now experience. Nature is "re-enchanted" when it is understood to be the realm of spiritual forces or vitality that modern science cannot explain or quantify. Ecofeminists gave the environmental critique of modern science a feminist twist by showing that modern science imaged nature (which modern science was to master and control) in feminine terms, which, in turn, exacerbated the marginalization, denigration, and control of women in modern society. The feminist critique of modern science and its negative impact on women will be discussed in this chapter, but the promoters of earth-based Judaism are not driven by the problematic relationship between religion and science. Some of the actors presented herein are aware of the negative impact of modern science on the environment and call for new environmental practices (for example, permaculture), but most of the individuals are interested not in the transformation of science but in the transformation of Judaism. It is for that purpose that they reclaim Goddess worship; reinterpret the symbolism of the Shekhinah, the feminine aspect of God in Jewish mysticism; and link worship of the Shekhinah to eco-Kashrut. Earth-based Judaism, then, reflects larger cultural trends as much as the specific concerns of Jews.

New Age, Neo-Paganism, and the Women Spirituality Movement

In chapters 1 and 2, we showed that the countercultural rebellion of the 1960s gave rise to New Age spirituality. As young people in the West critiqued materialistic capitalistic society, they looked for a radical break with modern conditions, seeking a different way of life and a new set of values associated with the mysticism, magic, and shamanism of the pre-modern world. "Self-spirituality," namely, the assumption that "the person is essentially spiritual" (Heelas 1996, 19) unites the various strands and practices of New Age spirituality. Rejecting thc authority of canonic texts, social institutions, or traditional rituals, New Agers regarded the inner realm as the source of authority, authentic vitality, creativity, love, tranquility, and wisdom.

Healing was the task of the New Age movement: healing of the Self, healing of others, and healing of the Earth. In the process of healing a special role is given to the shaman "who is able to mediate between the transpersonal of spirits and gods and the world of humans" (Hanegraaff 1998, 52). To cultivate Shamanic consciousness, one must view nature as sacred.

Recognizing the sacrality of nature generates practices that ensure that nature is respected and dictates how people live, work, and consume. New Age environmental spirituality expressed itself in various ways: some New Agers got involved in environmental organizations, others became involved with local exchanges and barter schemes or green consumption, and still others became eco-warriors, protesting the exploitation of natural resources or the destruction of natural habitats. The spiritual expression of New Age environmentalism is Neo-Paganism, a distinct strand or subculture within the New Age movement. Margot Adler, a journalist who has provided the most comprehensive information about Neo-Paganism in America, succinctly stated that Neo-Paganism is the "spirituality of the ecological movement" (Adler 2006 [1979], 155).

Neo-Paganism is a "nature religion" in the sense that it emphasizes the sacrality of nature and the presence of divinity in all processes of nature (Albanese 1990). Lacking authoritative dogmas or a central bureaucratic structure, Neo-Pagans hold a variety of worldviews, such as polytheism (i.e., they believe in plurality of gods and goddesses), animism (i.e., they believe that the material world is animated by a supernatural spiritual force), or pantheism (i.e., they believe that the natural world is identical with the deity) often endorsing two of these or even all three simultaneously. Neo-Paganism consists of various spiritual paths, all of which understand divinity as an immanent life force or creative energy. To tap into it, Neo-Pagans employ "magic," a range of techniques that mobilize the will, emotion, imagination, intention, and concentration to interact with the material world.

The primary way to practice Neo-Paganism is the festivals, namely, large public gatherings that include "workshops on astrology, tarot cards, mythology, magic, witchcraft, Native American medicine, African drumming, large rituals and late night-bonfires" (Pike 2004, xiii). Neo-Pagan festivals exemplify the re-enchantment of nature characteristic of New-Age ecological spirituality. Not only are Neo-Pagan festivals situated in natural environments (e.g., forests, natural parks, open fields, or deserts), Neo-Pagan rituals express the belief that "the divine is immanent: gods and goddesses live in the earth and trees and are more accessible in forests and farmlands removed from the distraction of urban life (Pike 2004, 22). Because the festival sites are set apart for specific ritual practice, they are considered "sacred," enabling the festivalgoers to connect with the deeper spiritual aspects of the human psyche as well as the spiritual energy of nature. Thus, nature is not merely the background for Neo-Pagan ritualized self-transformation but the very goal of the Neo-Pagan spiritual quest. Pike insightfully argues that Neo-Paganism is "not the triumph of secularization," but rather the "displacement of the sacred into spaces not considered 'religious'" (ibid.).

A prominent strand within contemporary Neo-Paganism is the Wicca cult. Wicca practitioners regard Witchcraft as the "Craft of the Wise" and their wisdom is decidedly experiential and embodied rather than propositional and theoretical. Echoing Jungian psychology, practitioners of Wicca or

Witchcraft consider the gods and the goddesses of ancient mythologies "to represent archetypes, real potencies and potentialities deep within the psyche, which, when allowed to flower, permit us to be more human" (Adler 2006, 26). In this regard, Wicca is a practice that concerns human enhancement and empowerment through communication with the power of nature. While feminist scholars debate the history of Wicca cult, feminist academic scholarship contributed to the intellectual credibility and popular appeal of Goddess religion in the United States (Eller 1995).

Working within the American Academy of Religion, feminist theologians (e.g., Mary Daly, Rosemary Reuther, and Carol P. Christ) argued that women need to adopt Goddess worship because it liberates women from the paradigms of oppressive patriarchy, empowers women to fight for social justice, and legitimizes female power, the female body, and female will. Referred to as "thealogy," Goddess theology was inseparable from nature-religion because "the power of the Goddess is present in the earth and in all beings in the universe" (Christ, 1997, 44). The Goddess speaks to us "through nature, through communities, through dreams and visions expressing her desire to manifest life ever more fully in the world" (Christ 1997, 106). The Goddess is the symbolic expression of the web of life or nature, and the Earth is viewed as the body of God. In Chapter 3 we discussed Sallie McFague's theological metaphoric language of Earth as the body of God. Although a feminist, McFague did not refer to the Goddess because her theology is Christ-centered, and Christ was a divine incarnation in the body of a man. Carol P. Christ critiques McFague, claiming that "not insisting that the earth is (also) the body of the Goddess leaves us free to assume (as our culture has taught us to do) that 'the body of God' is somehow superior to female bodies" (Christ 1997, 90). In contrast, Christ argues that Goddess-language compels women "to keep our focus on the ground beneath our feet and the beings with whom we share this planet" (Christ 1997, 91) and avoid vacuous abstractions about the universe. In her later work (Christ 2003), Christ softened her position by referring to divinity as God/Goddess and by explaining how the belief of the Goddess is philosophically cogent if it is understood in the context of Charles Hartshorne's Process philosophy, whom we mentioned in Chapter 3.

Carol Christ's thealogy is panentheistic: "the earth is holy because it is part of the body of the Goddess/God, but the body of Goddess/God is vaster than our earth" (Christ 2003, 63). She goes on to explain that Goddess/God "is the soul or consciousness and organizing principle of the world body. As the world body changes in large or small ways, so Goddess/God must also experience change" (Christ 2003, 64). The Goddess/God is not an external objective reality, "out there," but a dimension of the human self. According to Christ,

> Goddess/God is 'in' us in the sense of participating in our experiences. We are 'in' Goddess/God insofar as we are part of Goddess/God's body.

> Because we are in Goddess/God body, we can feel Goddess/God as a power within deepest self.
>
> (Christ 2003, 88)

Conversely, Goddess/God feels our pains and delights in our joys. Christ's feminist rendering of Process theology took its cue from the dialogical philosophy of Martin Buber (Christ 1997 1–2, 102, 108; Christ 2003 31–32, 88) to which she was introduced during her graduate studies at Yale by her Jewish friend and later collaborator, Judith Plaskow. We shall return to this point below when we discuss Judith Plaskow's indebtedness to Buber's relational philosophy. For now, let us note that the entry of women into the academic discipline of Religious Studies gave rise not only to systematic feminist theology but also to the fusion of academic feminism and ecological feminism, or ecofeminism.

Ecofeminism emerged in the early 1970s alongside the women's movement and the environmental movement as an internal critique of male-centered environmentalism. Within the environmental movement women felt that not only leadership positions are held exclusively by men; theorizing about nature ignored the distinctive connection between women and nature. In Western culture, women have been identified with "nature" and declared inferior to men, and conversely, nature has been imaged as a "female" and therefore human mastery of nature was taken to be morally justified and even necessary (S. Griffin 1978; Merchant 1980). The representation of nature as a passive, weak, and controllable female is especially strong in scientific literature that has played a normative role in modern society. Ecofeminists argued that nature and women could be liberated if we recognize Gaia as both the earth and the female aspect of the godhead and if we become aware that patriarchy construed women as the Other and men as inherently superior to women.

Ecofeminism is the "movement that sees a connection between the exploitation and degradation of the natural world and the subordination and oppression of women" (Mellor 1997, 1). Today ecofeminism is an international movement, an interdisciplinary academic discourse, and a vision for the future that encompasses both theory and activism. Intellectually diverse and multifaceted, ecofeminism comprises several strands that "reflect differently on the relationship between women and the natural world, and between misogyny and the ecological crisis" (Eaton 2005, 109). Despite their disagreement about the diagnosis of the ecological crisis and the ways to address it, ecofeminism offers a distinct spirituality that perceives and experiences the Earth as sacred, shifts from transcendence to immanence, and reintegrates the sacred and secular dimensions of private and political life. Like the New Age movement and Neo-Paganism, ecofeminism stresses the interconnectedness and interdependence of all existents, the mutually enhancing patterns of existence through the web of life, and the subjectivity of life itself. Because feminism began as a political movement that seeks to change women's social roles, ecofeminism too has a strong political dimension (Spretnak 1982). However, in the past two decades a second generation of ecofeminists has critiqued feminist Goddess worship, encouraging women concerned with sustainability to become more involved in politics (e.g., Sandilands 1999; MacGregor 2006).

Where do Jews and Judaism fit into the cultural landscape of New Age, Neo-Paganism, and ecofeminism? If Judaism emerged in antiquity as a critique of Paganism, can Jews today be Neo-Pagans? Isn't Neo-Paganism a form of idolatry that is theologically unacceptable to Jews? Can Jews be Wiccan Witches? And since Judaism is a patriarchal religion, does Jewish feminism entail a rejection of Judaism or its internal reformation? We will answer these questions by looking first at Jewish women who became most influential in the women's spirituality movement, promoting the revival of the Goddess religion and ecological feminism, or ecofeminism. While these women did not deny their Jewishness, neither did they anchor their ecofeminist sensibilities in Judaism. By contrast, we will examine Jewish feminists who wished to transform Judaism from within, and we will examine the role that nature plays in their reconstruction of Judaic imagery, symbolism, and rituals. We will conclude the chapter by looking at Jewish environmental activists, both women and men, and assess their similarities and differences.

Jews and the Worship of the Goddess

Starhawk: A Neo-Pagan, Feminist Jew

Starhawk is "the single most important reason for this growth of Witchcraft and paganism and for the spread of pagan ideas and practices in the US" (Salomonsen 2002, 9). Starhawk (née Miriam Simos) is a Jew who grew up in a traditional Jewish home in Los Angeles. Starhawk is the most celebrated High Priestess of Wicca and the author of the best-seller *The Spiral Dance: A Rebirth of the Ancient Religion of the Great Goddess*, which has sold over 300,000 copies. Later books sold fewer copies but they remained exceedingly popular worldwide. In addition to many publications, Starhawk founded hundreds of Wicca covens, and she is the leader of Reclaiming, a feminist collective of Wicca practitioners in San Francisco. Consisting of women-identified women and feminist women and men, Reclaiming is one of the many Neo-Pagan groups/communities in the Bay Area that produces its own journal, *Reclaiming Quarterly*. In 1997, after years of discussion, experimentation, and consensus building, the group articulated "Reclaiming Principles of Unity," which express the Reclaiming version of Witchcraft (Starhawk 1999 [1979], 6–7). Starhawk's writings, activities, rituals, poetry, and counseling offer a compelling blend of feminist spirituality, Neo-Pagan Witchcraft, environmentalism, psychotherapy, healing, shamanism, and community building, all of it as powerful rejection of mechanistic, masculinist, and oppressive worldview.

Starhawk is a self-defined Jewish Witch. She acknowledges her Jewish ethnic identity, upbringing, and belonging to the Jewish people, while constructing, articulating, and practicing Witchcraft. Starhawk's Jewishness is often glossed over because it is assumed that Neo-Paganism and Judaism are mutually exclusive. Judaism is monotheistic, whereas paganism is polytheistic, and Judaism regards paganism as idolatry. But the story of Starhawk complicates that assumption. Her Jewish identity was unambiguous. In fact, it was her identification with the historical experience of the Jewish people that led her to feminism, to liberation theology, and to political activism (Starhawk 1987).

In the Introduction to the tenth anniversary of *The Spiral Dance*, Starhawk speaks specifically about her Jewishness:

> Raised Jewish, I had been very religious as a child and had pursued my Jewish education to an advanced level. But as I reached young womanhood in the late sixties, something seemed lacking. The feminist movement had not yet entered its period of resurgence, and I had never heard the word *patriarchy*, but I sensed that the tradition as it stood then was somehow lacking in models for me as a woman and in avenues for the development of female spiritual power ... The Goddess tradition opened up new possibilities. Now my body, in all its femaleness, its breasts, vulva, womb, and menstrual flow, was sacred. The wild power of nature, the intense pleasure of sexual intimacy, took center stage as paths to the sacred instead of being denied, denigrated, or seen as peripheral.
>
> (Starhawk 1999 [1979], 14)

Like many other Jewish women in the late 1960s and early 1970s, Starhawk found traditional Judaism oppressive and unjust to women and was understandably attracted to the egalitarianism, empowerment of women, and freedom heralded by feminism. Some of these Jewish women became the public face of the women's movement and numerous others were in the forefront of academic feminism and the theorizing of feminist spirituality. But Starhawk was unique among these feminists in linking Goddess spirituality and Neo-Pagan rituals to environmentalism and political activism. In the mid-1970s she was deeply involved in the anti-nuclear movement, organizing the protests against building the Diablo Canyon nuclear power plant (Starhawk 1987, 152). Over the years she would "integrate more closely the worship of nature with knowledge that comes from the observation of nature, and ... infuse science, ecological design, and environmental activism with the deep connectedness that comes from acknowledging the sacred" (Starhawk 1999, 9). Starhawk best illustrates how a born Jewess integrates New Age spirituality, Neo-Paganism, feminist spirituality, Goddess religion, Witchcraft, and ecofeminism into a coherent whole.

Starhawk considers nature "the heart of my spirituality," but she grew up "a city girl in San Fernando Valley of Los Angeles" (Starhawk 2004, 4). Her

Jewish identity was forged in Jewish summer camps that "stressed studying Hebrew and saying prayers rather than learning woodcraft" (Starhawk, 2004, 4). Amid the countercultural revolution of 1968 she spent most of the summer "hitchhiking up and down the California coast and camping on beaches" (Starhawk 1999 [1979], 14) and that was her first experience of living in nature and feeling connected to a larger force. At the end of the summer, she enrolled in UCLA and there she began to experiment with witchcraft as a special project in an anthropology course, even though she knew very little about it. She was introduced to Witchcraft when a group of Witches came to her UCLA sorority house to read the so-called "Charge of the Goddess" written by Doreen Valiente in 1950. Starhawk, who already defined herself as a feminist, felt empowered by the Goddess religion and sensed "a natural connection between a movement to empower women and a spiritual tradition based on the Goddess" (Starhawk 1989, 3). It was in Los Angeles that Starhawk, who still went by her given name, Miriam Simos, met Z. Budapest, a lesbian separatist who claimed to inherit Witchcraft traditions from her Hungarian mother (Budapest 1976), and she introduced Starhawk to Witchcraft.

In 1979 Starhawk published her bestselling book, *The Spiral Dance*, and in that year she and a Jewish friend (Diane Baker) started the Reclaiming community, as a neopagan, feminist collective. *The Spiral Dance* was not just a bestseller; it was a catalyst for a movement. From a small feminist collective of about twenty people, Reclaiming would eventually become an international movement with chapters in the United States, Canada, Western Europe, and Australia with many Witch camps. About a thousand people per year are trained in these summer camps, without becoming identifiable members of the San Francisco Reclaiming community (Salomonsen 2002, 43). Although Starhawk is no longer associated with Jewish life, her "enchanted feminism" is part of the story of American Jewry. Her decision to establish a school and summer camp to disseminate the Goddess religion, her preoccupation with the transformation of the Self through therapy, and her focus on embodiment, ritual, and community building all resonate with Jewish sensibilities and are reflective of the Jewish experience in America. Starhawk's earth-based spirituality is indeed Neo-Pagan, but her earth-based spirituality is attractive both to Jewish feminists who are secular humanists as well as to feminists who are steeped in the Jewish tradition.

Covens are meant to empower their members. But what does "empowerment" mean? Starhawk's analysis of power distinguishes between "power-over," "power-from-within," and "power with" as distinct types of consciousness. "Power-over" is the consciousness that "sees the world as an object made up of many separate, isolated parts that have no intrinsic life, awareness, or value. Consciousness is fragmented, disconnected" (Starhawk 1987, 14). By contrast, the consciousness of "power-from-within" "sees the world itself as a living being, made up of dynamic aspects, a world where one thing shape-shifts into another, where there are no solid separations and no simple causes and effects" constitutes "power-from-within" (Starhawk 1987, 15). The bridging of "power-over" and "power-from-within" is "power with" that "embodies a particular consciousness ... that sees the world as a pattern of relationships, but its interest is in how that pattern can be shaped, molded, shifted" (Starhawk 1987, 15).

Starhawk's discussion of power is an insight of a spiritual visionary who wants to heal the world from the damage humans have inflicted on it because of the misguided consciousness of "power over." "Power-from-within" and "power with" are grounded in another source, akin not to violence but to spirit, but contrary to conventional understanding, spirit "is not severed from the material word and from the world of real political and economic struggle" (Starhawk 1987, 16). Contrary to patriarchal religion that "locates God and the sacred outside the world of form and earth and flesh and allows exploitation and destruction of human beings and the earth's resources," spirit is

> a presence immanent in the world, in nature, in the body, in the human community and all its creations ... 'Spirit' can be another way of saying 'immanent value.' When matter is sacred, there is no split, no severing of value from the here and now. Immanent spirit is the ground of the European Goddess tradition, as it is the ground of Native American, African, and other tribal traditions of shamanic practices everywhere.
> (Starhawk 1987, 16–17)

By locating spirit *in* the world and making it synonymous with immanent value, Starhawk spiritualizes the physical world as much as she makes mental processes into forces that can transform nature.

In a more recent publication, *The Earth Path: Grounding Your Spirit in the Rhythms of Nature* (2004) Starhawk reflects on her life's work and integrates the themes that have engaged her for nearly four decades. A mature expression of her "enchanted feminism," this book shows how Goddess religion, postmodern science, feminism, and environmentalism coalesce and cross-fertilize. The book teaches "nature-based rituals to communicate to the living beings around us," and provides practical ways "that we can transform our ways of living to be more in harmony with the earth and to have a sound basis for actions in defense of the earth" (Starhawk 2004, 13). Presumably, the reader who absorbs the book's teachings can conclude that "when science and spirit are reconciled, the world becomes re-enchanted, full of wonder and magic" (ibid).

Starhawk's earth-based spirituality takes nature as its framework and looks to natural systems as a model. A healthy ecosystem is characterized by cooperative and interdependent relationships among its members, and it is diverse and complex enough to be resilient to maintain itself in the face of change. The world is not comprised of physical entities but is rather (an/the) interplay of consciousness, energy, matter, or form; energy is transformed into matter and matter affects consciousness. Magic is the practice that teaches that consciousness can direct energy in both overt and subtle ways and that energy-flow sets the patterns that result in manifestations of form. Energy, Starhawk tells us, "moves in cycles, circles, spirals, vortexes, whirls, pulsations, waves and rhythms" (Starhawk 2004, 33) and these are self-constraining or self-regulating. Gaia is the "planetary organism that includes self-regulating circles" and observing, understanding and using these patterns can help us direct energy more effectively and create healthier systems. A Gaian interpretation of the theory of evolution highlights cooperation and complexity rather than randomness, competition, and predation, and is compatible with creation-centered spirituality articulated by Matthew Fox (1988) or with the earth-centered rituals of Judaism promoted by Arthur Waskow (Starhawk 2004, 43). Thus earth-based spirituality is fully compatible with monotheistic traditions, even though Starhawk regards herself to be a polytheist who worships a plurality of natural forces. She maintains that when people become informed about nature, their earth-based spirituality results in political activism, protecting air, water, or soil and cultivating earth-honoring practices such as permaculture, composting, soil regeneration, and organic agriculture. Ever a teacher, Starhawk offers the reader a bounty of practices on how to practice magic by tapping into the elements of air, fire, water, and earth, as well as how to practice permaculture in order to heal the earth.

Riane Eisler: Goddess Religion and Cultural Transformation

No less an influential ecofeminist is Riane Eisler (b. 1931), a social scientist, human rights attorney, educator, and children's advocate whose first book, *The Chalice and The Blade: Our History, Our Future* (1987), went through twenty-five printings! Eisler too illustrates the complex relationship between Jewishness, Goddess religion, and women's spirituality, but her life's work captures another important aspect of the women spirituality movement: the focus on care and caring. Eisler and Starhawk had much in common (e.g., they were both educated in UCLA, and both promoted the women spirituality movement and the worship of the Goddess) but these two Jewish ecofeminists are quite different from each other. Starhawk grew up in America, Eisler grew up in Europe; Starhawk replaced her Jewish traditional upbringing with Neo-Paganism and Witchcraft, but Eisler continues to pass on the Jewish heritage to her children and practices Judaism at home and in the synagogue (Eisler 2015, 66). Eisler's earth-based spirituality and endorsement of Goddess religion is inseparable from the trauma of the Holocaust.

Eisler's Jewish identity was forged in Vienna, where she witnessed the Nazi conquest of Austria in March 1938 and the destruction of Jewish life and businesses in Kristallnacht. When the Gestapo men came to arrest her father on Nov. 10, 1938, her mother courageously prevented his incarceration in a concentration camp by the force of her personality and the use of bribery

(Eisler 1987, xii; Eisler 1996, 393–94; Eisler 2015, 66). The family fled from Austria to Cuba and from there eventually migrated and settled in the United States. The Holocaust and her own refugee experience led her to ponder these questions:

> why do we hunt and persecute each other? Why is our world so full of man's infamous inhumanity to man—and to woman? How can human beings be so brutal to their own kind? What is it that chronically tilts us toward cruelty rather than kindness, toward war rather than peace, toward destruction rather than actualization?.
>
> (Eisler 1987, xiii)

The answers came from the women's spirituality movement and the vast feminist scholarship in history, archeology, anthropology, and sociology that recovered the ancient Goddess religion.

Eisler's goal has been to bring about cultural transformation: from a society based on domination to a society based on partnership. Eisler's feminist analysis of the human condition is rooted in the distinction between the "dominator model" and the "partnership model" of social organization. The "dominator model" is not part of "human nature" but a relatively recent type of social organization. The vast feminist scholarship on prehistoric societies shows that in the Paleolithic and Neolithic periods, prior to the Indo-European conquests, many societies in Europe, the Balkans, the Near East, and Anatolia were based on the partnership model. As a feminist, Eisler reclaims the prehistoric past in which Goddess worship offers an alternative vision for humanity, one in which "diversity is not equated with either inferiority or superiority" (Eisler 1987, xvii). For Eisler, New Age signals the embrace of partnership, equality, and care, especially care for the most marginal population in the society—women and children (Eisler 2000; 2003; 2008).

The feminist critique of the dominator modelled Eisler to critique ancient Israelite religion for its violence, hierarchy, and authoritarianism. She argues that "the Judeo-Christian Bible is a network of myths and laws designed to impose, maintain and perpetuate a dominator system of social and economic organization" (Eisler 1987, 94). Eisler knows enough about Judaism to concede that "Judaism goes far beyond the Old Testament in its conception of the deity and morality and [that] in the mystical tradition of the Shekhina it actually retained many of the elements of the old worship of the Goddess" (Eisler 1987, 95). Indeed, it was her embrace of ancient Goddess worship that helped her to return to the traditional spirituality of her mother who every Friday eve welcomed the Sabbath by lighting candles, reciting the kabbalistic hymn, Lecha Dodi, and preparing a festive meal that begins with a blessing over challah bread.

Understanding the history of Goddess worship enabled Eisler to reaffirm her Jewishness despite the feminist critique of patriarchal Judaism. As we shall see below, the kabbalistic symbolism of the Shekhinah was central to the spirituality of Jewish feminists, but whether and how this symbolism can support earth-based spirituality is a matter of debate. Eisler is not well versed in postbiblical literary sources of Judaism, including Kabbalah, but it is instructive that the Shekhinah symbolism enabled her to integrate her feminist embrace of Goddess worship with the traditional Judaism of her mother.

Moreover, understanding the history of Goddess worship, so Eisler believes, will bring about the cultural transformation she advocates: the shift from domination to partnership. Unlike Starhawk, Eisler is not an environmental activist, but as an ecofeminist she highlights the causal connection between the rule of men and the exploitation of nature which led to the ecological crisis. For Eisler, prehistoric societies that worshipped the Goddess offer the right model for social organization as well as the right attitude toward nature. She does not romanticize the ancient past and she admits that "many peoples past and present living close to nature have all too often been blindly destructive to their environment" (Eisler 1990, 31). Her "Ecofeminist Manifesto" rejects the binary dualism of "nature" versus "culture" or of technology versus spirituality. Our male-dominated society adopted "scientific objectivity" as the "substitution of detached measuring for an inquiry designed to enhance and advance human evolution" (Eisler 1990, 33). Yet the ethos of scientific detachment has led to indifference to human suffering and to a dualistic mindset that justifies "the domination of culture over nature, of man over woman, of technology over life, and of high priests and other so-called spiritual leaders over 'common' women and men" (Eisler 1990, 33). By contrast, societies based on the "partnership model" place empathy and caring as their key paradigms, replacing insensitivity, cruelty, exploitation, maiming, and killing. A partnership society will emerge only if we devote our energies to the education of children, inculcating the value of care and caring for self, for others, and for Mother Earth.

Eisler ecofeminist spirituality has much in common with feminist ethics of care that promotes compassion, nonviolence, and caring.

Irene Diamond and Gloria Feman Orenstein: Ecofeminist Jewish Academics

Starhawk and Riane Eisler are both academically trained, but their fame and social impact lie outside academia. Within the academy the newly established disciplines of women and gender studies, Jewish studies, and religious studies would play an important role in the feminist rethinking of the academy as well as in the feminist rethinking of Judaism. Gloria Feman Orenstein (b. 1938), an art critic and professor of comparative literature at the University of Southern California, and Irene Diamond (b. 1947), a political scientist who taught at the University of Oregon, co-edited one of the early anthologies of ecofeminist scholarship, *Reweaving the World: The Emergence of*

Ecofeminism (1990), which featured early ecofeminists including Starhawk and Riane Eisler.

Ecofeminism is "the convergence of feminist and ecological movements and a logical extension of each" (Eaton 2005, 30). Ecofeminism is variegated, encompassing many kinds of feminist and ecological viewpoints. In the amalgam of perspectives and strategies, spirituality may be "central, peripheral or irrelevant for individuals and groups" (ibid., 31), and ecological issues may be either "light green" (i.e., still within the anthropocentric paradigm) or "dark green" (i.e., concerned with the well-being of the whole earth and the flourishing of all life forms in interconnected ecosystems). Orenstein and Diamond's edited volume illustrated how ecofeminists sought to bring about "changes in consciousness, political action and spiritual practice" (Diamond and Orenstein 1990, vii) and promoted Goddess religion as a critique of contemporary science and technology and the capitalistic system that made them possible. Yet readers would never guess that both co-editors are Jews, since their Jewish identity was not stated and the analysis is secular, with no appeal to Jewish sources. Over time, however, these two scholars made deliberate efforts to integrate their ecofeminist and Jewish identities. This is especially clear in the case of Irene Diamond, an influential ecofeminist who criticized contemporary science and technology's attempts to control women's bodies.

In *Fertile Ground: Women, Earth, and Ecofeminism* (1994), to which Starhawk wrote an enthusiastic Foreword, Diamond drew on the Foucauldian analysis of power/knowledge and ecological feminist work to challenge the feminist assumption that "women's freedom resides in our gaining control over our bodies and sexuality" (Diamond, 2004, 11). Criticizing radical feminists who hailed the use of assisted reproductive technologies and genetic engineering, Diamond claimed that "from an ecofeminist perspective, this emphasis on technological mastery is particularly disturbing" (Diamond 1994, 92). This critique put Diamond at odds with her friends in the environmental and feminist movements who viewed the world through the concept of population control. But Diamond was unaware that ecofeminist critique is relevant to her Jewishness. As she attests: "until a Holocaust Remembrance Day in 1995 I had no sense that my views had anything to do with my Jewish background … I had no knowledge that Nazi law had attempted to prevent all Jewish women from birthing" (Diamond 2004, 11). It was that commemorative ceremony that made her reassert her Jewishness and begin to integrate her Jewish and ecofeminist identities.

By the mid-1990s Jewish environmental movement began to be present in organized Jewish life after the Coalition for Environment and Jewish Life (COEJL) was founded in 1993. In 1995 Diamond participated in two small conferences—one retreat organized by the COEJL held at the B'nai Brith Pearlman Camp in northeastern Pennsylvania and another at an ecofeminist encampment she co-organized with Carolyn Merchant, a leading ecofeminist, at an intentional community in the forests of southwestern Oregon. In her report of these events Diamond documents her encounter with

> a new generation of rabbinical students, rabbis, and Jewish educators who were interested in developing a methodology for Jewish ecological work that would go beyond selectively picking out the texts and holidays within the tradition that could be used to support a Jewish ecological sensibility. Several of the male rabbinic students specifically said that they felt it would be necessary to develop a Jewish ecological methodology akin to feminist methodology.
>
> (Diamond 1995, 86–87)

However, she was surprised that the event did not include Jewish women scholars interested in feminism or Jewish feminism. The discussion of Judaism in the first retreat led her "to see Jews as indigenous people that have managed to survive into the modern world," and made her feel that she "returned to my tribe" (Diamond, 1995, 87). In the second retreat, "participants who identified with their Jewish backgrounds learned that it was possible to practice and think openly as Jews without creating sharp lines of demarcation between religious and secular activities or having to shield observant activities from view" (Diamond 1995, 88). In 1997 Diamond decided to "to open up a Jewish discussion of human desiring bodies on relations to the earth by exploring the possibilities of Jewish ecofeminist practice" (Diamond 2004, 8). Two years later, she collaborated with David Seidenberg, a leading Jewish eco-theologian and environmental activist we discussed in Chapter 3 (Diamond and Seidenberg 1999). In 2003 they presented a joint paper at the Association for Jewish Studies in a session on ecofeminism. It was very poorly attended, indicating that scholars of Judaism are uninterested in ecofeminism.

For Diamond the integration of Jewishness and ecofeminism entails rejecting the "mechanistic understandings of the natural world that obscure the novelty and wonder of continual creation intrinsic to life." What she cares about most is respect toward "ancient cosmologies and emergent forms of non-mechanistic science that attest to the dynamic, non-linear flux, abundance and exuberance of the natural world, it foregrounds the errors of modernist approaches to history, politics and science" (Diamond 2004, 13). As a socially engaged, left-leaning political theorist, Diamond absorbs the Jews, to which she refers "a strange and wandering tribe," into the category of indigenous tribes whose culture we must protect from extinction. Like other indigenous cultures, Jewish observance "acknowledges the power of birth, magic, and a living and unfolding cosmos, and the belief in human agency, compassion for the oppressed and continual learning and questioning that permitted the flourishing of Western science, tolerance of human rights." For her, Shabbat observance, the Shekhinah symbolism, and Rosh Hodesh rituals enable ecofeminist Jews to honor human embodiment and create the ecological awareness necessary to protect the earth from environmental degradation.

Diamond's co-editor, Gloria Orenstein, also promotes the veneration of the Goddess, but with a focus on art, literature, psychology, and shamanism. In her *The Reflowering of the Goddess* (1990) Orenstein focused on

> he suppression of the Goddess and the impact of this on the creative imaginations of women in the arts historically … how the stifling of the female image of a cosmic Creator contributed to the madness and suicide of many women artists and writers.
>
> (Orenstein 1990, xvii)

Citing extensive feminist literature on the pre-patriarchal Goddess worship she sought "to reclaim and revalue the original image of the Creator as female and to reconceive the function of art as shamanic and magical" (Orenstein 1990, 2). Since the Goddess has always symbolized creation, the Goddess, by extension, also represents artistic creation. In the Goddess imagery

> there is no distinction drawn between woman as creator of nature and woman as creator of culture. The symbol of a woman Creator bringing forth both life and art, nature and culture, seems to represent the most poignant, powerful and concise way possible, the point at which many of the themes espoused by contemporary feminism converge.
>
> (Orenstein 1990, 6)

Literature and art produced by feminist matristic artists seek to transform the world, correcting patriarchy's abuses and injustices.

Relying on the work of Monica Sjöö and Barbara Mor (1987), Orenstein promotes a new global,

> organic spirituality that belongs innately to all of us, as children of the earth (Sjöö and Mor 1987, 421). While sharing the outlook of other ecofeminists who reclaim reverence of the Goddess, Orenstein admits that ecofeminists do not agree among themselves and do not necessarily require either a Goddess as a symbol or hearken back to the Neolithic for inspiration. For Orenstein, the key insight of earth-based ecofeminist spirituality is the notion that "spirit does not exist apart from the entire cosmos but rather that the Earth itself is sacred and that since we are an interdependent part of it, we are sacred too".
>
> (Orenstein 1990, 24)

Orenstein's analysis of Gertrud Stein's work (Orenstein 1990, 50–56), and her discussion of female Jewish artists in the United States and in Israel (Orenstein 2007), show how Jewish women reclaimed ancient female wisdom, female cosmogonic myths, and nature spirituality in order to celebrate female creativity more than heal the damaged Earth.

The Jewish ecofeminists discussed above were at the forefront of the women's spirituality movement and ecofeminism in America. While all four revere the Goddess, they did so in different ways. It is difficult to interpret these Jewish ecofeminists using the simplistic dichotomy between "secular" and "religious" domains. They do not deny their Jewish identity and their connection to Jewish culture, but (except for Irene Diamond) they are not interested in transforming Jewish communal life. To fully grasp the feminist variant of earth-based Judaism, we need to focus on Jewish feminists who deliberately seek to reform or reconstruct Judaism by highlighting nature imagery in God-talk, reclaiming the Shekhinah as a Jewish version of the Goddess, by retelling the sacred narratives of Judaism and constructing new nature-based rituals, and by resurrecting the priestess, an ancient model of female leadership.

Jewish Feminist Environmental Spirituality

Although Jewish women have contributed to ecofeminism, the vast theoretical literature generated by ecofeminists is barely cited by Jewish environmentalists and the revival of the Goddess religion has not been universally endorsed by Jewish environmental activists or by scholars of Judaism, notwithstanding Irene Diamond's efforts. Nonetheless, the Goddess discourse, the celebration of the connection between women and nature, and the reinterpretation of Judaism through the lens of earth-based spirituality have made a difference in contemporary Jewish life because of Jewish feminism.

Jewish feminism has profoundly transformed the "leadership, organizations, practices and beliefs of American Jews" (Prell 2007, 1). Emerging as an outgrowth of the countercultural revolution and Second Wave feminism of the 1960s, in which Jews were highly represented as both leaders and rank and file, Jewish feminism exhibits two complimentary trajectories: on the one hand, Jewish feminism adopted the claims, sensibilities, and demands of feminism for inclusion, justice, and equality for women, applying them to the beliefs and practices of Judaism. On the other hand, Jewish feminism recognized that feminism and the women's spirituality movement do not address their distinctive Jewishness; in fact, Jewish feminists have argued that the feminist critique of patriarchy is often tinged with anti-Semitism or anti-Judaism (Parker 1979; Pogrebin 1982; Plaskow 1991a). Instead of addressing their feminist needs within the framework of a global feminist movement, Jewish feminism sought to reinterpret Judaism in egalitarian, non-hierarchical, and just ways to transform the status of women in Jewish society and culture.

Feminism was crucial to the emergence of earth-based Judaism among contemporary Jews. The women's movement created a unique context for collaboration between Jewish and non-Jewish women who were united in their critique of patriarchy and who devoted their life to the liberation of women from male oppression. The collaboration between Jewish and Christian feminist intellectuals took place in academic settings that witnessed the proliferation of women and gender studies, Jewish studies, and religious studies. These academic sub-disciplines cross-fertilized, transforming religious practices of Jews and Christians. Outside the academy, earth-based Judaism was expressed in feminist and non-feminist ways.

Jewish feminist spirituality was most attentive to the ecological crisis and highlighted the connection between the oppression of women in Jewish patriarchy and the massive environmental destruction of life on earth. Jewish feminists sought to anchor Judaism in nature in four main ways: by changing the male-centered God-language of traditional Judaism; by worshipping the Shekhinah, the feminine aspect of God who is accessed through numerous nature symbols; by celebrating female embodiment in new rituals; and by reviving the priestess as a model for female leadership. Three Jewish feminists—Judith Plaskow Rabbi Lynn Gottlieb, and Rabbi Jill Hammer—illustrate how Jewish feminist environmental spirituality is rooted in the religious sources, rituals, and rhythms of Judaism while seeking to transform the secular world of contemporary Jews.

Judith Plaskow: Feminist Jewish Theology and Ecological Consciousness

Judith Plaskow (b. 1947) is a key figure in the emergence of Jewish feminist spirituality and systematic feminist Jewish theology. Brought up in a Reform congregation on Long Island, Plaskow was profoundly transformed by feminism and the women spirituality movement. She entered Yale Divinity School in 1968, where she studied systematic (Christian) theology and earned a PhD in 1975 with a dissertation on Reinhold Niebuhr and Paul Tillich (Plaskow 1979). At Yale she started her life-long friendship with Carol P. Christ, a leader of the women spirituality movement and a feminist ecotheologian, as we have mentioned above. In retrospect Plaskow refers to her encounter with feminism as a "conversion," an intense personal experience that shifts the way one seeks one's place in the world. Married at the time to Robert Goldenberg, a rabbi and scholar of rabbinic Judaism, she realized that as a woman she occupies a secondary, marginalized place within Judaism.

Feminist self-awareness brought her to realize that in traditional synagogue services "her presence was irrelevant to the purpose for which we have gathered" (Plaskow 1991, xi). As a woman she was not counted in the prayer quorum (*minyan*), she was not called to the Torah, and she was not involved in the interpretation of the received tradition. Refusing to give up Judaism for feminism or live as a woman whose identity is split between the two, Plaskow would gradually seek to affirm herself as a "Jewish feminist and a feminist Jew" (ibid.). That project took place within the contours of the discipline of religious studies and its leading professional organization, the American Academy of Religion, eventually becoming the President of the organization (1997–1998). However, the more Plaskow engaged other feminist academics, both Christian and post-Christian, the more she became aware of her own Jewishness and her need to reconcile her Jewish and feminist identities. In part that awareness grew out of the realization of the "persistence of anti-Semitic stereotypes in feminist literature" and the need to confront "Christian anti-Judaism" (Plaskow 2005, 100–101), but it also grew out of the need to fully express herself.

Plaskow forged her feminist interpretation of Judaism in a small "feminist collective," B'not Esh (Daughters of Fire). Founded in 1981, the initial group of sixteen women was a women-only Havurah that studied, prayed, and talked together, offering a model for the reconstruction of Judaism (Ackles-berg 1986), although in form the group was not so different from a Neo-Pagan coven or a feminist consciousness-raising group. Members of B'not Esh contributed greatly to Jewish feminism in America and to the impact of feminism on the discipline of Jewish Studies. The lived experience in the feminist collective led to personal life changes when she came out as a lesbian and her marriage ended in divorce in 1984. Since then, Plaskow's theological reconstruction of Judaism moved in three distinctive but complementary directions: first, she articulated egalitarian God-language rooted in women's experience; second, she critiqued the injustice imposed on gay, lesbian, bisexual, and transgender people within traditional Judaism, which practices compulsory heterosexuality; and third, she applied the critique of power not only to traditional Judaism but also to the policies of the State of Israel, especially in regard to the Palestinians. Like other members of B'not Esh, which still exists, Plaskow is aligned with the Jewish Liberal Left and endorses its social and political agenda.

In her path-breaking and most influential *Standing Again at Sinai* (1991), Plaskow offered systematic feminist Jewish theology that explicitly connects Jewish feminism with concern for the well-being of earth. She boldly asserts that "the devastation of the earth" and "the ugliness and suffering in society" are the direct outcome of our devaluation of human embodiment, especially the bodies of women: "To value our bodies means to value and care for the earth of which they are part; otherwise, this valuing has no relation to a material base that supports and sustains it"(Plaskow 1991, 231).

She adopts the "hermeneutics of suspicion" that subjects all expressions of patriarchal culture, especially canonic texts, to close analysis in order to show that even when these texts claim to speak truth about God, the world, and human beings, they in fact speak only about men and for men, because these truth claims were made by men. She exposes the exclusion of women from the formation of the tradition and instead demands inclusion by offering alternative readings of the canonic texts. To cultivate ecological consciousness and politics, as "an active awareness of, and responsibility to, the complex web of life" (Plaskow 1991, 231), it is necessary to liberate ourselves from patterns of domination that govern our God-language. The Abrahamic traditions use hierarchical God-language referring to God as "king," "lord," "father," or "husband." Androcentric God-language is wrong because its understanding of divine power as "power over," inevitably marginalizes women and leads to their oppression (Plaskow 1991, 130). Whether divine power is asserted over nature, over history, over other gods, or over humans, this understanding of power is itself the cause of injustice because it is perpetuated in human "schemes of dominance." To construct a just and egalitarian Judaism, Plaskow insists that we develop non-hierarchical, egalitarian, and relational God-language that expresses fluidity, multiplicity, and movement in God as the source of all being. To liberate women from male domination and liberate the earth from human oppression we must liberate God-language from patterns of domination.

Plaskow's feminist liberation theology follows three paths. First, she endorsed the practice of substituting anthropocentric and androcentric to the gender-neutral terms derived from nature, using nature-based metaphors that "teach us the intrinsic value of this wider web of being in which we dwell" (Plaskow 1991, 165). While nature metaphors are theologically necessary, as Sallie McFague has argued, they cannot address the emotional need of humans to think about God anthropomorphically since we are commanded to imitate God's ways.

Plaskow's second remedy is to retain anthropomorphic language but select egalitarian metaphors, speaking about God as "lover, friend, companion, and cocreator" (Plaskow 1991, 164). These egalitarian metaphors reflect Plaskow's indebtedness to the dialogical philosophy of Martin Buber, who insists that divine-human relations are not governed by law. Law belongs to the functional and instrumental I-It relations whereas true dialogue between the human "I" and the "Eternal Thou" is direct, unmediated, and in principle not amenable to systematization. When we think about divine–human relations in dialogical terms so that God becomes "partner, nourisher, and goad," we realize that God "is with us instead of over us, a partner in dialogue who ever and again summons us to responsible action" (Plaskow 1991, 164).

Agreeing that women need Goddess language to overthrow hierarchical ways of thinking, Plaskow argues that when Jews begin to refer to God as "She," "Queen of the Universe," or "Goddess" they will be more aware of the inadequacy of human language to express or fathom the infinite depth and complexity of divine reality. Plaskow supports other Jewish feminists who express their feminist spirituality by worshipping the Shekhinah, the feminine aspect of God. She also encouraged her feminist cohorts to create their own homilies (i.e., midrashim), construct new rituals for women, and struggle for the rights of the oppressed.

Plaskow's egalitarian, communal, embodied and ecological vision of Judaism links the lives of women, the fate of the earth, and the pursuit of justice to each other, urging us to struggle to act responsively and responsibility in history and in the community. Her inclusive social vision seeks to carve space for sexually marginalized, oppressed, or denigrated groups in Jewish society. She has been at the forefront of the fight to admit gays and lesbians to rabbinic seminaries but going beyond it, she has urged Jews to see human sexuality as more diverse, fluid, and porous than the binary distinctions between "heterosexual" and "homosexuals" that many men and women allow for (Plaskow 2005, 178–192). Her commitment to social justice is complemented by her commitment to ecological justice. Plaskow endorses the expanded concept of eco-Kashrut, stating that "concern for protecting our bodies might take the form of prohibiting foods that are grown with pesticides or that contain carcinogens or hormones" (Plaskow 1991, 236). Plaskow's ecologically minded feminist Judaism is a program for "the repair of the world" (*Tikkun Olam*) but as an academic she did not put it into action. Other Jewish feminists, especially those ordained as rabbis, followed the theological path blazed by Plaskow and translated her ideas into a feminist environmental spirituality to live by.

Lynn Gottlieb: Shekhinah Worship and Eco-Kashrut

Rabbi Lynn Gottlieb (b. 1950) is a Jewish feminist whose creative work as a storyteller, percussionist, ceremonial/visual artist, human rights advocate, peace educator, and author fuses feminism, Goddess worship, Judaism, and environmentalism. Growing up in a Zionist home in Bethlehem, Pennsylvania, and even spending one year of high school (1966) in Israel, she was at the forefront of the Jewish feminist movement in the United States. Like Plaskow she was associated with leaders of the women spirituality movement and was deeply engaged with the early phases of Jewish feminism. Determined to become a rabbi, she began her studies at the Hebrew Union College (HUC) in 1972, soon after the Reform Movement ordained the first female rabbi, Sally Priesand. Gottlieb eventually left the program at HUC but could not continue her rabbinic training at the Jewish Theological Seminary (JTS) because the Conservative Movement did not ordain women. However, she would go on to study privately with several Orthodox, Conservative, and Reform rabbis and scholars.

In 1973 Gottlieb became the spiritual leader of Temple Beth Or of the Deaf in Queens, while she was still a student at JTS. In 1977 Rabbi Wolfe Kelman, the executive vice president for the Conservative Movement's Rabbinical Assembly announced that he would ordain her privately if JTS continues to deny women ordination. This statement might have accelerated the debate about rabbinic ordination in the Conservative Movement, which finally approved the ordination of women in 1983. Prior to that, in 1981, Gottlieb was privately ordained by rabbis Schachter-Shalomi, Gendler, and Shlomo Carlebach, becoming the first ordained rabbi of the Jewish Renewal Movement. She would carry out her unique spiritual work as the spiritual leader of Nahalat Shalom in Albuquerque, New Mexico. Over the years, Rabbi Gottlieb would become a well-known voice of Jewish spiritual progressives. Her first book, *She Who Dwells Within: A Feminist Vision of Renewal Judaism* (1995) illustrates the fusion of Judaism, Goddess religion, and Jewish environmentalism.

All Jewish feminists were keenly aware that rabbinic Judaism refers to the divine presence of God as Shekhinah, meaning "dwelling." The term Shekhinah comes from the root *sh-kh-n*, meaning "to dwell" or "to abide," and it appears first in Targum Onkelos, the Aramaic translation of the Bible, where it is used to refer to God in non-anthropomorphic language, alternating with the terms *kavod* (glory) or *memra* (i.e., Word or Logos). Rabbinic sources use the term Shekhinah interchangeably with YHWH and Elohim to convey closeness to God or divine immanence (Abelson 1969 [1912]). Although the Hebrew word is grammatically feminine, rabbinic sources do not characterize the Shekhinah as female, but the qualities of God that are expressed in the concept of the Shekhinah, such as "love, charity, compassion, fairness, equity, healing, justice, power, sharing and connectedness" (Novick 2008, 7) are qualities that ancient civilizations associated with Goddesses. The full feminization of the Shekhinah took place in medieval Kabbalah and continued in modern Hasidism (Seidenberg 2019).

The kabbalistic symbolism of the Shekhinah introduced two novel interpretations of divine immanence: the Shekhinah was treated as a female deity and was identified with the national symbol of the Congregation of Israel (Knesset Israel). Scholars of Kabbalah have long debated the meaning of the Shekhinah symbolism in Kabbalah and their differences cannot be easily reconciled. Scholarly debates notwithstanding, Jewish feminists found Shekhinah symbolism especially attractive because it offered a ready-made female God-language that could anchor the project of reconstructing an egalitarian Judaism. Kabbalah provided a reservoir of images for a Jewish Goddess, if one was looking for it, whether one was versed in the intricacies of Kabbalah or not.

The Jewish Renewal Movement (both the leaders as well as the rank and file) was particularly open to reinterpret Shekhinah symbolism for the sake of reconstructing contemporary Judaism. Schachter-Shalomi has promoted the Shekhinah as "eternal feminine," a "mythical figure expressing a feminine essence that is noncognitive and mysterious" (Weissler 2007, 61). Similarly, his close friend, Arthur Green, has maintained that the Shekhinah "who is described as daughter, bride, mother, moon, sea, faith, wisdom, speech, and a myriad of other figures," could anchor "a truly feminine, a truly Jewish spirituality" (Green 1983 [1995], 259). And Arthur Waskow has similarly taught that "there are feminine fragments in Jewish tradition that could transform our view of Torah and society by uncovering women's spiritual experience" (Waskow 1983).

Within the Jewish Renewal Movement (and to a lesser extent in Reform Judaism) the worship of the Shekhinah led feminists to compose new prayers, reinterpret biblical and rabbinic sources, and propose new rituals that accentuate the power of the Shekhinah and, by extension, empower the women who connect with Her. Since the 1970s Rosh Hodesh groups created new communities for women, that celebrated the link between women and the cycles of nature and the biological dissimilarity between men and women (Myers 2007). The centrality of study in these activities also enabled Jewish feminists to study rabbinic texts and reinterpret them in more egalitarian ways or offer new liturgies to express women spirituality. By celebrating their own distinctive biology and linking it to the web of life more broadly, Jewish feminists found ways to engage in spiritual activity from which they were previously excluded presumably because of their inherent carnality.

Most relevant to our inquiry, Gottlieb translates the Shekhinah symbolism into a full program of eco-Kashrut, which for her means "loving the Earth's body" (Gottlieb 1995, 171). Following the footsteps of her two main mentors, Rabbis Schachter-Shalomi and Gendler, she boldly states: "Our growing environmental awareness has added a new dictate: Thou shalt be ecological." Rabbi Gottlieb warns us: "if we do not honor the complexity and diversity of nature, we are doomed to destroy nature and ourselves" (Gottlieb 1995, 173). Eco-Kashrut, Gottlieb explains, can be adopted by Jews as well as by non-Jews, and it consists of seven "sacred obligations, or mitzvot": "Honor the Mother; Offer Hospitality; Save Life; Do not be Wasteful; Be Kind to Living Creatures; Do not Oppress Others Economically; Keep the Sabbath" (Gottlieb 1995, 173).

If one follows this program of eco-Kashrut, one is committed to saving all forms of life, and not only human life. Eco-Kashrut makes any product that endangers life of earth; that contributes to deforestation; that depletes the ozone layer; poisons the water, the air, and the soil; or that abets the proliferation of war and violence against humans and animals unfit for use (Gottlieb 1995, 181). Commitment to eco-Kashrut requires that we avoid economic oppression, especially of women and children, and that we "combine our resources and focus our energies to serve those women who are marginalized by the mainstream community and require financial assistance" (Gottlieb 1995, 189). All the work on behalf of the earth and all its inhabitants is brought to rest on the Sabbath "that enjoins us to release people, animals and the earth from their utilitarian tasks in the world of labor and commerce and to allow all beings leisure time" (Gottlieb 1995, 190).

Her exposition of eco-Kashrut is feminist environmental spirituality par-excellence. It is shared by the community of Jewish spiritual progressives, associated with *Tikkun Magazine*, as well as by rabbis and spiritual leaders associated with the Jewish Renewal Movement. ALEPH: Alliance for Jewish Renewal holds biennial weeklong gatherings of Renewal Jews, known as Kallah, in which feminism, environmentalism, and Goddess worship are creatively intertwined. In 1990 two rabbis associated with ALEPH: Alliance of Jewish Renewal, Jeff Roth and Joanna Katz, established Elat Chayyim (literally, Goddess of Life, and as noted in Chapter 2, later to merge with Hazon and which today is part of Adamah: People, Planet, Purpose) to offer weekly spiritual retreats in the Catskill mountains of New York in which classes combined vegetarianism, yoga and meditation, and unique prayer experience.

Jill Hammer: Goddess Worship and Female Empowerment

Jewish feminism has greatly evolved since the 1970s, touching the life of all Jews in America, be they liberal/progressive, traditional/Orthodox, religious, or secular. In particular, the ordination of women as rabbis, the large number of women in Jewish studies, and the inclusion of women in the study and practice of Jewish Law have led to the dissemination of feminist values, goals, and sensibilities among American Jews. Rabbi Jill Hammer, who was ordained by the Jewish Theological Seminary, illustrates how earth-based spirituality goes beyond Conservative Judaism.

With Taya Shere, Rabbi Hammer established the (now defunct) Kohenet Hebrew Priestess Institute, which was affiliated with the Jewish Renewal Movement. That project engaged in the reclaiming of the Hebrew Goddess as well as revitalizing the multifaceted role of the ancient Hebrew Priestess (*kohenet*) as the paradigm for Jewish female leadership in the 21st century. An author, storyteller, homilist, teacher, leader, and institution builder, Rabbi Hammer articulated women-identified, earth-based Jewish spirituality that focuses on the empowerment of women and the healing of the self, society, and nature. Through the website Tel Shemesh: Celebrating and Creating Earth-Based Traditions, Rabbi Hammer sought to integrate "Jewish faith and practice with earth-based beliefs and way of living" (Tel Shemesh n.d.). Tel Shemesh can be translated as "Hill of the Sun," but '*tel*' refers to a mound where remains of previous civilization can be excavated. Like the website Neo.Hasid.com of Rabbi David Seidenberg, the website Tel Shemesh is informed by Kabbalah and Neo-Hasidism, and the ethics of "care and concern for the health and well-being of our planet" is expressed primarily through rituals, poetry, and homilies. Tel Shemesh grounds the practice of Judaism in nature but it does not attend to specific environmental problems such as climate change, loss of biodiversity, or treatments of animals. This feminist embodied Jewish spirituality demonstrates ecological consciousness that may or may not lead to environmental activism.

Inspired by Starhawk's work, the Kohenet Hebrew Priestess Institute was convened in 2006 on the grounds of Elat Chayyim Retreat Center in Accord, New York (Rock-Singer 2020; 2022). The first class of priestesses graduated in 2009, and a West Coast branch was opened as well, indicating the growing interest in the vision. The style, values, and activities of the Kohenet Institute illustrate the life trajectories of the co-founders, Rabbi Jill Hammer and Taya Shere: whereas Hammer was fully grounded in traditional Judaism, Shere was a worshipper of the Goddess who lived in a women's collective house and who initially rejected the Judaism of her childhood in favor of Goddess religion. Shere describes her experience in the women's collective as a "living laboratory of ecofeminist spirituality," where she realized her need "to find a way to love Judaism" (Hammer and Shere 2015, 14). Their collaboration integrated Goddess worship, earth-based spirituality, Jewish feminism, and the rituals of Judaism. Other graduates of the Kohenet Institute, whose life stories are briefly presented in their book (Hammer and Shere 2015, 260–66) confirm paths like Shere's: they came to Kohenet Institute to affirm their Jewish identity as well as their reverence to the Goddess.

In our post-secular, post-modern age, hybridity is rampant, and Jews creatively fuse elements from various spiritual traditions.

Rabbi Hammer's fluid, nonbinary, embodied spirituality reflects LGBTQ sensibility as well as the sensibility of New-Age Judaism of the Jewish Renewal Movement. The Kohenet Institute is endorsed by male rabbis who offer alternative, nature-based and shamanic-oriented Jewish spirituality such as Rabbi Jay Michaelson (Michaelson 2006; 2009) and Rabbi Gershon Winkler (Winkler 2009). However, the fact that the Kohenet Institute is now defunct indicates that this eclectic blend, as creative as it may have been, has limited appeal within the Jewish community.

Women and Environmental Education

Ellen Bernstein: Biblically Inspired Environmental Spirituality

Feminism has clearly inspired Jewish Earth-based spirituality, and women have taken leading roles in articulating the environmental agenda, establishing new organizations, and spearheading environmental education. Although women could not have done so without the feminist revolution, not all forms of Earth-based Judaism are feminist. Rabbi Ellen Bernstein, who is lovingly referred to as "the birthmother of Jewish environmentalism," is a case in point. As the founder Shomrei Adamah and an author/editor of several key publications that articulated Jewish environmental spirituality, Bernstein clearly experienced feminism but her contribution to Jewish environmentalism is not inspired by the values of the women spirituality movement.

Ellen Bernstein grew up in Haverhill Mass, right on the New Hampshire border. A child of a Jewish middle-class family, Bernstein was educated in public schools and benefited from the economic prosperity, social acceptance, and cultural integration of Jews in America in the post-WWII years. Growing up in the beautiful environment of New England, Ellen Bernstein loved the outdoors, but already as a high school student (1967–1970), she became aware of and deeply concerned about environmental degradation, pollution, excessive development, and consumption. Bernstein graduated high school just after the responsibility of world religions to the environmental crisis was identified by Lynn White Jr. and ecotheology by Christian and Jewish theologians emerged in response to charge that the Bible shaped the exploitative attitudes that brought about the environmental crisis. She entered UC-Berkeley in 1971 and enrolled in the program in Conservation of Natural Resources, one of the first college-level programs in environmental science. During her college years she cultivated her love of nature through trips to the foothills of the Sierras where she and her like-minded friends discussed the works of nature writers, including Annie Dillard, Lewis Thomas, and Wendell Berry. The beauty, grandeur, complexity, and connectedness of the natural world fascinated and awed Ellen, nourishing her nature spirituality and encouraging her to impart that love to other people through teaching.

After graduating from UC-Berkeley in 1975 she earned an MA in biology and education from Southern Oregon State University and made a living by teaching biology to high school students and spending her summers as a river guide in Northern California. Nature was her first love, so to speak, but as she experienced nature, studied nature, and taught about nature, she realized that "stories, poetry, and literature can help shape our awareness of the environment and inspire a sense of responsibility for the earth" (Bernstein 2024, xv). Written late in her life and published posthumously, this statement expresses the insight that inspired her entire career: world and words, cosmos and logos, matter and spirit are closely intertwined. But which words or which literature should convey the voice of a Jewish biology teacher? Over time, the answer became clear to her–the Hebrew Bible, rather than the writings of American naturalists, became the literary prism through which she engaged the physical world that God had created.

Although Bernstein self-identified as a Jew, by her own admission, by the time she entered college, she "has long since abandoned her ancestral tradition" and verses of the Bible were no more than "faint childhood memories … likely buried in my consciousness" (Bernstein 2024, xv). This was quite common for Jewish baby boomers who acknowledged their Jewishness as an ethnic fact but were disengaged or even alienated from Judaism as a religious way of life. It was at UC-Berkeley that Bernstein began to take her Judaism seriously, affirming her Jewishness culturally, socially, and religiously. That meant learning Hebrew well enough to read the Bible, becoming affiliated with a Jewish community, and adopt the rhythms of Jewish religious life, especially celebrating the Sabbath and festivals.

Rekindling her Jewish commitment led her to relocate back East, settling in Mt. Airy in Philadelphia. The choice was no coincidence, since Mt. Airy in Philadelphia boasted a thriving Jewish countercultural community led by Zalman Schachter-Shalomi. His outreach to previously unaffiliated Jews inspired them to reaffirm their Jewishness by familiarizing themselves with the spiritual strands of Judaism, especially Kabbalah and Hasidism. Schachter-Shalomi called for an ecologically oriented Judaism based on the Jewish mystical tradition. In Mt. Airy during the 1980s, Ellen Bernstein forged her Jewish environmental spirituality, anchoring her love of nature in the sources of Judaism which she interpreted in the light of her scientific knowledge. While mining the Jewish textual tradition for its ecological insights began already in the 1970s, Bernstein was the first to identify the need for a national, environmental organization that will cross denominational divides and bridge the gap between rabbinic leadership and the "Jews in the pews," so to speak. We already mentioned that in 1988 she founded Shomrei Adamah (Keepers of the Earth) whose mission was "to help illuminate the ecological dimensions of Judaism which had so long been overlooked" (Bernstein 2024, vi). Working with other environmentally concerned Jews, she aspired "to illuminate the deep ecological culture at the heart of Judaism and the Hebrew Bible" (Bernstein n.d.) thus integrating her commitment to both environmentalism and Judaism. Shomrei Adamah inspired the emergence of other Jewish environmental organizations on local and regional levels all engaged in education, consciousness raising, and reconstruction of Judaism. Bernstein illustrated how a Jew who is imbued with environmental consciousness can adhere to the beliefs and practices of Judaism.

Education was the primary concern of Shomrei Adamah, as much as it is the main concern of most Jewish environmental organizations. Under Bernstein's leadership, the organization generated books and essays, composed curricula for Jewish schools, organized workshops and conferences, and organized retreats. With Rabbi Dan Fink, Bernstein co-authored *Let the Earth Teach You Torah: A Guide for Teaching Jewish Ecological Wisdom* (1992), which identified key sources that harbor Jewish ecological wisdom.

The book's title was intentionally intriguing. It told the reader that sacralization of material life is not simply a matter of following the commands of the Torah, but rather that Earth itself (in Hebrew, *adamah*) instructs us how to treat it so that life on Earth could flourish. When Genesis 2:15 teaches us "to till and protect" the Earth (Gen. 2:15) (in Hebrew: *le-ovdah u-leshomrah*), the imperative manifests the wisdom inherent in creation. To observe that imperative, Bernstein insists, we must use our scientific knowledge that enables us to see, appreciate, and cherish the complexity, interconnectedness, and fragility of the Earth. And when we do so we heal the rift between humanity and the Earth that signifies the brokenness from which we all suffer. As a scientifically oriented, religious Jew, Bernstein was hopeful that the brokenness could be repaired when we integrate science and religion, when we become attuned to the wisdom of Judaism in which *adam* and *adamah* are inextricably intertwined through language, speech, and action. That foundational link is asserted by other environmentalists, most notably Arthur Waskow (e.g., Waskow 2012).

Bernstein's later writings greatly contributed to educating American Jews and articulate a distinctively Jewish environmental spirituality. In 2000 she published *Ecology and Judaism: Where Nature and the Sacred Meet* (2000), an early harvest of Jewish environmental writings by rabbis, educators, activists, and scholars. In the late 1990s, when the field of religion and ecology established itself within Religious Studies, Judaism was often represented by Bernstein. In the first decade of the 21st century, one could offer college-level courses on religion and environment and Ellen hoped to benefit from this development, but to do so she had to enhance her academic credentials. She earned another MA degree in Jewish education from the non-denominational Hebrew College in Newton Massachusetts (2006), which was led by Rabbi Arthur Green, and a few years later she received rabbinic ordination from the Academy of Jewish Religion in New York (2012), where Rabbi Jill Hammer taught. In 2016 she accomplished her goal to enter academia when she was appointed an Advisor for Identity and Praxis at Hampshire College (2016–2020), returning to the region where her love of nature was first kindled.

Rabbi Bernstein made an indelible mark as a writer, public speaker, and influencer rather than as an academic. In her three books—*The Splendor of*

Creation: A Biblical Theology (2005), *The Promise of the Land: A Passover Haggadah* (2020), and the posthumously published *Toward a Holy Ecology: Reading the Song of Songs in the Age of Climate Crisis* (2024)—she taught Jews and non-Jews how to live Jewishly and environmentally. Although she was not a systematic theologian, she had a consistent theological message that grew out of her close reading of biblical, rabbinic, philosophic, and kabbalistic texts. She articulated an eco-hermeneutic that treats nature as a text to be explicated, highlights the agricultural roots of Jewish canonic texts, and appreciates the creative power of language.

Bernstein was not a systematic theologian, but she articulated consistent eco-theology rooted in the concept of creation rather than in eco-feminism. Like Lubarsky, Rabbi Bernstein focused on the centrality of beauty, because "beauty teaches us to love the world." For her, beauty is objective and not merely a subjective perception constructed in the eyes of the beholder. The beauty of the created world is manifested in its intricate unity within plurality, holistic connectedness within differentiation, stability and orderliness within dynamism and vivacity. Along with Lubarsky, Bernstein was convinced that when we become attuned to the beauty of the world, we become less prone to destroy, pollute, or squander the richness of creation.

Bernstein introduced environmental values through education and many other women followed in her footsteps (Krone 2019; Jewish Women Archives 2025) teaching Jews how to garden, be attuned to environmental issue, and become politically involved. As rabbis, lay leaders, and educators, women have been crucial to contemporary Jewish environmentalism, but most of them have not been interested in Goddess worship. Similarly, several Jewish men have promoted earth-based Judaism, which was not inspired by Goddess worship, although they are respectful of feminism and committed to egalitarianism. For example, Rabbi Mike Comins, a Reform rabbi, established Torah Trek: The Center for Wilderness Spirituality in 2004 (Comins 2007) to teach God how to find God in nature. Rabbi Zelig Golden, an environmental lawyer ordained by Rabbi Zalman Schachter-Shalomi, created Wilderness Torah in 2009 to teach Jews to reclaim and embrace the pagan roots of Jewish monotheism (Golden 2010; 2018). And Rabbi Natan Margalit, who teaches at the nondenominational Hebrew College, created Organic Torah: Torah in Nature, Nature in Torah as a web-based educational resource to disseminate a holistic vision of Judaism that bridges the gap between nature and culture (Margalit 2007; 2022). For these spiritual entrepreneurs and educators, the future of Judaism depends on the awareness that God revealed the Torah in the wilderness, that the three pilgrimage festivals of Judaism—Sukkot, Passover, and Shavuot—were agricultural celebrations linked to the cycles of the seasons, that Jewish spirituality is embodied and sensual, and that embodied spirituality is most fully experienced outdoors. To live as a religiously committed Jew does not entail alienation from nature but rather experiencing God and living the life of Torah in nature. In this vision, the natural world re-enchants the practice of Judaism, which has become arid, meaningless, or alienating to many Jews.

Through their efforts, the people discussed in this chapter have sought to revive contemporary Judaism by anchoring in the rhythms and vitality of the earth.

References

Abelson, Joshua. 1969 [1912]. *The Immanence of God in Rabbinic Literature.* London: Allen and Unwin.

Acklesberg, Martha A. 1986. "Spirituality, Community and Politics: B'not Esh and the Feminist Reconstruction of Judaism." *Journal of Feminist Studies in Religion* 2 (2): 109–120.

Adler, Margot. 2006 [1979]. *Drawing Down the Moon: Witches, Druids, Goddess-Worshippers and Other Pagans in America*, revised and updated edition. New York: Penguin.

Albanese, Catherine L. 1990. *Nature Religion in America: From the Algonkian Indians to the New Age.* Chicago and London: University of Chicago Press.

Bernstein, Ellen. n.d. Personal website. https://www.ellenbernstein.org/.

Bernstein, Ellen (ed.). 2000. *Ecology and the Jewish Spirit. Where Nature and the Sacred Meet.* Woodstock, VT: Jewish Lights Publishing.

Bernstein, Ellen (ed.). 2005. *The Splendor of Creation: A Biblical Theology.* Cleveland: Pilgrim Press.

Bernstein, Ellen (ed.). 2020. *The Promise of the Land: A Passover Haggadah.* Milburn, NJ: Behrman House.

Bernstein, Ellen (ed.). 2024. *Toward a Holy Ecology: Reading the Song of Songs in the Age of Climate Change.* Rhinebeck, NY: Monkfish Books Publishing Company.

Bernstein, Ellen and Dan Fink. 1992 (eds.). *Let the Earth Teach Your Torah: A Guide for Teaching Jewish Ecological Wisdom.* Shomrei Adamah.

Christ, Carol P. 1997. *Rebirth of the Goddess: Finding Meaning in Feminist Spirituality.* New York and London: Routledge.

Christ, Carol P. 2003. *She Who Changes: Re-Imagining the Divine in the World.* New York: Palgrave Macmillan.

Deutscher, Isaac. 1968. *The Non-Jewish Jew and Other Essays.* London, New York, and Toronto: Oxford University Press.

Diamond, Irene. 1990. "Babies, Heroic Experts, and Poisoned Earth." In *Reweaving the Word: The Emergence of Ecofeminism*, ed. Irene Diamond and Gloria Feman Orenstein, 201–211. San Francisco: Sierra Club Press.

Diamond, Irene. 1994. *Fertile Ground: Women, Earth, and the Limits of Control.* Boston: Beacon Press.

Diamond, Irene. 1995. "What, You're an Ecofeminist and You Don't Practice Mikvah?" *Bridges: Journal for Jewish Feminists and our Friends*, 5 (2): 85–88.

Diamond, Irene. 2004. "Toward a Cosmology of Continual Creation: From Ecofeminism to Feminine Ecology and Umbilical Ties." *Cross Currents* 54 (2) 7–15.

Diamond Irene and David Seidenberg. 1999. "Sensuous Minds and the Possibilities of a Jewish Feminist Practice." *Ethics and the Environment*, 4 (2): 185–195.

Eaton, Heather. 2005. *Introducing Ecofeminist Theologies.* London and New York: T & T International.

Eisler, Riane. 1987. *The Chalice and the Blade: Our History, Our Future.* San Francisco: HarperSanFrancisco.

Eisler, Riane. 1990. "The Gaia Tradition and the Partnership Future: An Ecofeminist Manifesto." In *Reweaving the World: The Emergence of Ecofeminism*, ed. Irene Diamond and Gloria Feman Orenstein, 23–34. San Francisco: Sierra Club Books.
Eisler, Riane. 1996. *Sacred Pleasure: Sex, Myth, and the Politics of the Body—New Paths to Power and Love*. New York: Harper Collins.
Eisler, Riane. 2000. *Tomorrow's Children*. Boulder, CO: Westview Press.
Eisler, Riane. 2003. *The Power of Partnership: Seven Relations That Will Change Your Life*. Novato, CA: New World Library.
Eisler, Riane. 2008. *The Real Wealth of Nations: Creating a Caring Economics*. Oakland, CA: Berrett-Koehler Publishers.
Eisler, Riane. 2015. "Women's Spirituality: From Domination to Partnership." In *Foremothers of the Women's Spirituality Movement: Elders and Visionaries*, ed. Miriam Robins Dexter and Vicki Noble, 65–73. Amherst, NY: Teneo Press.
Eller, Cynthia. 1995. *Living in the Lap of the Goddess: The Feminist Spirituality Movement in America*. Boston: Beacon Press.
Fox, Matthew. 1991. *Creation Spirituality: Liberating Gifts for the People of the Earth*. San Francisco: HarperCollins.
Golden, Zelig. 2010. "Earth-Based Judaism—Reclaiming Our Roots, Reconnecting to Nature." *Zeek*. Available at: http://zeek.forward.com/articles/116841.
Golden, Zelig. 2018. "*Earth-Based Judaism: Early Origins, Polemics, and a New Vision*." MA Thesis, Graduate Theological Union, Berkeley.
Gottlieb, Lynn. 1995. *She Who Dwells Within: A Feminist Vision of Renewal Judaism*. San Francisco: HarperOne.
Griffin, Susan. 1978. *Woman and Nature: The Roaring Inside Her*. New York: Harper and Row.
Hammer, Jill and Taya Shere. 2015. *The Hebrew Priestess: Ancient and New Vision of Jewish Women Spiritual Leadership*. New York: Ben Yehuda Press.
Hanegraaff, Wouter J. 1998. *The New Age Religion and Western Culture: Esotericism in the Mirror of Secular Thought*. Albany, NY: SUNY Press.
Heelas, Paul. 1996. *The New Age Movement*. Oxford: Blackwell.
Jewish Women Archive. 2025. "Jewish Women in Environmental Activism." https://jwa.org/discover/throughtheyear/january/environment.
Krone, Adrienne. 2019. "Jewish Environmentalism." *The Shalvi/Hyman Encyclopedia of Jewish Women*. Jewish Women Archive. https://jwa.org/encyclopedia/autho/krone-adrienne.
MacGregor, Sherilyn. 2006. *Beyond Mothering Earth: Ecological Citizenship and the Politics of Care*. Vancouver and Toronto: UBS Press.
Margalit, Natan. 2007. "Naturally Thinking: Judaism, Modes of Thought, and the Environment." In *Studies in Jewish Civilization*, vol. 20, "Jews and the Environment," ed. Leonard Greenspoon, 87–104. Omaha, NE: Creighton University Press.
Margalit, Natan. 2022. *The Pearl and the Flame: A Journey into Jewish Wisdom and Ecological Thinking*. Boulder, CO: Albion-Andalus.
Mellor, Mary. 1997. *Feminism and Ecology*. New York: New York University Press.
Merchant, Carolyn. 1980. *Death of Nature: Women, Ecology and the Scientific Revolution*. New York: Harer.
Michaelson, Jay. 2006. *God in Your Body: Kabbalah, Mindfulness, and Embodied Spiritual Practice*. Woodstock, VT: Jewish Lights Publishing.
Michaelson, Jay. 2009. *Everything Is God: The Radical Path of Nondual Judaism*. Shambhala/Trumpeter.

Myers, Jody. 2007. "Phasing In: Rosh Hodesh Ceremonies in American Jewish Life." In *Women Remaking American Judaism*, ed. Riv-Ellen Prell, 231–256. Detroit: Wayne State University Press.
Novick, Leah. 2008. *On the Wings of Shekhinah: Rediscovering Judaism's Divine Feminine*. New York: Quest Books.
Orenstein, Gloria Feman. 1990. *The Reflowering of the Goddess*. New York: Pergamon Press.
Orenstein, Gloria Feman. 2007. "Torah Study, Feminism and Spiritual Quest in the Work of Five American Jewish Women Artists." *Nashim: A Journal of Jewish Women's Studies and Gender Issues* 14: 97–130.
Parker, Rozsika. 1979. "Being Jewish: Anti-Semitism and Jewish Women," *Spare Rib* 79 (February): 27–31.
Pike, Sarah M. 2004. *New Age and Neopagan Religions*. New York: Columbia University Press.
Plaskow, Judith. 1979. *Sex, Sin, and Grace: Women's Experience and the Theologies of Reinhold Niebuhr and Paul Tillich*. Lanham, MD: University Press of America.
Plaskow, Judith. 1991. *Standing Again at Sinai: Judaism from a Feminist Perspective*. San Francisco: HarperSanFrancisco.
Plaskow, Judith. 1991a. "Feminist Anti-Judaism and the Christian God." *Journal of Feminist Studies in Religion* 7 (2): 99–108.
Plaskow, Judith. 2005. *The Coming of Lilith: Essays on Feminism, Judaism and Sexual Ethics, 1972–2003*. Boston: Beacon Press.
Pogrebin, Letty Cottin. 1982. "Anti-Semitism in the Women's Movement." *MS Magazine* (June): 45–49, 62–74.
Prell, Riv-Ellen (ed.). 2007. *Women Remaking American Judaism*. Detroit: Wayne State University Press.
Rock-Singer, Cara. 2020. "Milk Sisters: Forging Sisterhood at Kohenet Hebrew Priestesses Institute." *Nashim: A Journal of Jewish Women's Studies & Gender Issues* 37 (Fall): 87–114.
Rock-Singer, Cara. 2022. "The Kohenet Hebrew Priestess Institute and the Gendered Politics and Media of Jewish Renewal." In *Jewish Revival Inside Out: Remaking Jewishness in a Post-Secular Age*, ed. Rachel Werczberger and Daniel Monterescu, 245–268. Detroit: Wayne State University Press.
Salomonsen, Jone. 2002. *Enchanted Feminism: Ritual, Gender and Divinity Among the Reclaiming Witches of San Francisco*. New and London: Routledge.
Sandilands, Catriona. 1999. *The Good-Natured Feminist: Ecofeminism and the Quest for Democracy*. Minneapolis and London: University of Minnesota Press.
Sjöö, Monica and Barbara Mor. 1987. *The Great Cosmic Mother: Rediscovering the Religion of the Earth*. New York: HarperOne.
Spretnak, Charlene (ed.). 1982. *The Politics of Women's Spirituality: Essays on the Rise of Spiritual Power within the Feminist Movement*. New York: Knopf Doubleday Publishing Group.
Starhawk. 1987. *Truth or Dare: Encounter with Power, Authority and Mystery*. San Francisco: Harper & Row.
Starhawk. 1989. *Encounters with Power, Authority and Mystery*. San Francisco: HarperOne.
Starhawk. 1999 [1979]. *The Spiral Dance: A Rebirth of the Ancient Religion of the Great Goddess*. San Francisco: HarperOne, 20th Anniversary Edition.
Starhawk. 2004. *The Earth Path: Grounding Your Spirit in the Rhythms of Nature*. San Francisco: HarperOne.

Taylor, Bron. 2010. *Dark Green Religion: Nature Spirituality and the Planetary Future*. Berkeley: University of California Press.

Tel Shemesh: Celebrating and Creating Earth-Based Traditions. n.d. "*About Tel Shemesh*." http://telshemesh.org/about/#top.

Waskow, Arthur. 1983. "A Feminist Judaism: Restoration of the Moon." In *On Being a Jewish Feminist*, ed. Suzannah Heschel, 261–272. New York: Schocken.

Weissler, Chava. 2007. "Meaning of Shekhina in the 'Jewish Renewal' Movement." In *Women Remaking American Judaism*, ed. Riv-Ellen Prell, 51–81. Detroit: Wayne State University Press.

5 The Holy Land

Zionism and the Dialectic of Post-Secularism

Post-Secular Hybridity and Zionism

Judaism is a land-centered religion, but for most of their history the Jews have been a diasporic people, dispersed among all the world's nations. For over two millennia, Jews prayed for, speculated about, and dreamed of the Land of Israel, but only in the late 19th century did Zionism, a modern movement of Jewish nationalism, make the dream a reality. From the 1880s, Zionism concretized what religious Jews yearned for while living in Exile. Zionism brought Jews back to their ancestral homeland, reconstituted a Jewish polity—the State of Israel—and created a modern Hebraic culture. Whether the accomplishments of the Zionist movement were possible because Jews acted as (secular) historical agents without relying on God, or whether these accomplishments manifested God's hidden will to redeem the Jewish people is still debated, differentiating secular and religious interpretations of Zionism and of Judaism more broadly.

This chapter illustrates the dialectics of post-secularism by focusing on the attitudes toward nature in Zionism and the State of Israel. How did a land considered "holy," that is, a spiritual entity, become the vehicle for the physical rejuvenation of Jewish People? How did technoscience transform a small nation into a leading global power but at a very high cost to the well-being of the physical environment? How did environmentalism emerge to express a nonreligious, Western-oriented, Liberal identity only to generate a religious backlash whose veneration of the land came at the expense of the land's environmental well-being as well as the non-Jewish residents of the land? How did the canonic sources of Judaism inspire a faith-based environmentalism, on the one hand, as well as religious denunciation of environmentalism, on the other hand? In short, the story of Zionism and the State of Israel is filled with paradoxes that go to the heart of the post-secular condition in which the secular and the religious, the physical and spiritual, are dialectically intertwined.

The Secularization Thesis understands modernity as a story of "disenchantment" by means of secular technoscientific "progress." Zionism and the State of Israel complicate this framing because many of the features of the

DOI: 10.4324/9781351123945-6

"post-secular"—the emergence of spirituality as hybridization of the secular and the religious, the re-enchantment of the physical world, and the investment of embodiment with spiritual meaning—manifested in Zionism long before the term 'post-secular' appeared in the 1980s. Zionism and the State of Israel make clear that secularism, secularity, secularization, and post-secularism are historically situated and culturally dependent formations that bely generalizations. No model fits all cases; each must be understood on its own terms.

Hybridity is one of the markers of the post-secular age. As a theoretical meta-construction of social order, hybridity "works with and within the cultural design that reshapes our understanding of the interstices—social and psychic signs of cultural similitude with emergent signifiers of alterity" (Bhabha 2015 [1997], 1; Cf., Easthope 1998). In the post–Cold War era, the concept of hybridity expressed postmodern skepticism of all essentialized identities or binary dichotomies and was applied to various academic discourses including postcolonial studies, ethnic and racial studies, justice studies, and cultural studies. In all of them hybridity signifies a critique that challenges notions of purity, authenticity, and ownership while promoting instead creativity, syncretism, diffusion, marginality, and cultural fusion (Hutnyk 2006). In the case of Zionism, however, hybridity does not necessarily function subversively. Instead, it manifests in the paradoxical fusion of revolution and evolution; discontinuity and continuity; rejection and rejuvenation; destruction and reconstruction (Avineri 1981 [2017]). As Shlomo Avineri succinctly put it, Zionism "created a universe of discourse that, while revolutionary and innovative, was deeply steeped in the symbolical language of the Judaic tradition" (Avineri 1998, 5).

Four aspects of Zionism—land, nature, embodiment, and language—capture its unique paradoxicality. Zionism placed the Land of Israel, a geographical territory, at the center of its personal and national renaissance. In the Zionist analysis of Jewish history, the exile (*galut*) of the Jews from their ancestral homeland distorted Jewish life individually and collectively, creating social, economic, political, and spiritual anomalies, disabilities, and pathologies. The hatred of Jews, repeated persecutions, humiliating poverty, cultural backwardness, and social marginalization characterized exilic life, and these conditions could be overcome only if Jews leave their countries of residence *en masse* and settle in their ancestral homeland, the Land of Israel, where they would reconstitute Jewish political sovereignty, engage in agriculture and productive work, and recreate a modern Hebraic culture. That was the core message of the Zionist attempt to solve the "Jewish Problem" once and for all.

Zionism regarded the return of the Jews to the homeland as the return of the Jews to "nature" (*teva*). In the Zionist worldview, Nature was the antithesis of Exile and its inherent limitations. The Zionist return to nature was predicated on the reinterpretation of the Messianic idea. Traditionally, Jews waited for the Messiah, an individual chosen by God who was tasked with gathering the Jewish exiles in the Promised Land. Zionism departed from this view by insisting that the Jews themselves should initiate their own redemption through historical, political, and economic activities within the material world. Thus "redemption" (*geulah*) became an earthly event: redeeming the Land of Israel (*geulat ha-aretz*) by purchasing swaths of land for Jews to cultivate with their own hands. This transformation, in turn, would bring about the redemption of the Jewish people (*geulat ha-am*) from the humiliating conditions of Exile. The return to "nature" intended to normalize the Jews, making them a people like all others, a new, modern nation with its own territory, language, and culture, the markers of sovereign nations.

The normalization of the Jews signaled a new conceptualization of the Jewish body (Gilman 1991). In exile, so the Zionists insisted, Jews were physically weak, fearful of their gentile neighbors, engaged in nonproductive occupations, and disengaged from manual labor. Rejecting the debilitating conditions of exile, the "new Jews" in the Land of Israel would make a living in farming, road paving, or construction rather than in the traditional Jewish occupations of finance, commerce, and trade (Conforti 2009). The "new Jews" would not devote their life to the study of sacred texts but would be rooted in the rhythms of nature and willing to use physical force in their self-defense. Inspired by the Nietzschean vision for the "new European man," the antithesis of the European bourgeois paradigm (Mosse 1996, Golomb 2004; Peled 2002), Zionism sought to restore Jewish pride, overcoming centuries of humiliation and shame. Paradoxically, the Zionist desire to transform the Jews could become a reality only because the Zionist settlers believed that the very soil of the Land of Israel empowers, energizes, and rejuvenates those who return to it.

The Zionist construction of the modern Jewish nation in the Land of Israel was to be forged through the Hebrew language, yet another aspect of the paradoxicality of Zionism. If prior to Zionism, Hebrew was a "holy tongue" (*leshon ha-kodesh*), the language of sacred texts, prayer, and commentary, now Hebrew would become the language of everyday life, a vernacular for daily communication. How modern Hebrew should sound was itself a matter of considerable dispute among those who toiled in reviving modern Hebrew, but all agreed that the use of Hebrew was "an expression of national authenticity" (Segal 2010, 45). Modern Hebrew thus signified the Zionist rupture with Jewish exilic existence, on the one hand, and the reconnection with the remote, biblical past, on the other hand.

Knowledge of the Bible, which many Zionist pioneers possessed because of their traditional upbringing, now framed their encounter with the Land of Israel, concretizing biblical narratives and bringing them to life, so to speak. Written in modern Hebrew, Zionist prose and poetry both de-sacralized Jewish life by focusing on the experience of natural phenomena, while at the same time investing secular, natural life with spiritual meaning that arose from the linguistic associations of the Hebrew language itself. By means of language, Zionism articulated an embodied, secularized Jewish environmental spirituality that collapsed the boundary between Self and Nature, between interior life and physical environment. The creation of a modern, Hebraic nation linked Zionism back to the textual tradition that it critiqued or denounced, thus making the past forever present and making Zionism inevitably paradoxical. The paradoxicality of Zionism, and the dialectical relationship between religious traditionalism and modern secularism, between exile and homeland, between Torah and "nature," is exemplified in the very concept to the Holy Land

The Holy Land: Geographic Territory and Spiritual Reality

In the Bible, the land, not the Torah, was God's essential gift to Israel (Hoffman 1986). The land offered a wellspring of sustenance, because it was blessed by God. Nonetheless, the Bible does not refer to the Land of Israel as a Holy Land (*eretz ha-kodesh*). While the Bible indeed recognizes the sanctity of place (Exod. 3:5), it does not refer to the Land of Israel in terms of sanctity or sacrality (*kedushah*). One exception is Zechariah 2:16, where Jerusalem is a "land of holiness," but the Land of Israel's merit lies in the moral integrity of its inhabitants, not its inherent sanctity or sacrality. Protecting the Land's moral purity is behind many of the Bible's land-based commandments, especially during the Sabbatical Year (*Shemitah*), and it is no coincidence that this ancient practice has become of primary interest to contemporary Jewish environmentalists.

After the destruction of the Second Temple in 70 CE, rabbinic Judaism reinterpreted biblical religion, shifting the center for the Jewish sacred myth from the Land of Israel to the Torah. The Land of Israel retained its value and importance, but the rabbis also created a complex religious system enabling Jews to relate to God anywhere and everywhere. In rabbinic Judaism, liturgical acts such as food rituals, prayers, blessings, and Torah study became the modes of communication with God, enabling diasporic Jews, who once again became landless people, to live holy life outside the Land of Israel, even without observing the land-based commandments prescribed in the Torah. Although Jewish religious life did not depend on presence in the Land of Israel, rabbinic Judaism praised the intrinsic merits of the Land of Israel: only the Land of Israel "atones for sins and crimes," only the Land of Israel generates great wisdom, and only it serves as the locus of prophecy (Yisraeli 2006). And yet, the rabbis were quite restrained in terms of asserting the holiness of the Land of Israel in its entirety.

Endowing the Land of Israel with holiness as a metaphysical attribute emerged first among Judeo-Hellenistic Jews, especially Philo (d. ca. 50 CE), Josephus Flavius (d. ca. 100 CE), and the anonymous Jews who generated the Apocalyptic literature (Halpern-Amaru 1986). In contrast to Hellenistic Judaism of the Diaspora, the framers of rabbinic Judaism recognized the intrinsic merits of the Land of Israel, but the holiness of the Land is only one element in a complex system of a grid of holiness. In rabbinic Judaism the Torah gave cosmic meaningfulness of actions in everyday lives of ordinary men and women, thereby diminishing the centrality of the Land in daily Jewish religious practice. Rabbinic Judaism created a comprehensive religious system that sanctified all aspects of daily life, Jews now resided outside the Land of Israel, lost their cultic center in Jerusalem, and could no longer observe the land-based commandments of the Bible (*mitzvot ha-teluyot ba-aretz*). The life of religious observance as articulated by rabbinic Judaism made it possible for Jews to be Jewish outside the Land of Israel while continuing to aspire and yearn to return to the Land of Israel.

In the Middle Ages, Jewish philosophers and mystics debated the religious significance of the Land of Israel (Hallamish and Ravitzki 1991). For Yehuda Halevi, the entire Land of Israel (and not just Jerusalem) was holy because the divine Presence dwells in it. For Maimonides, the Land of Israel was not intrinsically sacred, but human performance of the commandments conferred sacrality on the land.

Rabbi Moses ben Nahman, Nahmanides (d. 1271), held that the very soil of the Land of Israel was inherently pure, and only Jews who perform the commandments correctly could preserve and protect the purity of the land. His essentialist interpretation of the land's sacrality was inseparable from his identity as a proponent of Kabbalah, which transformed the Land from a geographical territory to a mystical symbol laden with erotic meaning (Idel 1986). In the symbolic vocabulary of Kabbalah, "land" (*eretz*) is one symbol of the Shekhinah, the divine aspect of God. Hebrew is a grammatically gendered language, and the word "land" is feminine; since the land's fertility makes human life possible, it is easy to understand how the land became a symbol of the Shekhinah. As we saw in Chapter 3, in kabbalistic symbolism, the Shekhinah receives the divine efflux from the Sefirot; in turn, like a good, caring mother She bestows the divine efflux on all beings, particularly the People of Israel, the Chosen People. kabbalistic symbolic theosophy turned Jewish religious worship into a theurgical act—it is to the Shekhinah that the (male) kabbalist directs his mystical intention (*kavanah*) when he performs the commandments. Through his mystical worship, the (male) kabbalist unifies the feminine Shekhinah with her "Husband," Tif'eret. Unifying God's feminine and masculine aspects redeems all levels of reality (i.e., human, societal, cosmic, and metaphysical), and Exile is no more.

Mystifying the Land of Israel dominated traditional Jewish imagination throughout the early-modern period. Zionism rebelled by calling for the Land's demystification, exemplifying what Weber characterized as "disenchantment": the core of modern secularity. But the story is more complicated than a binary dichotomy suggests, since the demystification was accomplished by Zionists who grew up in traditional Jewish homes in which Hasidism and Kabbalah framed the way of being Jewish (Shapira 1996, 244; Sh. Almog 1998). It was precisely because the Zionist pioneers understood the Land of Israel as a (feminine) metaphysical entity with whom the (male) believer yearned to unite that they accomplished the difficult feat of settling the arid, uncultivated, malaria-stricken land. The very spiritualization of the Land of Israel in Jewish religious imagination made the Zionist settlement project possible, even though Zionism demanded the negation of spiritualization.

In calling for the secularization of Judaism, Jewish culture, and Jewish education, the Haskalah (Jewish Enlightenment) demystified the Land of Israel. It critiqued rabbinic obscurantism and the social backwardness of Jewish life, calling for the transformation of the Jews into a "productive" people. The Haskalah secularized Judaism by historicizing the Jewish past, thereby paving the way for Jewish nationalism, which in turn used history to support the Jewish right to settle in the Land of Israel. While many Maskilim endorsed a non-territorial solution to the Jewish problem (i.e., integrating Jews into Western society), the Haskalah also advanced the Zionist emphasis on the Land of Israel as the only viable territorial solution to the problems of the Jews.

Jewish Nationalism: Synthesizing Religion and Secularism

Zionism was a response to the profound crisis Jews experienced at the end of the 19th century. Emancipation, modernization, secularization, acculturation, and assimilation upended Jewish life, forcing Jews to answer three key questions: What is Judaism? Why should one remain Jewish? How to be Jewish? The identity crisis was especially acute for Eastern European Jews, who composed Zionism's rank and file.

Zionism emerged in Eastern Europe as a synthesis of religious traditionalism and Enlightenment ideals. In contrast to Western European Maskilim who defined Judaism as a *faith*, a confessional community, Zionism insisted that Jews are a *people*, an ethnic group with a distinctive cultural identity that includes religion. Traditionalism implied respecting religious observance and revering religious texts, values, ideals, and norms, but commitment to Enlightenment ideals implied interpreting the past in secular, historical categories that privilege human agency over divine miracles; modernizing an exilic life that stunted the Jewish spirit; and recognizing the impossibility of fusing religion and Enlightenment in the Diaspora. Yet, in focusing on the Land of Israel, Zionism fused 19th-century European nationalism with critique of the Haskalah "that moved away from a connection to the Holy Land" (Saposnik 2008, 14). Saposnik aptly characterizes Zionism as "a secularized revolt against the processes of Jewish secularization just as much as it was a revolt against traditional religiosity" (ibid.). Given its hybrid nature, Zionism could be interpreted in a variety of conflicting ways, all competing bitterly for the Jewish masses' support and for control of Zionism's institutions. Zionism was a movement of the national renaissance, but the precise nature of the national rebirth was vehemently debated.

The main strands of Zionist thought presented different interpretations of the Zionist idea and different programs for realizing it (Shimoni 1995).

Political Zionism and Revisionist Zionism focused on Jewish political sovereignty and attaining statehood. Cultural Zionism focused on creating in the Land of Israel a modern, Hebraic culture rooted in the Jewish literary past and interpreted in secularized and historicized categories. In critique of both Political Zionism and Cultural Zionism, Practical Zionism focused on colonization of the Land of Israel through land purchase and agricultural settlements. For Socialist/Labor Zionism, transforming the Jewish economy meant reorganizing society in large farming collectives, with communal life following the principles of justice and equality. And Religious Zionism spoke about divine redemption of the Jewish people, albeit in collaboration with non-observant, secular Jews. These diverse approaches resulted in bitter ideological debates in the newspapers and journals of the Jewish community of Pre-State Israel (Yishuv) and among Diaspora Zionists gathering in the World Zionist Organization's congresses since 1897. The debates were so vehement because they pertained to the core of Jewish identity: What constitutes Jewish collective identity? What are the political and cultural ramifications of the Jewish return to the ancestral homeland? If one no longer observes Judaism's precepts, is one still Jewish? What is secular Judaism? How does modern, secular Jewish culture relate to the Jewish tradition's values, norms, symbols, and tropes?

This chapter sheds light on these questions by looking at the connection between self, nature, and society in Zionism and in Israeli culture. Zionism's environmental spirituality pre-dates, by half a century, recognition of the environmental crisis and the subsequent environmental movement (of which Jewish environmentalism is but one example). This does not mean that environmentalism per se motivated Zionism, since the Zionist determination "to make the desert bloom" also inflicted considerable damage, recognized and corrected only when environmentalism began to take hold in Israel in the 1970s. The chapter argues that while Zionism succeeded because of its fusion of secularity and religiosity, the Six-Day War of 1967 pulled that synthesis asunder. The current turmoil in Israel, exacerbated by the horrific events of October 7, 2023, and the still ongoing Israel-Hamas war, manifest the perpetual debate about the meaning of Zionism, a debate that began with the very emergence of Zionism but intensified since 1967.

The Six-Day War of 1967 exposed Zionism's paradoxes, dividing the nation and Jews worldwide on the legal and political status of the territories. For secular Israelis on the left of the political spectrum, the physical environment's well-being (*eichut ha-sevivah*) is linked to quality of life (*eichut chayim*) in a democratic society where individuals have civic rights: the right to clean air, clean water, and open spaces. The young state would have to rely on environmental science to correct the environmental damage caused by high population density, rapid urbanization, and massive industrialization since its founding in 1948. Israel's environmental movement has been central to constructing an ostensibly secular national culture that secularized the Jewish religion, turning a certain variant of Zionism into the civil religion of the nascent state. In contrast, the messianic interpretation of the Six-Day War, which spawned the Settlement Movement on the right of the political spectrum justified retaining territories occupied in 1967 by framing them as a literal Holy Land, a land imbued with holiness. For the acutely messianic settlers, the land's holiness is irrelevant to environmental well-being; in fact, settlers have engaged in destructive environmental acts (e.g., uprooting olive trees) while attempting to "Judaize" the land and making it off-limits to non-Jews. Between the secular (environmentalist) camp and the messianic (non-environmentalist) camp is a small but growing trend that seeks to integrate Zionism, traditional Judaism, Western liberalism, and environmentalism. In Israel, land illustrates the dialectics of post-secularism.

Zionism as the Jewish Return to Nature

Secularized Messianism

Zionist national renewal secularized Jewish messianism by returning the Jews to "nature." For centuries the messianic idea nourished Jewish yearning for a better future while perpetuating Jewish passivity and political weakness. For traditional Jews, exile would end by miraculous divine intervention, a supernatural event attesting God's presence in human history.

Zionism rejected that view, calling for Jews to actively engage in their own redemption by settling the Land of Israel. The so-called "harbingers of Zionism," Rabbi Judah Alkalai (d. 1878) and Rabbi Zvi Hirsch Kalischer (d. 1874), reinterpreted the messianic idea by suggesting that human beings can instigate the messianic age by resettling the land. Thus in 1843 Rabbi Alkalai daringly asserted that "Redemption will begin with efforts by the Jews themselves; they must organize and unite; choose leaders and leave the lands of exile" (Hertzberg 1973, 106).

Redemption would be a natural, historical, political event, brought about by earthly human activity. Kalischer envisioned a new type of Jewish, this-worldly activism when he speculated that "an organization be established to encourage settlement in the Holy Land, for the purpose of purchasing and cultivating farms and vineyards" (Hertzberg 1973, 113). Secularizing the messianic impulse, he argued that "Jewish farming will be a spur to the ultimate Messianic Redemption. As we bring redemption to the land in a 'this-worldly' way, the rays of heavenly deliverance will gradually appear" (Hertzberg 1973, 114).

From its inception, then, Zionism fused religious and secular interpretations of Jewish messianism, enabling Zionism to take root among East-European Jews. In the 1880s the European rabbinic establishment rejected the doctrine of natural worldly redemption and continued to endorse the passive, miracle-centered view of Jewish messianism. Thus, Rabbi Jacob Reines stated that "redemption in itself is above nature and human effort ... It is only in the heights of miracle and wonder that it is to be established" (cited in

Ravitzky 1996, 33). However, because Rabbi Reines differentiated traditional messianism and modern Zionism, he could establish in 1902 the movement of Religious Zionism (Mizrahi Movement) that joined the new efforts to resettle the Land of Israel. The differentiation allowed Orthodox Jews like Rabbi Reines to collaborate with secular Zionists who rebelled against Jewish traditionalism. Despite different motivations, the camps could cooperate because they agreed that resettling the Land of Israel was necessary for national renewal.

In response to the pogroms of 1881–1883, the Love of Zion movement (Hibbat Zion) began settling the Land of Israel as a project of self-liberation, or "self-emancipation," as Leon Pinsker called it. Settlement in the Land of Israel—not emancipation—would solve European Jews' problems. The "Aliyah" (meaning "ascent" or "ascension") comprised five immigration waves to the Land of Israel prior to 1948. The First Aliyah (1881–1899) brought 25,000 Jews, mostly from the Russian Empire, attracting traditional and stridently nontraditional Jews alike. They founded thirteen agricultural colonies based on viniculture and citrus orchards that were backed financially by Baron Edmond de Rothschild of the French branch of the Rothschild banking family. Since the Jewish settlers lacked farming experience, Rothschild invested in technological and social services to facilitate independent Jewish farming and regenerate the Jewish people. When Baron Rothschild removed his financial support at the beginning of the 20th century, the Zionist settlement project was taken over by the Jewish National Fund and the Jewish Colonial Trust, organs of the World Zionist Organization (WZO). Established in 1897 by Theodor Herzl, the WZO funded the Zionist settlement project, functioning as a statist institution even before Israel became a state.

Although the First Aliyah colonists changed the Jewish economy by shifting to agriculture, they did not transform the relationship between Jews and nature, since they did not engage in manual labor; that work was performed by local Arab laborers. It was the Second Aliyah (1903–1914) of about 35,000 Jews, mostly young, single, and secular, who articulated a new attitude toward the Land of Israel's physical environment. The Second Aliyah reinvented modern Jewish identity; established the Jewish farming collectives, the kibbutzim; and launched modern Hebraic culture. For the Second Aliyah pioneers, Zionism was a national and personal renaissance built on the identification of man (*adam*) and soil (*adamah*). As Eric Zakim correctly noted, the opening line of the Zionist folk song, "We came to the land to build and be built by/in it," summarizes the Second Aliyah's ethos as a project of individual and collective rebuilding (Zakim 2006).

Boaz Neumann brilliantly analyzed the Zionist pioneers' (*halutzim*) ethos as an erotic desire to unite with the land (Neumann 2011). These young men and women "lacked any defined occupation and had no formal or professional training. Most were Socialist by persuasion. The *halutzim* immigrated to Palestine with the express intention of becoming manual laborers"

(Neumann 2011, 5). The pioneers shaped the ethos, ideology, style, and culture of the pre-State Jewish community and even the first decades of the State of Israel. They concretized their passion for the Land of Israel through physical labor that transformed them physically and spiritually. As literal "lovers of Zion," their physical labor rooted them in the soil of the Land of Israel, thereby giving birth to the "New Jew" and regenerating the Jewish People. The Second Aliyah was a profound project of self-transformation, of being "born again" within Mother Earth (*imma adamah*), the land from which the Jews were uprooted two millennia earlier.

European Romanticism, Tolstoy's religious populism, and Nietzsche's *Lebensphilosophie* were some of the intellectual strands that framed the attitudes of the Second Aliyah pioneers to the Land of Israel. But no less important are the Jewish religious tropes that feminized and eroticized the Land. The motherland (*moledet*) substituted for the biological mother that the young pioneer left behind in the Diaspora. Returning to the motherland was the most radical expression of Zionism's "negation of the Diaspora" (*shelilat ha-golah*): the pioneers did not just liberate themselves from exile, they erased its existence by being born again in the bosom of the ideal perfect motherland. Many signified their rebirth by Hebraizing their surnames and even changing their chronological date of birth (Neumann 2011, 43–47). But the Zionist rebirth was not limited to individuals; it was no less the collective rebirth of the Jewish nation that for centuries suffered from persecution, humiliation, poor health, and backwardness. The nation had to be reinvented by returning the Jews to nature.

A.D. Gordon: Zionist Ecological Spirituality

The thinker who best articulated Zionism as a return to nature was Aaron David Gordon (d. 1922). Gordon emigrated to the Land of Israel in 1904 and stood out in the Second Aliyah: unlike the young, single, and secular pioneers, Gordon was old, married, and religious. Admired and revered by many, he emerged as the spiritual leader of the pioneering society and was even fictionalized in a short story by Joseph Hayim Brenner, his close friend and a leading Zionist intellectual. The pioneering ethos became known as "religion of labor" (*dat ha-avodah*), a phrase misattributed to Gordon. For Gordon, labor was a way to worship God within the corporeal world in an embodied, nonverbal, and noncerebral way. Ron Margolin has argued for characterizing the outlook of A.D. Gordon, Franz Rosenzweig (d. 1922), Martin Buber (d. 1965), and Hans Jonas (d. 1993) as "secular religiosity," in which "rational scientific thought combined with elements of religious existentialism. It is a style of thought and behavior not identifiable with religions and tradition nor yet wholly atheistic" (Margolin 2008, 233). These thinkers "chose not to live a fully halakhic or Orthodox life … because they could not accept the traditional and heteronomic belief in the divine revelation of the Law" (Margolin 2008, 235). Whether we characterize their thought as "secular religiosity" or as "religious secularism," a phrase Donald S. Moore (1996) applied to Martin Buber, it is important to see it as a way of being in the world, an expression of lived experience, not an ideology.

Gordon's ideas gave Labor Zionism its intellectual depth and inspired thousands of Zionist pioneers, even though other leaders of the Labor movement (e.g., Berl Katznelson and David Ben Gurion) would eclipse him. Gordon was the first Jewish ecological thinker who critiqued humans' destructive control of nature, long before environmentalism, religious environmentalism, and Jewish environmentalism appeared in the second half of the 20th century (Jobani 2023). In 1909 Gordon had already published his philosophical essay, "Man and Nature" (*Ha-Adam Ve-ha-Teva*), in which he critiqued the human exploitation of nature and showed it to be the result of alienation from the sources of vitality and creativity—God (Margolin and Jobani 2020; Jobani 2023). For Gordon, Zionism would be the *avant garde* that showed humanity how to heal its rupture with nature and how to root universal humanism in national particularism. Gordon's "secular religiosity" sheds light on the hybrid nature of post-secularism by showing that it is always historically situated and culturally dependent. Finally, Gordon influenced 20th-century Jewish thinkers (e.g., Martin Buber, Mordecai Kaplan, and Abraham Joshua Heschel) who inspired contemporary Jewish environmentalism in Israel and in the Diaspora. His views resonate with the teachings of the Jewish Renewal Movement and with the view of the Spiritual Progressives. Yet, Gordon's ecological outlook emerged out of his own idiosyncratic biography.

Gordon was born in 1856 into a traditional Jewish family in the Ukraine province of Podolia. As relatives of the Baron Joseph Ginzburg, a wealthy industrialist in Russia, Gordon's family enjoyed economic security and were also open to the Haskalah. Tutored privately at home, Gordon was first introduced to traditional Jewish learning and at 14 was sent to study in a yeshivah, but poor health forced him to return home and continue his private education. As a teenager he mastered European languages and began to immerse himself in European literature, philosophy, and science. The lack of formal academic training probably contributed to Gordon's idiosyncratic synthesizing of diverse viewpoints without feeling beholden to any of them. As an autodidact, Gordon studied the works of Spinoza, Kant, Hegel, Marx, Schopenhauer, Tolstoy, Soloviev, Nietzsche, and Bergson and engaged them critically from his own Judaic perspective (Turner 2020, 39–46). A traditional East European Maskil, Gordon was also open to Zionism, since the Maskilim were the first to criticize the social, economic, and cultural ills of Jewish exilic existence. As a passionate member of the Hibbat Zion (Love of Zion) movement, Gordon explored Zionist ideas in the synagogue and welcomed other Zionist preachers, to the consternation of Hasidic Jews in town. Although Gordon was imbued with Hasidic ideas, he disliked the Hasidic communities' authoritarian leadership style.

In 1903 the relative stability of Gordon's life ended when Baron Ginzburg's lease on the Mohilna estate expired, and the estate was sold to a Russian nobleman who fired Gordon. Unemployed for the first time in his life, Gordon soon after had to cope with the death of both his parents within several months of each other. The year 1903 was consequential year also for Jewish life in the Pale of Settlement. The notorious pogrom of Kishinev (Zipperstein 2018) laid bare Jews' profound inability to defend themselves against Anti-Semitism. The existential crisis compelled Gordon to consider emigration, but unlike many other Jews from the Pale of Settlement who emigrated to America, Gordon, the Zionist, emigrated to Palestine in 1904.

Disease, hunger, despair, and suicide were the main challenges confronting the Second Aliyah pioneers; only 10 percent of them remained. The harsh conditions were even more difficult for Gordon, who was much older than the pioneers. Gordon insisted on making a living as a day laborer and worked the fields owned by the First Aliyah settlers. Making a living in farming presented several challenges to the pioneers. First was the physical difficulty and utter lack of experience in manual labor. No less challenging was the competition with the Arab peasants, who were more skilled and much cheaper. Since the Arabs understood the long-term implications of Zionism perhaps even better than the Zionists, they were unsurprisingly hostile to the Zionist project and periodically attacked the pioneers' life and property, requiring them to organize their own self-defense. Manual labor and self-defense were the two intertwined aspects of the Second Aliyah's attempt to generate the "new Jew," and both aspects consisted of conflict with the local Arab population. Zionist colonization did not start as a military conquest but as a legal land purchase, but the pioneers referred to their project as "conquest" (*kibbush*). They had to "conquer" labor (*kibbush ha-avodah*) from the Arab laborers and "conquer" their (exilic) self through willed discipline and self-control, psychological processes that facilitated the emergence of an authentic self through "self-realization" (*hagshamah atzmit*). The Zionist return to nature through manual labor was a therapeutic project of self-healing as much as a project of national and cultural rebirth.

Within the pioneering community Gordon emerged as an intellectual and a spiritual leader, a father figure and a guru of sorts. A member of the Socialist, non-Marxist Young Worker party (*Ha-Poel Ha-Tza'ir*), Gordon published essays in its weekly newspaper on Hebrew labor, self-defense, Jewish education, Jewish Socialism, and the relationship between the Yishuv and the Jewish Diaspora. A decade after he settled in Palestine, his reputation extended to Zionist circles in Europe, when Martin Buber published his public appeals to Diaspora Jews in 1917 in the Zionist journal *Der Jude,* a monthly magazine that Buber founded with Salman Schocken and edited from 1916–1928. In 1920 Gordon represented the Ha-Poel Ha-Tza'ir party in the Zionist conference in Prague and delivered a lecture that greatly impressed Martin Buber, who was present, because Gordon's message resonated with his own life experience (Ratzabi 2011, 22). In 1919, a year prior to that trip, Gordon joined Deganya, the first collective agricultural settlement, and there he remained until his death in 1922 from esophageal cancer. Despite his short-lived residence in Deganya, where Gordon's daughter, Yael, remained a lifelong member, Gordon became the symbol of the kibbutz movement by best articulating Labor Zionism's ideals and by joining the first kibbutz.

Gordon's thought is of relevance to this study for three reasons: first, he thought in environmental terms long before others; second, his Zionism blended religiosity and secularity characteristic of post-secularism, and partly for this reason, third, Gordon's legacy has attracted new interest in Israel (e.g., Schweid 1970; 2014; Shapira 1996; Strassberg-Dayan 1995; Ramon 2007; Shamir 2018; Fuehrer 2019; Jobani 2023).

Zionism was a response to the identity crisis of European Jews at the end of the 19th century. For Gordon, the Jews' problematic existence was a most acute crisis of modernity, brought about by capitalism, urbanization, industrialization, rationalism, technoscience, individualism, and atheism. He saw modernity as distorting human existence by disengaging individuals from the sources of vitality and creativity—Nature—which for Gordon was synonymous with God. According to Gordon, modern instrumental rationality, which yielded the technoscientific achievements of modern civilization, required humanity to alienate from nature, which in turn resulted in human exploitation of nature and of other humans. Gordon imaged the modern Self as a tortoise (*tzav*) whose thick shell of analytic, discursive, instrumental rationality cuts it off from the creative energy of life. The modern person's excessive individualism simultaneously results in loneliness and in glorifying power and control (Ramon 2007, 90). Gordon's critique of modern life echoes German social theory at the beginning of the 20th century, especially Georg Simmel's 1903 essay, "The Metropolis and Mental Life" (Zakim 2006, 157).

Philosophically, Gordon's worldview is often misleadingly labelled as "pantheism." Gordon shrank from any dogmatic "-ism" that he believed only "hypnotizes" its adherents, preventing them from thinking freely and being fully present. Scholars have traced the pantheistic tendencies of Gordon's outlook to various intellectual sources, including Spinoza, Hasidism, Schopenhauer, Darwinism, and Zen Buddhism. The autodidact Gordon was indeed familiar with them, but his outlook's coherence emerged from the depth of his personality and lived experience rather than from the systematic presentation of his ideas. Gordon applied his distinctive outlook to contemporary social and political issues and to the personal struggles of his correspondents, who were also engaged in the birth of the "new Jew," a process of self-realization by means of manual labor. For Gordon, self-realization involved seeing human beings embedded in nature.

In Gordon's organismic worldview, reality is an organic whole of dynamic processes, "an organic manifold of matter and spirit" (Turner 2020, 40) that can only be experienced but not conceptualized. In his poetic, emotionally charged style, he likens cosmic reality to "an ocean of motion, an ocean that is perpetually boiling (*retiha olamit*), and all beings (*nimtzaim*) and all of visions (*hezyonot*) are like bubbles of this cosmic boil" (Gordon 1951, 42). Using kabbalistic terminology, Gordon refers to the totality of the cosmic whole as *havayah* (Being), a word that comprises but rearranges the four letters of the ineffable name, YHVH. The totality of Being is God, a living reality to which Gordon refers as "nature" (*teva*), "cosmic Being" (*havayah olamit*); "infinite being" (*havaya einsofit*), "infinite motion" (*tenu`ah einsofit*), "infinite ocean" (*yam einsofi*), and "eternal life" (*hayey olam*). Being is not a static but dynamic reality of unending, ever-changing, organic processes of perpetual creativity that bring new things into existence. The source of the perpetual creativity is the Ein-Sof, the Infinite, which is both transcendent and immanent, simultaneously beyond and within the created world.

How does humanity fit into this organismic conception of God/Nature/Being? Gordon tells us that each atom in the human body derives its vitality from "infinite nature" (*ha-teva ha-einsofi*), and nature in its entirety [*kulo*] "suckles from the Infinite (*yonek mi-ha-Ein-Sof*)" (Gordon 1951, 42). The metaphor of "suckling" can also be traced to kabbalistic and Hasidic texts and conveys the organic connection between humanity and nature as well as total dependency of humanity on nature, the source of life. As long as the human is connected to the flow of life (*zerem ha-chayyim*), the creative principle, which is immanent in all organisms and is responsible for growth, change, and adaptation, enables humanity to thrive. The crisis of modernity arises from humans' having been cut from the infinite source of vitality by excessive rationality, thus alienating humanity from lived experience.

Human alienation from God/Nature/Being is rooted in the very nature of human consciousness. Humans can apprehend the world in two ways. Rational cognition (*hakarah*) conceptualizes the world in analytic, discursive, and causal terms that differentiate between subject and object, between the "I" and the not-I. It enables humans to observe nature, study its processes, and generate science and technology. Yet, rational cognition also distances us from the spontaneous, pre-conceptual consciousness that Gordon called *chavayah*, a neologism that combined *havayah* (i.e., Being) and *chayyim* (i.e., life). *Chavaya* is lived experience (in German, *Erlebnis*), it is Being that has reached the level of self-awareness, becoming aware of life in its entirety. In lived experience, the human has access to the "image of God" (*tzelem Elohim*), the invisible, non-conceptual "hidden intelligence" (*sekhel ne'elam*), which only humans share with God. The "image of God" is embedded in the human. Those who aspire to live in accordance with the image of God need not seek to transcend themselves or live according to an external ideal; instead, they must allow the true human essence to deepen and spontaneously expand into "cosmic life" (*chayyim olamiyim*). The "image of God" is the categorical imperative whose realization makes individuals realize their most authentic Self (Ratzabi 2011, 188).

These two types of human consciousness—*hakarah* and *chavayah*—determine human orientation to the world: the former distances the Self from the world, allowing for the rise of culture, civilization, science, and technology that constitute a "mechanistic" existence hidden behind a cognitive "shell." By contrast, the latter is an intuitive, unmediated, spontaneous, holistic experience of life's vitality that cannot be conceptualized. Using kabbalistic/Hasidic terminology, Gordon characterizes these "contraction" (*tzimtzum*) of the Self and "expansion" (*hitpashtut*), respectively. Both are necessary for human functioning, but excessive rational cognition cuts individuals off from the sources of creativity and vitality (namely, Nature/God), positioning them opposite Nature, which they seek to know, control, or exploit. Only if they retain the pre-conceptual experience of lived experience can humans "unite with infinite wholeness (*ha-kol ha-einsofi*) and live within the infinite" (*chai be-khol ha-einsofi*) (Ratzabi 2011, 190). What Gordon wrestled to express using somewhat cumbersome and obscure language, Buber would articulate as the "I-It" and "I-Thou" orientations: a formulation that became world famous and influence many discourses, including religious environmentalism.

For Gordon, excessive instrumental rationality caused the crisis of modernity: "the more 'advanced' (*mitkadem*) the human becomes, the more social life concentrates in large cities … erected at the expense of small towns and villages," the more disengaged the human becomes from creativity and vitality of natural life (Gordon 1951, 46). To heal the existential wound, humans must be open to the "flow of life" (*zerem ha-chayyim*) and be totally present to each other in their social relations.

Modernity exacerbated Jewish alienation from the sources of vitality because of the rational skepticism, critique of religion and unbelief (*kefirah*) that led many to renounce traditional lifestyles and become secular Jews. More than any other critic of religion, Nietzsche exerted profound influence on Zionist intellectuals and on "marginal Jews," namely assimilated Jewish intellectuals, who were deeply conflicted about their Jewish identity but could not bring themselves to convert to Christianity (Golomb 1997). But Gordon rejected Nietzsche's destructive nihilism, which only leads to despair, and adopted his positive, life-affirming message to which he gave a Judaic twist, so to speak. For Jews the only way to live authentically and heal the crisis of modernity was to settle in the Land of Israel and create Hebraic culture that concretized the human ideal of AM-ADAM, namely, of universal humanity expressed in national particularity. Through manual labor in the Land of Israel, Jews could show humanity how "turtle-like" individualism could be overcome by opening the Self to God's presence in the natural world.

Gordon held that the human ideal is not the isolated, "buffered" individual of modern Liberalism but the "humane nation" (AM-ADAM), namely, a nation in which the "image of God" is realized in individuals' organic lives. The nation is a creative, living organism comprising families, within which individuals absorb the nation's values, norms, and expectations through language. Mechanistic society disrupted the human connection with nature, destroying the natural basis of national existence. The only way to revive the nation is to return to manual labor to recreate individuals who live in harmony with nature and are critical of mechanistic social formations. Gordon disliked politics and was critical of Political Zionism, which aimed to establish a Jewish state. Instead, Gordon's vision of organic Jewish society in the Land of Israel begins with the individual and culminates in harmonious humanity. The vision did not ignore Arabs' existence in the Land of Israel. Gordon believed in harmonious Arab and Jewish coexistence because Zionism was an apolitical settlement movement aimed at creating an organic society alongside local Arabs, not at their expense.

Gordon's view inspired Buber and other Jewish intellectuals to establish Brit Shalom in 1925, a study circle devoted to sponsoring informed and responsible discussion of the Arab question. As Scholem stated years later, the group's founders believed that "the land of Israel belongs to two peoples, and these peoples need to find a way to live together … and to work for a common future" (Mendes-Flohr 1983, 72). Buber, a member of the German chapter of the organization, was the most vocal promoter of the notion that a bi-national state is the best solution to the conflict between Zionism and Arab nationalism. But this view was much vilified within the Zionist movement. Instead, the pioneers of the Third Aliyah (1919–1924) took a more dichotomous, militaristic and materialistic approach. Under the sway of the 1917 Bolshevik revolution, they believed that the "country's fate depended on the degree to which Palestine's Jews could successfully deploy military forces" (Engel 2013 [1998], 133). The new pioneers were therefore organized in the Labor Battalion (*gedud ha-avodah*), lending the "conquest of labor"—no longer merely a matter of self-exertion or psychological self-transformation—a more militaristic flavor.

In 1917, Great Britain's Balfour Declaration recognized Jews' right to a "national home" in Palestine, which Britain had conquered. Under the British Mandate the Zionist project of rebirth and renewal was no longer a quasi-mystical fusion of "Man and Nature" but a large-scale construction project as the Zionist settlers transformed the natural landscape by drying wetlands, planting trees, building roads, establishing large collective farms (*kibbutzim*), and building towns and cities. Given Arab resistance to Zionist settlement, the kibbutzim took an increasingly strategic function as "the kibbutz was the village designated to define and maintain borders in the face of this continuing danger" (Troen 2003, 64). As attacks on Jewish rural settlements increased during the 1930s, the kibbutz became the preferred mode of settlement, superseding the moshav (i.e., the private agricultural settlement), and Tower and Stockade (*Homah u-Migdal*) design characterized the rural settlements, surrounding them with defensive walls. In the 1930s the Yishuv also witnessed growing urbanization and industrialization that changed how the Land was imagined, designed, and built (Troen 2003, 112–40).

When Jewish architects, engineers, and city planners from Nazi Germany settled in Palestine in the Fifth Aliyah (1933–1939), the modernist architecture of the Bauhaus movement began to dominate the aesthetics of public spaces of the Yishuv, with Tel Aviv as the model. The rationalist, industrialized approach of modernism characterized the style of many cities and towns. The Zionist dialectical articulation of "to build and be built" no longer meant the "back-to-the-land" movement's romantic pastoralism but a new attempt "to synthesize artifice and nature in the very act of building the country" (Zakim 2006, 152; Cf. Chowers 2002). The leaders of the Yishuv understood that modern technoscience and engineering would determine their survival in a hostile geopolitical environment. Zionism was as much a love affair with technoscience as with the Land of Israel's physical environment. Since the Third Aliyah, the "eros of the land" and the "eros of science" were intertwined (Efron 2011); this fusion persisted after the founding of the State of Israel in 1948 and accounts for much of Israel's technoscientific success.

State Building, Technoscience, and Environmentalism

Zionism, Technoscience, and Technocracy

For Herzl technoscience was essential to the goal of Political Zionism: creating a state in which Jews are the majority. From the start, Herzl conceived the WZO as a state-building institution and, while not a socialist, supported linking the national settlement project to cooperative settlements. After Herzl's death in 1904, his vision of a technoscientific utopia was translated into action by "a group of technical experts and settlement activists who possessed the technical managerial skills needed in order to realize this goal" (Penslar 1991, 1). Together these German and German-trained East-European Jewish engineers provided the technocratic and technological expertise that ensured the Zionist settlement's success.

The focus on rural settlement was crucial. As Penslar explains, "The settlement engineers argued that a technologically advanced agricultural sector would ensure a stable and healthy national economy, and they conceived a program of social engineering that gave priority to agricultural colonization" (Penslar 1991, 2). Technology was the means to creating a healthy national economy that would turn the Jews into "productive" people. However, the technological transformation of the backward Middle East was to be carried out by and for Jews; Palestine's local Arab population was left out of the plans. The settlement engineers regarded Arab animosity as one of many obstacles that "careful planning could clear away" (ibid., 4). Only Arthur Ruppin, who headed the WZO Palestine Office, appreciated the strength of Arab nationalist resistance to Zionism and would later be among the founders of Brith Shalom who called for a bi-national state. Other settlement engineers believed that the Arabs' goodwill could be won once technoscience's benefits to the region became apparent. In historical hindsight, Ruppin was closer to the truth.

Faith in technoscience's transformative capacity reflected Zionism's secular strand. Beginning with Herzl, many German and Central European Zionists who promoted technoscience were assimilated Jews removed from Jewish tradition. Yet it was East European engineers such as Moshe Berman and Yitzhak Wilkansky—who knew Hebrew, came from traditional Jewish homes, and managed the training farms in Palestine that taught agricultural skills to the pioneers—who successfully implemented technoscience in the Zionist settlement project. Marrying technoscience and Labor Zionism's "back to nature" idealism allowed the pioneers "to wrench from arid Palestine better yields than had been witnessed since Joshua and the spies observed that this was a land of milk and honey" (Efron 2011, 421; Cf. Efron 2014, 64–93). Science was "synonymous with progress, enlightenment, and rationality, traits that many Zionists, who were weary of being viewed in Europe as primitives, greatly admired" (Efron 2014, 91). However, reverence for science did not conflict with the Romantic passion for the land but facilitated many innovations constituting Israel's so-called "agricultural miracle."

David Ben Gurion (1886–1973), the leader of Labor Zionism and the first Prime Minister of Israel, perpetuated Herzl's marriage of Political Zionism and technoscience. In the 1930s and 1940s Ben Gurion recognized the importance of science and technology for the success of Political Zionism and the making of a modern nation-state. In 1942 Ben Gurion could not have known the full destructive power of modern technology that the Nazis used against the Jewish people, but he was fully aware of the existential threat to European Jewry. In 1942–1943 Ben Gurion proposed quickly transferring one million Jews from Europe to Palestine. The "Million Plan" expressed the modernist, technoscientific outlook of Ben Gurion's secularized political theology. In October 1943, on the eve of establishing the Planning Committee to carry out the Million Plan, Ben Gurion declared:

> We need engineers and nature researchers who will plan on the basis of our experience in Palestine and the experience of other lands, restoring the ruins and making the desert bloom, on the sale of a national state, so that we can do to all the available and infertile land what we have done with the six percent we have conquered so far. [...] Let the sages of Israel—the sages of earth and rain and plants—reveal to the sages of the nations the new truth from Eretz Israel: the truth of a desolate land waiting for redemption.
>
> (cited in Barell and Ohana 2014, 18)

Technoscience would save Jews from extinction.

Under Ben Gurion's leadership, the Land of Israel was reimagined as a place of unlimited absorptive capacity that could dramatically increase its population, regardless of the Land's small size and limited natural resources. He believed that the outstanding "human factor," with its innovative and creative energy, would create a model state, a secular "light to the nations." The Fourth and Fifth Aliyot established urbanization prior to 1948, as architects, engineers, and designers worked diligently on building "new suburbs, housing projects, stores, hospitals, factories, public buildings and even sumptuous villas for the wealthy" (Troen 1988, 22). Ben Gurion and the Labor movement leaders instituted national plans for absorbing the mass migration of Jews. These plans, drawn together with British civil servants of the Mandatory administration, pertained to industry, transportation, communication, social services, education, and leisure no less than to agriculture, but even agriculture had to be modernized by technoscience. After 1950 Israel would take for granted technoscience—manifested in industrialized agriculture, massive infrastructure projects, desalinization plants, nuclear power reactors, and other technoscientific achievements—as the key to the Jewish state's survival and thriving.

The "Greening" of Zionism: Israeli Environmentalism Emerges

Large-scale technoscientific and engineering projects changed the landscape of the Land of Israel in the 1950s and 1960s, and ironically enough, they would give rise to the emergence of Israeli environmentalism. Concern for the environment's well-being had already emerged in Israel in the 1950s. The Israeli environmental movement began with the Society for the Protection of Nature (SPNI), founded in 1953 following the campaign to oppose plans by Israel's government and the Jewish National Fund (JNF) to drain the Huleh Valley wetlands (Tal 2002, 113–38). Founded in 1901 in the Fifth Zionist Congress, JNF has played a key role in the environmental story of Israel as the leading land acquisition agency of the World Zionist Organization. Committed to the goal of "conquering the land," JNF raised funds from Jews all over the world for land purchase, land reclamation, and land improvement, making thousands of dunams (a measure of land area in the Ottoman Empire which is equal to about 900 square meters) into arable land or laying the foundation for urban development, thus enhancing the bond between Diaspora Jews, Zionism, and the Land of Israel. Realizing Herzl's idea that the Land of Israel should be owned by a public trust rather than private owners, JNF purchased land on behalf of the Jewish people, serving "as the Jewish People's trustee of the land." It leased land long-term rather than selling it to individuals. During the British Mandate (1920–48) JNF was mainly involved in afforestation, working closely with the British civil administration, sometimes introducing species of trees that were not suited to the environment (Tal 2013, 30–55). In 1953 the JNF was dissolved and reorganized as an Israeli company (JNF-KKL) and in 1961 JNF and the Israeli government merged their land holdings for the purpose of joint management under a single Israel Land Authority. Since 1961 JNF has sole authority to conduct afforestation activities in Israel and still owns 13 percent of Israel's public land. Together with the Israeli government, 90 percent of land in Israel is public real estate. As Alon Tal notes, land ownership in Israeli public policy illustrates "the Earth-is-the-Lord's approach" (Tal 2008, 295), another example of Zionist secularization of religious values (intentional or not). That approach and its biblical source (Ps. 24:1) are cited by Jews in the Diaspora to justify religious environmentalism.

JNF's afforestation and land reclamation projects changed the look of Israel's landscape but also acted as a catalyst of environmental consciousness (Tal 2002, 79–85). The draining of the Huleh Valley wetlands is a case in point. Although the campaign against the project failed, the opposition to it galvanized "a core of Green activists and set the stage for the establishment of the Society for the Protection of Nature in Israel" (ibid., 117). In its first decade (1953–63) SPNI focused on nature conservation and "public educational activism," which led to establishing government bodies responsible for protecting nature: the Nature Reserves Authority and the National Parks Authority. In the early 1960s, a budding environmental awareness began to "green" Zionism, namely, give rise to a new environmental code of ethics that defined relations between people and environment in Israel. From the mid-1960s to the end of the 1980s, SPNI emphasized environmental education and hiking, and since the 1990s has added a sustainability focus. As environmentalism came to the fore of SPNI's agenda, the organization has led the struggle to keep open spaces, preserve the built and natural environment, and include the public in planning future projects to be more sustainable. SPNI now boasts 45,000 full-time members and over 600 personnel, an annual budget of over $196 million, and 400,000 annual participants in walking tours, nature hikes, trips to archeological sites, and field schools.

Through its field schools and organized nature hikes, SPNI inculcated a love of the land, its topography, flora, and fauna in scores of native-born "new Jews," whose symbol became the prickly pear cactus, the Sabra (O. Almog 2000). A major instrument of socialization, the nature hike (*tiyul*) has enabled Israelis to express their emotional attachment to the land in ritualized behavior that was no less than "secular worship of the land" (Aran and Gurevitz 1991, 31). Now physical contact between the human body and the earth was not accomplished through manual labor in agriculture or construction, as in Zionism at the beginning of the 20th century, but through "conquering" challenging terrains. The Sabra culture venerated the land not as an abstract metaphysical reality or a religious aspiration, but as object of intimate, embodied knowledge. Orit Ben David, who analyzed SPNI's nature hikes, succinctly defined the experience as a "component of Israeli civil religion" (Ben David 1997, 131). Going even further, the sociologist Oz Almog speaks of "the Zionist religion" presenting the Sabra as a "secular yeshivah student" (O. Almog 2000, 18–22). By secularizing religious concepts such as "redemption," "covenant," and "sacrifice," Zionism merged religion and nationalism; the "pioneer was a devout believer who observed the precepts of patriotism with physical and spiritual valor" (O. Almog 2000, 20).

The kibbutzim, the communal settlements that functioned as the intellectual and cultural *avant garde* of the Labor movement, spearheaded the secularization of Judaism in modern Israeli culture. In the kibbutzim, the Jewish holidays of Passover, Hanukkah, Lag Ba-Omer, Shavuot, Sukkot, and

Purim were "refashioned in a way that reduced, and sometimes even eliminated, the religious observances attached to them, while emphasizing their links to the agricultural cycles of planting and harvest" (O. Almog 2000, 21). This was particularly evident in the celebration of Passover as members of kibbutzim wrote new Haggadot in which they celebrated nature, human agency, freedom, and political renewal. The modernist art of kibbutz culture links nature's vitality and fecundity to the renewal of Jewish national life in the ancient homeland, a historic homecoming.

Zionism replaced Judaism's traditional myths with its own foundational myths that functioned like sacred narratives but glorified human agency and heroism, past and present. The myth of new beginning, of fighting to the end, of the few against the many, and of voluntary sacrifice forged the Sabra's self-image and justified the Jewish historical right to the Land of Israel (O. Almog 2000, 35–45; Zerubavel 1995). These myths and the public ceremonies they generated shaped Israeli collective memory, national ethos, and cultural sensibilities as a secularized Judaism rooted in no text but the Bible.

Although we can trace the rise of environmental consciousness to the 1950s, in the state's first three decades, environmentalism had low priority in the national agenda, which privileged development, construction, absorption of immigrants, and security. Until the 1960s, the first generation of Israeli environmentalists focused on protecting nature from development, providing nature education to strengthen the attachment to the land, and preserving wildlife and habitats as part of larger nature recreation. With growing industrialization during the 1960s, the environmental movement increasingly focused on distinctly environmental issues, including "population growth, the environmental impact of technology, air pollution, safe drinking water, hazardous waste management and pesticide use" (Orenstein and Silverman 2013, 358).

Since 1977, late capitalism, globalization, postmodernism, technoscience, and post-secularism have profoundly transformed Israel's politics, economy, society, and culture. As Israel became more ethnically diverse and socio-economically stratified, post-Zionists leveled scathing critiques at Zionist narratives, exposing disparities, inequalities, and injustices either to the indigenous Arab population of Palestine or within the Jewish population of the State of Israel (Nimhi 2003). Toward the end of the 20th century, higher standards of living and technoscientific successes clearly made Israel a "developed" rather than "developing" country. Precisely because it was emerging as a developed country, Israel had to pay attention to environmental issues that impact its citizens' quality of life. Environmentalism marked Israel as an economically and techno-scientifically advanced liberal democracy.

In the 1980s public health emerged as the overarching theme of Israeli environmentalism. New environmental organizations emerged—such as MALRAZ: The Public Council against Noise and Air Pollution, *Adam Teva*

V'Din: The Israeli Union of Environmental Defense, the Coalition for Public Health, and the Green Course (Tal 2002)—which focused on air quality, waste disposal, water quality and quantity, health risks, and public transportation. Since the 1990s and especially after 2000, Israeli environmentalism shifted again, "as the environmental movement's agenda has been converging with that of the social justice movement" (Orenstein and Silverman 2013, 364). Concerns for social justice emerged from growing economic disparities between the haves and have-nots, the Sephardim's resentment toward the Ashkenazic privileged elites, and awareness of the plight of marginalized groups (e.g., the Arab sector, the Bedouins, the disabled, or senior citizens). By promoting the notion of human rights to address social disparities, the environmental movement translated commitment to social justice to commitment to the environment. A more integrative, place-centered approach, which views humans as integral to the natural world, has emerged, according to which "what is good for nature is good for people and vice versa" (E. Schwartz 2009). The converging justice and environmental discourses engendered a nationwide network striving to protect the environment, conserve nature, address public health issues, and promote sustainable development. As the environmental movement became a social movement, it appealed to both science and technology and the moral and ethical values of the Jewish tradition to address specific challenges. This gave rise to Judaic environmentalism, which we will discuss below.

Israel has no shortage of environmental challenges. Growing attention to environmental problems led to establishing the Ministry for Environment Quality in 1984 (later to change its name to Ministry of Environmental Protection), though it remains the least funded governmental ministry. In the 1980s international environmental organizations (e.g., GreenPeace and Friends of the Earth) opened branches in Israel, making the Israeli environmental movement part of a larger global discourse and increasingly professional. Israeli media also began covering environmental issues, especially as environmental organizations got involved in legal action against development projects that they claimed would endanger the environment. Some of the successes convinced environmental activists that non-governmental organizations can make a difference in the national agenda. Public advocacy led to establishing parks and natural preserves, saving species from extinction, passing environmental laws, and fueling resistance to development projects that reflect economic interests or security concerns at the expense of ecological integrity.

The environmental movement has also organized itself politically to make an impact on environmental legislation, albeit with limited success. The first environmental party, The Greens (*Ha-Yerukim*), was founded in 1997, but has yet to gain seats in the Israeli parliament (the Knesset), despite a 2008 merger with a rival green party, the Green Movement (*Tenuah Yerukah*). In the first decade of the 2000s the Israeli environmental movement learned how to become more effective while working within the parliamentary process and the rough-and-tumble of Israeli politics (Karassin 2013). Although the green parties failed to enter the Knesset, they won seats in municipal elections, indicating the growing importance of environmentalism for Israeli citizens. The most interesting development was the emergence of a liberal religious party, Meimad, which focused on education, peace, and religious pluralism. Under the leadership of the Danish-born Rabbi Michael Melchior, Meimad indicated that environmentalism is a religious issue and that Judaism has much to contribute to the greening of Israel; environmentalism need not be the banner of secular Israelis. Meimad did not last long as a political party but fusing traditional Judaism with environmental consciousness has inspired a new, faith-based civic consciousness, to which we will return below.

Israel has a robust environmental movement that cannot be discussed here in detail. The works of Alon Tal (e.g., 2002; 2008; 2011; Tal et al. 2013), Itay Greenspan (2015), and Benny Furst (2016) provide overviews of the movement's accomplishments and challenges. Israeli environmental activism's main concerns are climate change, renewable energy, water issues, public health, ecological economics, environmental planning, gray water, communal gardens, and organic agriculture. Organizations for preventing animal cruelty were not included in the survey, but that is yet another concern of environmental activism. Many organizations share concern for sustainability, environmental planning, democracy and public inclusion, preservation of open spaces, public health, protection of nature and biodiversity, air pollution, and hazardous materials; desertification, overpopulation, soil erosion, and fair trade receive much less attention, even though environmental activists recognize their importance.

One nonprofit NGO—Hayyim U-Sevivah (Life and Environment: The Israeli Union of Environmental NGOs)—provides a broad framework to foster cooperation and offer professional assistance among Israeli environmental organizations. Its civically important work; however, it makes no reference to Judaism, Jewish values, or Jewish literary sources to legitimate its social agenda. Environmentalism therefore is rightly perceived as an expression of Israeli secularism.

Despite Life and Environment's efforts, the environmental movement is quite fragmented, and tensions exist among the various organizations. On some issues (e.g., globalization) the organizations diverge, since small organizations are more concerned about the environmental and social implications of globalization than the large organizations. Smaller organizations promote sustainable development, protection of nature, social justice, integrative economy, environmental technology, public health, anti-globalization, and love of the Land of Israel.

Just as all environmental organizations in the United States face the major challenge of fundraising, in Israel many environmental projects operate without a formal budget, and private philanthropies, civic organizations, and local municipalities are now the main sources of funding. Since Israel has adopted Neo-liberal, market economics, it is no surprise that environmental organizations must rely on private funding. Diaspora-based philanthropies such as the New Israel Fund, the Abraham Fund, the Nathan Cummings Foundation, and the Jim Joseph Foundation, which have also supported American environmental organizations, have underwritten environmental initiatives and projects in Israel.

The Israeli environmental movement has questioned the earlier Zionist embrace of technoscience and shown the environment's relevance to the quality of life in a democratic Israel. As the movement has matured, environmental activists have increasingly turned environmentalism into a social movement that invokes values such as fairness, justice, equality, pluralism, and diversity. These values came not from Western (secular) liberalism but from the Jewish tradition. They reflect Judaism's growing presence in the Israeli public sphere—in other words, the emergence of Israel as a post-secular society. This transformation began after the Six-Day War of 1967, which deepened the conflict among the various strands of Zionism precisely because it pertained to the Holy Land, its boundaries, and spiritual meaning.

Nature, Religion, and Politics in Israeli Post-Secular Society

Political Theology and the Settlement Movement

The Six-Day War in June 1967 was a watershed in the history of Zionism and the State of Israel. Exposing Zionism's inherent paradoxicality but not resolving it, Israel's remarkable victory over the Egyptian, Jordanian, and Syrian armies proved Israel could defend itself against its enemies while creating an insoluble conundrum, a "catch '67," as Micha Goodman insightfully called it, that has torn Israel apart from within (Goodman 2017). The war brought the West Bank, the Golan Heights, the Gaza Strip, and the Sinai Peninsula under Israel's control, but what should be the status and future of these new territories?

For many secular Israelis, the conquered territories would best be negotiated in a peace agreement with Arab nations as the territorial basis of a Palestinian State. Placing the territories under Israeli sovereignty would undermine Israel's identity as a democratic state in which Jews are the majority. In contrast, for many religious Jews, 1967 was not just a war for Israel's survival but a miraculous, divine intervention that brought Jews and the entire world closer to redemption. For them, the newly conquered territories were not assets in political negotiations but sacred land that had to remain in Jewish control for Jewish settlement only. This was particularly true of Judea and Samaria, the terrain of much of biblical history. The Gaza Strip, Golan Heights, and Sinai Peninsula were also claimed to be part of the Greater Israel (*Eretz Yisrael Ha-Shlemah*) that had to be settled by Jews, but they lacked the same level of sacrality. The secular and religious visions were bound for collision.

The fact that Israel's swift victory in the Six-Day War came two decades after the Holocaust explains the euphoria that engulfed Israel and Jews all over the world as well as the messianic significance attributed to the war (Sagi and Schwartz 2019, 16–19). With support from Diaspora Jewry, the young state proved that it could not be defeated, bringing great pride to Jews worldwide and unleashing renewed commitment to Zionism. In the immediate months after the astonishing victory, it was difficult to heed the lone voices (e.g., Yeshayahu Leibowitz) who called for returning the conquered territories in a comprehensive peace settlement with Israel's Arab neighbors. Instead, the emerging Settlement Movement received support from the Labor government, which saw it as continuation of the Zionist project.

Gush Emunim (The Block of the Faithful), founded in February 1974 and entering Israeli political life in December 1975, spearheaded the Settlement Movement. A group of restless young politicians, agitating for change within the Religious National Party (RNP) and responding to the humiliation of the Yom Kippur War of 1973, organized Gush Emunim. It was imbued with messianic consciousness that questioned existing political institutions' legitimacy and operated under the ideology of illegalism.

From its inception, Gush Emunim activists presented the Settlement Movement as the "real Zionists" and true heirs of the Zionist pioneers who created the collective settlements in the 1920s and 1930s that had since grown lax in their commitment to settling the Land of Israel. Adopting the tactics of the "Tower and Stockade" campaign from the 1930s—in which Jewish settlements were established overnight as elevated observation points surrounded by fences and walls to establish "facts on the ground"—the Settlement Movement appropriated the mantle of Zionist activism, erecting settlements in the heart of the West Bank Arab population. Unsurprisingly, Labor politicians at the forefront of the Zionist settlement project gave their initial blessing to the Settlement Movement, even though the latter was deeply critical of the Labor movement's hegemony and keen to expose "the impotence and flaws of democracy" and "show the Jewish people the light" (Zertal and Eldar 2007, 185). The "light" would come from the mystical teachings of Abraham Isaac Kook (d. 1935), which his son, Rabbi Zvi Yehuda Kook (d. 1982), translated into full-fledged political theology.

Abraham Isaac Kook: The Holy Land and Humanity's Religious Perfection

Known as Rav Kook, Rabbi Abraham Isaac Kook is considered "the greatest Jewish mystical thinker of the twentieth century" (Fine 1995, 23). A towering Talmudic scholar, religious existentialist, poet, and spiritual leader, Kook was crucial to Yishuv history as a figure seeking to bridge the secular and religious camps. Like A.D. Gordon, Kook settled in Palestine in 1904, where he was first the Chief Rabbi of Jaffa and the surrounding colonies (1904–1919) and later the Chief Ashkenazi Rabbi of the Yishuv (1919–1935). Intellectually and spiritually, Kook and Gordon had much in common, integrating Talmudic learning, Kabbalah, Hasidism, Haskalah, and

European thought into a coherent worldview that guided their involvement in Yishuv public affairs (Strassberg-Dayan 1995). But whereas Gordon was an autodidact, Kook was the product of elite Talmudic educational institutions; whereas Gordon lived with the secular pioneers and eventually became a non-observant Jew, Kook was an Ultra-Orthodox Jew living by Halakhic strictures, for whom Judaism had a universal vocation: bringing humanity to religious perfection. Interpreting the Zionist national renewal in messianic categories, Kook was remarkably open to the secular pioneers, whom he regarded as active contributors to Judaism's messianic task, although they themselves failed to understand it. He wanted the Orthodox camp to respect the Zionist pioneers and not reject them as "sinners" and "heretics," and conversely, he wanted the Zionist secularists to understand the settlement project's religious and even cosmic significance. This fusion of the secular and the religious dimensions makes Kook, no less than Gordon, a thinker who anticipated post-secularism and whose ideas could support ecological spirituality.

Born in 1865 in Grieva, Latvia, into an ultra-Orthodox family suffused with Chabad Hasidism, Kook was considered a child prodigy. At fifteen he went to Liutsin (Ludza) to study with a Hasidic-minded Talmudist while also becoming familiar with Haskalah literature in the library of his cousin, a self-professed Maskil who regularly gathered youthful disciples. Like other proponents of the Haskalah movement, the young Kook became proficient in modern Hebrew and familiar with European philosophy and science, although he continued to express "critical views of modernization and secular university education" (Mirsky 2014, 14). He set to study at the famous yeshivah of Volozhin, which attracted young men from all over Eastern Europe. There he became the disciple of the Naftali Zvi Yehudah Berlin (d. 1893), who integrated orthodoxy with knowledge of philosophy and science (Perl 2013).

In the last quarter of the 19th century, as we have noted, Imperial Russia's Jews experienced profound crisis. In addition to frequent pogroms, poverty, growing political insecurity, and mass emigration, Jews abandoned their traditional way of life and joined secular revolutionary movements. The profound crisis of faith convulsed Eastern Europe's Orthodox academies (yeshivot). Kook was most attentive to the turmoil and understood the depth of the crisis that led many Orthodox Jews to abandon traditional belief and practice to join revolutionary movements, become adherents of European Enlightenment, or endorse Zionism.

Familiar with Haskalah writings and sympathetic to the Love of Zion movement, Kook was deeply concerned about Political Zionism because it defined itself in secular terms. For Kook, only Jewish messianism legitimates Zionism. In 1904 Kook agreed to settle in Palestine as the Rabbi of Jaffa, joining his father-in-law, who two years earlier became the deputy Chief Rabbi of Jerusalem. Kook's spiritual mission was protecting Judaism against modern unbelief, on the one hand, while framing Zionism as part of Jewish messianism, on the other. His efforts to bridge the Yishuv's secular and religious camps often brought him into conflict with other leaders of the Orthodox community. The Sabbatical year controversy was a case in point.

Colonizing the Land of Israel raised religious challenges for observant Jews. Precisely because the Land of Israel was the Holy Land, how should it be farmed? The First Aliyah settlers were observant Jews for whom returning to the Land of Israel made possible observance of Judaism's land-based commandments, impossible to observe in the Diaspora. But how do you apply the ancient commandments to the reality of modern agriculture? The issue was acute, because the laws of the Sabbatical year (*Shemitah*) required the land to lay fallow every seventh year. As early as the 1880s, settlers of the early colonies appealed to the Lithuanian rabbinate. Three Lithuanian rabbis met in Vilna and designed a *hetter mechirah*, a rabbinic legal device that allowed the settlers to sell their land to non-Jews for the duration of the Sabbatical year, during which the settlers could continue cultivating the land, since the Sabbatical year laws do not apply to land owned by non-Jews. This ensured that the young agricultural colonies would not collapse. Rabbi Yitzchak Elchanan Spektor, Chief Rabbi of Kovno and pre-eminent Halakhic authority, approved the sale permit. Settlers relied on these permits during the Sabbatical years of 1889, 1896 and 1903 but the matter was by no means resolved, since highly respected rabbinic authorities vociferously opposed the lenient ruling.

With 1909–1910 a Sabbatical year, Kook had to address the issue just five years after settling in the Land of Israel. He took a forceful position defending the Halakhic validity of the sale permit and published a treatise, *Shabbat Ha-Aretz* (The Sabbath of the Land), to explicate his legal reasoning along with detailed discussion of the Sabbatical year laws. According to Kook, the sale of the Holy Land to non-Jews is halakhically valid and "the categories of work that are biblically forbidden can be permitted only if done by Gentiles" (Nehorai 1995, 129). This position was lenient in permitting land cultivation during the Sabbatical year and indicated Kook's historical awareness, stating that "in our days, when the basis of the Yishuv is commercial agriculture, and preventing this commerce would destroy all livelihood and future standing ... it is downright obligatory to maintain the permissive annulment through sale" (Mirsky 2014, 74). However, Kook departed from the views of previous legal authorities, including his own revered teacher, in regarding permission of land sale to non-Jews not merely as a temporary concession but because the Zionist settlement project already possessed a redemptive quality. Therefore, the Zionist settler would

> seek through his action and his influence, to hasten the flourishing of our salvation and the revelation of the light of the Messiah. For the lights will be effective in the abundance of their holiness even if, Heaven forbid, his actions, in practice, were unsuccessful.
>
> (Kook, *Iggerot* II:196, cited in Nehorai 1995, 130)

Because he invested Zionism with messianic import, Kook held that "the freethinking rebels play an effective role in the sequence of events, helping to move matters along and even struggling toward a certain end without grasping the inner logic of the occurrences, their true meaning or real consequences" (Zertal and Eldar 2007, 194). He visited the agricultural settlements and developed personal relations with A.D. Gordon—his spiritual kin—and other leaders of the pioneering community. In the grand scheme of universal, sacred history, Zionism is but the first step in the messianic process of redeeming humanity.

Rav Kook believed in Judaism's unique role in cosmic evolution. The universal striving toward perfection, toward the full realization of the absolute good in the Ein-Sof, is made possible by the Torah of Israel. It is born by the People of Israel and requires their presence in the Land of Israel. In this religious interpretation of Zionism, the physical revitalization of the Jewish people through Zionism laid the foundation for spiritual perfection. This too required the Land of Israel. Settling the Land of Israel makes possible the observance of the Torah's land-based commandments, enabling observant Jews to unite life's material and spiritual dimensions. The Land's essential, spiritual quality—its holiness—comes to light when Jews not only dwell in the land and build it into a homeland for the people of Israel but when they fully grasp its unique, spiritual quality.

Rav Kook exhorts all Jews to see Zionist settlement as a religious, messianic project. In his *Orot Ha-Kodesh*, he states: "We are commanded to bite deeply into the delightful sweetness of the land of Israel's glorious, invigorating holiness. [...] Come back, come back, you cast-away, dispersed children, to your mother's bosom" (cited in Schweid 1985, 184).

Kook's holistic and organic worldview grounds the sacred in the mundane, investing it with holiness, and conversely, the mundane, or the corporeal, concretizes and enlivens the sacred. But how does one translate this into politics? Rav Kook died before the State of Israel's founding, but he was very involved in the Yishuv's political life and in Jewish Diaspora life. Jews could not find redemption by participating in the political life of ordinary states, because the state is not "the supreme happiness of man." But the future State of Israel would differ as "the foundation of God's throne in the world." The phrase "State of Israel" was not commonly used when Rav Kook was alive, but he envisioned the Yishuv's achieving statehood. That state, he held, would be unique, governed by Halakhah: a theocracy. It was Rav Kook's son, Rabbi Zvi Yehuda Kook, who turned the vague vision into the political theology that sacralized the State of Israel and gave theological coherence to the post-1967 Settlement Movement.

Re-Mystifying the Land of Israel

The Six-Day War unleashed messianic fervor that presented the new State of Israel as a divine miracle in response to the catastrophe of the Holocaust.

After the Yom Kippur War, the messianic rhetoric became even more pronounced, and Gush Emunim was founded under the leadership of graduates from Mercaz Ha-Rav Yeshiva (Hellinger, Hershkowitz, and Susser 2018, 32–37). Rabbi Zvi Yehuda Kook's disciples believed that enormous messianic progress had occurred which in turn demanded that they move beyond focusing on Torah study for its own sake to become public figures who shape Israeli society and culture (Inbari 2012, 35). They insisted that redeeming the Land of Israel is the core element of both Judaism and Zionism and that the People of Israel, the Land of Israel, and the State of Israel were all participants in the eschatological drama led by Gush Emunim. The nation, the land, the State, and the Torah were integrated into a single messianic vision (Aran 2005; 2013).

Since the Land of Israel is "the soul of the Jewish people," Gush Emunim members were prepared to disobey the law of the State should it hamper their settlement effort. They saw,

> our right to settle everywhere in the land [as an] elementary Jewish right. This right has not expired with the establishment of the state and every directive that limits it is neither legal nor moral—it is imperative to revoke it.
>
> (cited by Sprinzak 1991 [1986], 125)

Civil disobedience became the mark of the Settlement Movement, which believed that the commandment to settle the Land of Israel is equal in worth to the entire Torah. In their mind, only when Halakhah governs the State of Israel will Israel become "the essential throne of God in the world," as Rav Kook envisioned. In his messianic vision, Israel becomes a Halakhic state, that is, a theocracy, and human history becomes sacred history.

In Rabbi Zvi Yehuda Kook's political theology, all the land conquered in 1967 is Holy Land and part of Greater Israel (*Eretz Yisrael Ha-Shelemah*). Inspired by the teachings of Nachmanides, who interpreted Numbers 33:53 ("Take possession of the land and settle in it, for I have given you the land to possess") to mean that we "are commanded to take possession of the land and not leave it in the hands of other peoples or allow it to become a wasteland," Rabbi Kook held that this commandment applies to all generations and compels everyone in Israel. He further insisted that one must sacrifice life for every single part of the Land of Israel and that no part of it should be transferred to non-Jews. Furthermore, according to Rabbi Kook, the land's sacredness corresponds to the State of Israel's sacredness. He declared that

> The State of Israel is divine. Not only must there be no retreat from [a single] kilometer of the Land of Israel, God forbid, but on the contrary, we shall conquer and liberate more and more … In our divine, world-encompassing undertaking, there is no room for retreat.
>
> (cited in Zertal and Eldar 2007, 212)

When the Likud Party came to power in 1977, sacralizing Greater Israel became public policy. The Settlement Movement grew by leaps and bounds through its fervent adherents' eagerness to live the holy life and the Likud government's heavily subsidizing mortgages and investing enormous sums in upscale homes. Between 1967 and 1977 there were 6,000 settlers; between 1977 and 1981 thirty-five new settlements were established with 17,000 residents; by 1992, there were over 100,000 settlers in the West Bank. The Camp David accord of September 1978 and the peace agreement with Egypt in March 1979 required full retreat from the Sinai Peninsula, including sixteen settlements and the town of Yamit. The evacuation of Yamit in 1982 (a few months after Rabbi Zvi Yehuda Kook's death) pitted the Likud government of Menachem Begin (who signed the peace treaty with Sadat) against religious Zionism's leadership (who supported Gush Emunim) and Yamit's secular settlers against the religious Zionist activists (Hirschhorn 2017, 58–97). In 2005 Israel disengaged from the Gaza Strip and dismantled all the settlements of Gush Katif, a process led by Ariel Sharon, who as Minister of Agriculture was instrumental to the Settlement Movement's success. Retreating from the Sinai and disengaging from Gaza made Judea and Samaria (i.e., the West Bank) even more crucial for the Settlement Movement and not subject to negotiation. Neither the Oslo Peace Process in the 1990s, under the Labor government of Yitzhak Rabin, nor the continued rebuke of successive American governments could halt the Settlement Movement's growth. After the 1995 assassination of Rabin by a Gush Emunim supporter, Yigal Amir, and Netanyahu's 1996 rise to power, the Settlement Movement created "facts on the ground," undermining any attempt to establish a Palestinian State in the West Bank.

Today there are about 465,000 settlers in the West Bank and another 220,000 in East Jerusalem, and the government of Benjamin Netanyahu now considers annexing these territories to Israel. The settlers' commitment to Greater Israel's holiness had little to do with environmentalism and everything to do with the Jewish political sovereignty in what they define as Greater Israel. From their perspective, Arab resistance permits violence, since the Holy Land must not be in the gentiles' possession. Radicalizing Zvi Yehuda Kook's teachings, Rabbi Yitzchak Ginsburgh and his disciples articulated a uniquely ethnocentric, messianic blend of kabbalistic-Hasidic ideas justifying the killing of non-Jews during peacetime and in war. This ideology supports violent vigilante behavior by the marginal settler group Tag Mechir (Price Tag), "loose congeries of disaffected, frustrated, often psychologically troubled youth who see it as their duty to take revenge upon Israel's enemies, to exact 'payment' for their perfidy" (Hellinger, Hershkowitz, and Susser 2018, 248).

Hilltop Youth is a radical offshoot of the Settlement Movement that positions itself in various outposts unauthorized by the government, aware that they may be forcibly removed by the Israeli army (IDF). Whether the army would do so depended on the government in power. Some several hundred youths coalesced around a hard core of violent activists who establish illegal or disputed outposts outside existing settlements. Several thousand people from within the Settlement Movement support their actions and ideology. The militant Hilltop Youth (most of whom are no longer young) see themselves as people who "return to nature" by consuming only natural foods, riding horses, and eschewing the trappings of Israeli consumerist society. Their lifestyle derives inspiration from the early Zionists' agrarian romanticism, but they are much more bellicose than the early Zionists; these isolated settlers have no qualms about harassing their Arab neighbors by preventing the olive harvests, "Judaizing" springs and making them off-limits to Palestinian shepherds and their herds, destroying property, and employing physical abuse. The justification for these actions came from the chauvinist teachings of Meir Kahane, who urged "deportation, revenge, and annihilation of Gentiles that posed a threat to the people of Israel" (Pedahzur 2012, 135–36). The militant actions received mystical justification from Rabbi Ginsburgh and his followers, who consider violence integral to returning to the soil. Indeed, the soil itself is endowed with metaphysical import that excludes gentiles' presence in the Holy Land (Hellinger, Hershkowitz, and Susser, 252). In 2014 the UN Office for the Coordination of Humanitarian Affairs documented 330 hate crimes in the West Bank, the direct outcome of Settlers' aggression. The IDF does not interfere in preventing violence, and the police are highly ineffective in these remote areas.

While Hilltop Youth theologize and mystify the Land of Israel, they have no interest in protecting and conserving nature. Their disregard for environmentalism also pertains to the massive urban settlements in the West Bank that attracted many American Jews, about 15 percent of the total settler population. The Six-Day War rekindled Zionist passion among middle-class, college-educated American Jews who were disenchanted with Jewish life in America, either because it failed to address their spiritual quest or disappointed their expectations following the civil rights movement (Hirschhorn, 2017). The turmoil of the 1960s generated diverse responses among American Jews, including the return to Orthodoxy (i.e., the Ba`alei Teshuvah movement discussed in Chapter 2), Jewish feminism and Neopaganism (Chapter 4), the flowering of Jewish Studies, and the making of Zionism a primary commitment for American Jews.

The public confrontation between green organizations and the Settlement Movement took place in 2007, but it pertained to settlement within the Green Line rather than in the West Bank. Israel's unilateral disengagement from the Gaza Strip in 2005 was a traumatic event for the Gush Katif and North Samaria settlers, and it raised the question of where to resettle them. After prolonged deliberation, the government authorized settling the evacuees, mainly associated with Gush Emunim and the National Religious Party, in the Lachish region bordering the Hebron mountain's southern slopes. This decision encountered vociferous opposition from environmental organizations, led by SPNI, who argued that the region is ecologically sensitive and that settling 350 families in the region would undermine its ecological integrity. The clash between the (religious) evacuees and the (secular) green organizations highlighted different understandings of nature within the framework of Zionism. The green organizations appealed to environmental sciences to support their claim that human settlement would harm the region's birds and vegetation, whereas the evacuees argued that by settling the area they are fulfilling Zionism's religious and national goals, marking the Land of Israel as Jewish space. The bitter legal struggle reached Israel's Supreme Court, which sided with the evacuees and allowed the settlement of Lachish to proceed.

Closely examining the case shows how important nature was to both sides and how each framed nature differently, misperceiving the opposing side's meaning (Shani 2009). The evacuees claimed the greens privilege "nature" over "Man," whereas the greens claimed the evacuees are beholden to an old but mistaken Zionist view according to which love of the Land of Israel is manifested in settlement, planting trees, and nature hikes. The conflict juxtaposed the greens' interest in the "environment" (*sevivah*) against the religious Israelis' loyalty to "land" (*adamah*), indicating that the rhetoric of nature is always culturally embedded and inseparable from politics. The settlers believed that attachment to the land determines the Land of Israel's political boundaries, whereas environmental organizations were acutely aware of the environmental cost of politicizing nature. These perspectives clashed in the Lachish region and even more in the debate about the West Bank barrier fence, built in 2005 and causing profound ecological damage, social disruption, and human rights violations for the resident Palestinians (Sadeh 2010). Holiness could not prevent ecological damage to the highly politicized land.

Judaic Environmentalism and the Betterment of Society

The complex story this chapter presents makes clear that sanctifying the land does not necessarily entail committing to environmentalism and conversely that Jews' attending to environmentalism does not necessarily mean committing to Jewish religious values. However, since the 1990s there have been deliberate attempts to integrate Judaism, Zionism, and environmentalism into a coherent vision of Israel as a democratic, pluralistic, progressive, technoscientific society that addresses its environmental challenges by infusing knowledge of environmental sciences with civic-mindedness and concern for social justice. Abraham Joshua Heschel, the spiritual leader, peace and civil rights activist, religious philosopher, and eco-theologian inspires this vision. Rising to meet these needs is the Heschel Center for Sustainability, Environmental Leadership and Learning, which links the environmental value of sustainability with the Jewish value of *Tikkun Olam*.

Founded in 1998 by Dr. Eilon Schwartz and Dr. Jeremy Benstein, "the Heschel Center works to promote a sustainable Israel: a just and cohesive society, a robust and democratic economy, and a healthy and productive environment for all its residents, now and in the future" (Heschel Center n.d.). The organization's strategy is "to attract change-makers from all sectors of Israeli society and provide the knowledge, ideas, and tools, as well as the support networks, to create innovative projects and platforms for change and to inspire others to action" (ibid.). In the Heschel Center, environmental activism, advocacy, and education combine with concern for social justice, civic mindedness, and Jewish ethics to offer Israelis an ethically informed environmental spirituality.

For the Heschel Center, sustainability is the key to Israel's future. Instead of approaching environmentalism as a reactive and confrontational niche issue, the Heschel Center promotes "a broad proactive agenda, integrating economic, political and social change, and training and partnership with key players in all those realms." The organization presents Heschel as a person who "believed that humankind could not fulfill its spiritual potential without the struggle for justice, the eradication of poverty and racism, tireless action for peace, and a renewed relationship to the natural world" (Heschel Center n.d.).

Over the past two decades, the Heschel Center has translated environmental spirituality into an impactful array of initiatives and educational programs. These include its Center for Sustainability Studies, Community Place-Based Education, Environmental Fellows Program, Center for Local Sustainability, Green School Network, and Local Sustainability Network. Like other Israeli environmental organizations, it receives funding from private Israeli and Diaspora foundations and from international organizations and individuals donating to Friends of the Heschel Center. But what makes this environmental organization specifically religious and distinctly Jewish? The answer is made clear in one of its initiatives, "Siach: An Environment and Social Justice Conversation," a forum whose goals are conversation, connection, and cooperation (Siach n.d.).

Working with Hazon: Jewish Lab of Sustainability (United States), the Jewish Social Action Forum (UK), and Bema'aglei Tzedek (Israel), Siach is a platform that "creates global visions for Jewishly inspired social change, visions that have the power to inspire and galvanize the international Jewish community and the world at large." This conversation foregrounds Jewish religious values such as justice (*tzedek*) and holiness (*kedushah*). Rabbi Schwartz acknowledges tension between justice and holiness: justice requires full engagement with the world and concerns for the rights of all people (i.e., universalism); holiness affirms Jewish particularism, an age-old tension in Judaism that can be traced to biblical prophets. While all traditionally grounded Jews refer to the ideal of holiness, the Heschel Center's understanding is clearly radically different from the Settlement Movement's ethnocentric and exclusivist understanding of holiness.

The Heschel Center has been most successful in disseminating the value of sustainability in Israeli society. In 2015 it published "The Optimism Report," covering 60 Israeli initiatives, projects, and organizations committed to sustainability in areas including sustainable management of resources; food and agriculture; community and local economy, youth, education, and training; practical sustainability; democratic culture; business innovation; and sustainable cities (Ettinger et al. 2015). The report refers to sustainability not as a utopian dream or professional occupation but as a "a lens or a compass through which we examine, orient our actions, from daily personal decisions, through business and social activities to public, governmental and global policy" (ibid., 10). The report promotes "appreciative inquiry" that highlights "the goodness in people, their organizations, and the world around them." Positive Thinking (a term coined by Abraham Maslow, which Martin Seligman turned into the psychological approach of Positive Psychology) inspires people to innovate and build a better future for positive change. Inspiration comes from "the willingness of humanity to restore, repair, reform, rescue, reimagine and rethink … this is neither liberal nor conservative activity; it is holy activity" (Ettinger et al. 2015, 10). The Heschel Center's post-secular nature could not be stated more forcefully.

If the Heschel Center illustrates how religious environmentalism in the Diaspora has impacted environmentalism in Israel, mobilizing Diaspora Jewry to support the environment in Israel has been the goal of The Green Zionism Alliance, founded in 2001 by Alon Tal, Eilon Schwartz, and Rabbi Michael Cohen. Schwarz, Tal, and Cohen were graduates of the American Young Judea movement, and all three environmental activists lived for a while in Kibbutz Keturah in the Arava Valley, which is affiliated with the American Young Judea movement. In Kibbutz Keturah, Alon Tal, the environmental scholar-activist, founded the Arava Institute for Environmental Studies in 1996 as an academic research institute that trains leaders to address environmental challenges of the Middle East. Rooted in the conviction that "nature knows no political borders," the Arava Institute convenes Israelis, Palestinians, Jordanians, and students from all over the world to study and collaborate so that they could lead the region and the world in solving today's most pressing environmental challenges. Still in existence, the Arava Institute works across sectors to tackle water, energy, and food security issues—work that requires cooperation and collaboration among groups that are commonly viewed as enemies—but that work receives very little attention from the media.

The Jewish American environmentalists who settled in Israel have been committed to bridging between Diaspora Jews and Israeli Jews. The Green Zionist Alliance, which was located in New York, changed its name to Aytzim: Ecological Judaism, whose goal is to mobilize Diaspora Jewry to support the Israeli environment and makes Zionist institutions such as the World Zionist Congress attentive to environmentalism. If the Heschel Center focuses on environmental education and capacity building, Aytzim illustrates how environmentalists work within existing political institutions. Aytzim was the first environmental party within the World Zionist Congress and for nearly two decades has utilized the World Zionist Congress to obtain seats on KKL-JNF's board, where its representatives have pushed KKL-JNF to become more environmentally sustainable, participate in the UN Framework Convention on Climate Change, treat Jews and non-Jews fairly, and protect the Jerusalem Forest. Aytzim's religious dimension has recently become even more explicit. Its most distinctly Judaic project is developing seven-year environmental plans to reduce greenhouse gas emissions, inspired by the biblical practice of the Sabbatical Year. Indeed, the Sabbatical Year has become the main religious concept inspiring Israeli Judaic environmentalism, as we shall see with Teva Ivri.

Despite impressive advances and a more evident post-secular identity, Aytzim failed to get on the World Zionist Congress ballot. Could it be that non-environmental interests favoring development over conservation have prevailed? Perhaps. In Israel too, the sister organization Hatenuah Ha-Yerukah, Israel's largest environmental party, also failed to run a slate in Israel's election for the Knesset in 2015. It seems that while environmentalism and the ideal of sustainability have successfully transformed Israeli civil society, accomplishing environmental goals through politics may have reached a limit. Yet, there is no doubt that fusing Judaism and environmentalism has been gaining popularity in Israel in the last two decades. The best example is Teva Ivri, an Israeli NGO promoting Jewish sustainability based on Israeli local culture.

Translated literally, Teva Ivri (founded in 2007) means "Hebrew nature," echoing the Zionist commitment to creating modern Hebraic culture. However, the organization's English name is "Jewish Nature," which locates the organization within the Jewish tradition and therefore resonates better with Diaspora Jews. Its founder and director, Einat Kramer, is an Israeli-born Orthodox woman who graduated from the Heschel Center's Environmental Fellows Program and from Tehuda, a professional Jewish leadership program. Kramer, another excellent example of female leadership in environmental organization, believes the "environment in Israel can reflect both a Judaism that is relevant and an environmentalism that is uniquely Jewish" (Kramer, Feb. 25, 2014), thus illustrating the post-secular character of Judaic environmentalism in Israel. As a member of the Israeli delegation to the 2012 Rio+20 Earth Summit in Brazil, Kramer helped promote a notion of "sustainable development" anchored in the sources of Judaism (Kramer 2012). A Jewishly inspired environmentalism means that "humans are commanded to work the land, that is, to develop civilization, but that they are entrusted with the responsibility to protect the land from destruction and overutilization." The official statement of the Israeli delegation refers to the standard sources of Jewish religious environmentalism including creation care, prohibition against wanton destruction (*Bal Tashchit*), and concern for animal welfare (*Tza`ar Ba`aley Hayyim*).

Two distinctly Jewish religious practices—the Sabbath and the Sabbatical Year—anchor the fusion of Judaism and environmentalism. Teva Ivri makes them the focal point of its educational initiatives. In Teva Ivri's environmental interpretation, the Sabbath, which offers freedom from production and consumption, is a practice that creates an openness to the spiritual values that consumerist society marginalizes (Kramer May 7, 2014). All Jewish environmental organizations and faith-based non-Jewish organizations recognize the Sabbath's environmental significance. As we have already noted, concern for the Land of Israel's well-being and future is one area that can establish common ground across the ideological and social spectrum, leading to the collaboration between Israeli environmentalists, educators, and thinkers (Kramer March 18, 2015). Deliberately bridging the polarized secular and religious camps gives rise to educational institutions and social platforms in which secular and religious Israelis study together, learn from each other, and create new cultural hybrids. The Israeli Sabbath is a successful new practice that seeks to bridge religious and secular Israelis by offering new ways to integrate humanity and nature.

To some scholars Teva Ivri and its initiatives demonstrate New Age spirituality's growth in Israel (Ruah-Midbar 2012). While integrating feminism, Neo-Paganism, shamanism, and Judaism does occur in Israel, Teva Ivri does not exemplify this type of post-secular hybridity. Rather, Teva Ivri is closer to another organization also founded in 2007, HaShomer HaChadash (the New Watchman). Inspired by the early 20th century Zionist organization HaShomer (The Watchman), HaShomer HaChadash sees itself as an evolution of Practical Zionism, allowing Israelis to engage in voluntarism to protect agricultural and open spaces in Israel. Teva Ivri suggests that Israelis are seeking new ways to bridge the religious-secular divide and that they are open to creative, alternative ways to celebrate and experience ancient rituals such as the Sabbath, the Sabbatical year, and nature-based Jewish festivals.

Today, "Jewish sustainability"—environmentally responsible conduct rooted in the sources of Judaism—informs many Israeli grassroot social organizations, educational initiatives, and community projects. The Biblical creation narrative (Gen. 2:15) according to which the first human was placed in the Garden of Eden "to till and protect the Earth" provides the title of these initiatives that go by the name "LeOvdah u-LeShomrah," all of which inculcate the value of sustainability, nature conservation and preservation, and environmental responsibility. We find educational programs anchoring the love of nature and concern for future generations in Jewish religious sources across the Israeli religious spectrum (from Reform to Ultra-Orthodox). That Judaism cares about nature as much as it cares about social justice is no longer a radical idea. Similarly, environmentalism is now taught in Israel's Jewish, Christian, and Muslim seminaries. The "Report on Faith and Ecology Instruction in Holy Land Seminaries" documents how "a number of seminaries have embarked on a host of greening initiatives on their campuses" (Ben Arie and Neril 2016, 5). Courses on the bible, theology, and ethics engage ecological values, and environmental issues frame the informal dialogue and seminaries' ethos. For example, the Reform rabbinical seminary, Hebrew Union College in Jerusalem, introduced a course on "The Environmental Crisis and Jewish Education"; the Schechter Institute of the Masorti (Conservative) Movement introduced a new course on "Jewish Ecology and Spiritual Sustainability"; and Midreshet Ein HaNatziv, a religious seminary for women, includes a new course on "Jewish Sustainability." The institution's leader, Rabbi Yitzchak ben David, teaches a weekly lesson addressing connections between environmental issues and economics. Whether a Reform, Conservative, or Orthodox curriculum, attentiveness to environmental issues in religious education is evident, although environmental subjects do not dominate. Most importantly, concern for the Holy Land's environmental integrity and recognition that the Abrahamic traditions preach the ethics of care and responsibility create the common ground for collaboration among the faith communities cohabiting the Holy Land.

The story of Zionism and the State of Israel enriches our understanding of the post-secular moment, characterized by hybrid formations fusing life's "secular" and "religious" spheres. Zionism was not a secular rebellion against traditional Judaism but a variegated process of secularizing Judaism by reconstructing religious myths, ethical values, habits of mind and heart, social institutions, and public rituals into modern, Hebraic, cultural forms. Jewish attachment to the Land of Israel—geographic location, physical territory, metaphysical reality, and object of passionate love—facilitated the Zionist return to nature in the beginning of the 20th century as much as it has facilitated Zionism's continued growth and renewal in the early 21st century. If in the early decades of the nascent state, environmentalism was the mark of secular Zionism, Israel has emerged in the 21st century as a post-secular society in which Judaic environmentalism came to the fore, imbuing environmental activism with traditional Judaism's ethical values. Today a "renewed and uncommitted yet genuine interest in God and spiritual matters occupies 'middle ground' between tradition, religion, spirituality and modernity" (Keshet and Liberman 2014, 93). Judaic environmentalism is but one expression of Israeli post-secularism, along with the prevalence of folk religion, rising authority of rabbis, veneration of saints and pilgrimages to their gravesites, membership in New Age cults, popularity of Neo-Kabbalah and Neo-Hasidism, proliferation of alternative study halls, ubiquity of Eastern healing practices, and spiritual counseling and coaching (Garb 2005; Huss 2014; Werczberger 2016). An array of spiritualities saturates Israel's public sphere, signally its creative vitality, social and ethnic diversity, and commitment to "improving" the world. While "spirituality" is often juxtaposed with "religion," this chapter shows that environmental spirituality is embedded in Judaism's ethical values and sacred texts, deepening Israelis' and Jews' emotional ties to their land and ensuring Zionism and Judaism's perpetual renewal.

Israelis and Diaspora Jews interpret Zionism in numerous, often conflicting ways. There is no reason to believe that the spirited debate about Israel's future and the meaning of living in the Holy Land will end soon. In fact, the tragic events of October 7, 2023, have only exacerbated the debate within Israel between conflicting visions, making Israeli ever more polarized and a larger segment of American Jewry ever-more emotionally disconnected from Israel. The following chapter returns the focus to the United States to trace the dialectical interplay of secularization and re-sacralization and the role of science in these processes. In America, which Jewish migrants in the 19th and 20th centuries regarded as "the New Promised Land," the relationship between self, nature, and society will take different forms than in the Land of Israel.

References

Almog, Oz. 2000. *The Sabra: The Creation of the New Jew.* Berkeley and Los Angeles: University of California Press.

Almog, Shmuel. 1998. "The Role of Religious Values in the Second Aliyah." In *Zionism and Religion*, ed. Shmuel Almog, Yehuda Reinharz, and Anita Shapira, 237–250. Hanover, NH: University of New England Press and Brandeis University.

Aran, Gideon, 2005. "Religious and Ideological Dimensions of the Israeli Settlement Issue: Reframing the Narrative." *Negotiation Journal* 21 (2): 177–191.

Aran, Gideon, 2013. *Settlers Politics as God's Playing Field.* Jerusalem: Carmel Publishing House.

Aran, Gideon and Zali Gurevitz. 1991. "About Place (Israeli Anthropology)" (in Hebrew). *Alpayyim: A Multidisciplinary Publication for Contemporary Thought and Literature* 4: 9–44.

Avineri, Shlomo. 1981 [2017]. *The Making of Modern Zionism.* New York: Basic Books.

Avineri, Shlomo. 1998. "Zionism and the Jewish Religious Tradition: The Dialectics of Redemption and Secularization." In *Zionism and Religion*, ed. Shmuel Almog, Yehuda Reinharz, and Anita Shapira, 1–9. Hanover and London: The University Press of New England and Brandeis University.

Barell, Ari and David Ohana. 2014. "'The Million Plan': Zionism, Political Theology and Scientific Utopianism." *Politics, Religion & Ideology* 15 (1): 1–22.

Ben Arie, Ari and Jonathan Neril. 2016. "*Report on Faith and Ecology: Teaching in Holy Land Seminaries.*" The Konrad Adenauer Stiftung and the Interfaith Center for Sustainable Development. Available at: https://interfaithsustain.com.

Ben David, Orit. 1997. "Tiyul (Hike) as an Act of Consecration of Space." In *Grasping Land: Space and Place in Contemporary Israeli Discourse and Experience*, ed. Eyal Ben-Ari and Yoram Bilo, 129–146. Albany, NY: SUNY Press.

Bhabha, Homi K. 2015 [1997]. "*Forward.*" In *Debating Cultural Hybridity: Multi-cultural Identities and the Politics of Anti-Racism*, ed. Pnina Werbner and Tariq Modood, ix–xiii. London: Zed Books.

Chowers, Eyal. 2002. "The End of Building: Zionism and the Politics of Concrete." *The Review of Politics* 64 (4): 599–626.

Conforti, Yitzhak. 2009. "The New Jew in Zionist Thought: Nationalism, Ideology and Historiography." *Israel* 16: 63–96.

Easthope, Antony. 1998. "Bhabha, Hybridity and Identity." *Textual Practice* 12(2): 341–348.

Efron, Noah. 2011. "Zionism and the Eros of Science and Technology." *Zygon* 46 (2): 413–428.

Efron, Noah. 2014. *A Chosen Calling: Jews in Science in the Twentieth Century.* Baltimore: Johns Hopkins University Press.

Engel, David. 2013 [1998]. *Zionism.* London and New York: Routledge.

Etkes, Immanuel. 2005. *The Besht: Magician, Mystic, and Leader*, trans. Saadya Sternberg. Waltham, MA: Brandeis University Press.

Ettinger, Lia, Jeremy Benstein, Meira Hensen, Amit Lerner, Rina Magen, Orli Ronen. 2015. *The Optimism Report: 2014* (in Hebrew). Tel Aviv: Heschel Center.

Fine, Lawrence. 1995. "Rav Abraham Isaac Kook and the Jewish Mystical Tradition." In *Rabbi Abraham Isaac Kook and Jewish Spirituality*, ed. Lawrence J. Kaplan and David Shatz, 23–40. New York and London: New York University Press.

Fuehrer, Ehud. 2019. *New Man, in A Jewish Form: A New Reading in A.D. Gordon's Philosophy* (in Hebrew). Ramat Gan: Bar Ilan University Press.

Furst, Benny. 2016. "Ecology, Environment, Sustainability: The Development of the Environmental Movement in Israel." *Cultural and Religious Studies* 4 (4): 238–253.

Garb, Jonathan. 2009. *The Chosen Will Become Herds: Studies in Twentieth Century Kabbalah.* New Haven: Yale University Press.

Gilman, Sander. 1991. *The Jew's Body.* New York and London: Routledge.

Golomb, Jacob. 2004. *Nietzsche and Zion.* Ithaca and London: Cornell University Press.

Golomb, Jacob (ed.). 1997. *Nietzsche and Jewish Culture.* London and New York: Routledge.

Goodman, Micha. 2017. *Catch-67: The Left, the Right, and the Legacy of the Six-Day War* (in Hebrew). Tel Aviv: Zemora Bitan.

Greenspan, Itay. 2015. "Processes of Change and Development in the Environmental Movement in Israel: From Political Orientation to Civil Society and Partnership" (in Hebrew). *Bitahon Soziali* 98: 157–193.

Gordon, Aharon David. 1951. *Ha-Adam Ve-Ha-Teva*, ed. Samuel Hugo Berman. Jerusalem: Ha-Sifryah Ha-Tzionit.

Halpern-Amaru, Betsy. 1986. "Land Theology in Philo and Josephus." In *The Land of Israel: Jewish Perspectives*, edited by Lawrence Hoffman. 65–93. South Bend, IN: University of Notre Dame Press.

Hallamish, Moshe and Aviezer Ravitzky (eds.). 1991. *The Land of Israel in Medieval Jewish Thought* (in Hebrew). Jerusalem: Yad Itzhak Ben Zvi.

Hellinger, Moshe, Isaac Hershkowitz, and Bernard Susser. 2018. *Religious Zionism and the Settlement Project: Ideology, Politics and Civil Disobedience.* Albany, NY: SUNY Press.

Hertzberg, Arthur. 1973. *The Zionist Idea.* New York: Schocken Books. Heschel Center for Sustainability. n.d. https://heschel.org.il/en/.

Hirschhorn, Sara Yael. 2017. *City on a Hilltop: American Jews and the Israeli Settler Movement.* Cambridge, MA and London: Harvard University Press.

Hoffman, Lawrence A. 1986. "Introduction: Land of Blessing and 'Blessings of the Land'" in *The Land of Israel: Jewish Perspectives*, ed. Lawrence A. Hoffman, 1–23. South Bend, IN: University of Notre Dame Press.

Huss, Boaz. 2014. "'Spirituality': The Emergence of a New Cultural Category and Its Challenge to the Religious and the Secular." *Journal of Contemporary Religion* 29: 47–60.

Hutnyk, John. 2006. "Hybridity." *Journal of Ethnic and Racial Studies*, 28 (1): 79–102.

Idel, Moshe. 1986. "The Land of Israel in Medieval Kabbalah." In *The Land of Israel: Jewish Perspectives*, ed. Lawrence A. Hoffman, 170–187. South Bend, IN: Notre Dame University Press.
Inbari, Motti. 2012. *Messianic Religious Zionism Confronts Israel Territorial Compromises*. Cambridge: Cambridge University Press.
Jobani, Yuval. 2023. *The First Jewish Environmentalist: The Green Philosophy of A.D. Gordon*. New York: Oxford University Press.
Karassin, Orr. 2013. "The Battle of the 'True Believers': Environmentalism in Israeli Party Politics." In *Between Ruin and Restoration: An Environmental History of Israel*, ed. Daniel E. Orenstein, Alon Talk, and Char Miller. Pittsburgh: University of Pittsburgh Press.
Keshet, Yael and Ido Liberman. 2014. "Seeking Empowerment and Spirituality in the Secular Age: Secular and Traditionalist Israelis Consulting Rabbis." *Sociology* 48 (1): 92–110.
Kramer, Einat. 2012. *Sustainable Growth According to Jewish Sources: A Jewish-Israeli Position; Prepared for the International Summit on Sustainable Development; Rio de Janeiro*, June 2012.
Kramer, Einat. May 7, 2014. "Shabbat Behar: Ready, set, LET GO!" *Times of Israel*. http://blogs.timesofisrael.com/shabbat-behar-ready-set-let-go.
Kramer, Einat. February 25, 2015. "The Next Chapter in the Social Change Movement—Israeli Shmita." *Times of Israel*. http://blogs.timesofisrael.com/the-next-chapter-in-the-social-change-movement-israeli-shmitta.
Kramer, Einat. *March* 18, 2015. "Do We Choose Well? My Personal Campaign." *Times of Israel*. http://blogs.timesofisrael.com/will-we-choose-well-my-personal-campaign.
Margolin, Ron. 2008. "Hans Jonas and Secular Religiosity." In *The Legacy of Hans Jonas: Judaism and the Phenomenon of Life*, ed. Hava Tirosh-Samuelson and Christian Wiese, 231–258. Leiden: Brill.
Margolin, Ron and Juval Jobani (eds.). 2020. *A.D. Gordon: Man and Nature: Meditations and Dreams of a Radical* (in Hebrew). Jerusalem: Magnes Press.
Mendes-Flohr, Paul (ed.) 1983. *A Land of Two Peoples: Martin Buber on Jews and Arabs*. Chicago: University of Chicago Press.
Mirsky, Yehuda. 2014. *Rav Kook: Mystic in a Time of Revolution*. New Haven: Yale University Press.
Moore, Donald J. 1996. *Martin Buber: Prophet of Religious Secularism*. New York: Fordham University Press.
Mosse, George L. 1996. *The Image of Man: The Creation of Modern Masculinity*. New York and Oxford: Oxford University Press.
Nehorai, Michael Z. 1995. "Halakah, Meta-Halakhah, and the Redemption of Israel: Reflections on the Rabbinic Rulings of Rav Kook." In *Rabbi Abraham Isaac Kook and Jewish Spirituality*, ed. Lawrence J. Kaplan and David Shatz, 120, 156. New York and London: New York University Press.
Neumann, Boaz. 2011. *Land and Desire in Early Zionism*. Translated by Haim Watzman. Waltham, MA: Brandeis University Press.
Nimhi, Ephraim (ed.). 2003. *The Challenge of Post-Zionism: Alternatives to Fundamentalist Politics in Israel*. London: Zed Books.
Orenstein, Daniel E. and Emily Silverman. 2013. "The Future and History of Israeli Environmental Movement: Is a Major Paradigm Shift under Way?" In *Between Ruin and Restoration: An Environmental History of Israel*, ed. Daniel E. Orenstein, Alon Tal, and Char Miller, 357–382. Pittsburgh: University of Pittsburgh Press.

Pedahzur, Ami. 2012. *The Triumph of Israel's Radical Right*. New York and Oxford: Oxford University Press,

Peled, Rina. 2002. "*'The New Man' of the Zionist Revolution: Hashomer Hazair and His European Roots*" (in Hebrew). Tel Aviv: Am Oved.

Penslar, Derek J. 1991. *Zionism and Technocracy: The Engineering of Jewish Settlement in Palestine, 1870–1918*. Bloomington and Indianapolis: Indiana University Press.

Perl, Gil S. 2013. *The Pillar of Volozhin: Rabbi Naftali Zvi Yehuda Berlin and the World of Nineteenth-Century Lithuanian Torah Scholarship*. Boston: Academic Studies Press.

Ramon, Einat. 2007. *A New Life: Religion, Motherhood and Supreme Love in the World of Aharon David Gordon*. In Hebrew. Jerusalem: Karmel.

Ratzabi, Shalom. 2011. *Anarchy in Zion: Between Martin Buber and A.D. Gordon*. In Hebrew. Tel Aviv: Am Oved.

Ravitzky, Aviezer. 1996. *Messianism, Zionism, and Jewish Religious Radicalism*. Chicago: University of Chicago Press.

Ruah-Midbar, Marianna. 2012. "Current Jewish Spiritualities in Israel: A New Age." *Modern Judaism* 32(1): 102–124.

Sadeh, Shahar. 2010. "Evergreen: Environmental Organizations and the West Bank Barrier Project" (in Hebrew). *Theory and Criticism* 37: 184–211.

Sagi, Avi and Dov Schwartz. 2019. *Religious Zionism and the Six-Day War: From Realism to Messianism*. London and New York: Routledge.

Saposnik, Arieh Bruce. 2008. *Becoming Hebrew: The Creation of a Jewish National Culture in Ottoman Palestine*. Oxford: Oxford University Press.

Schwartz, Eilon. 2009. *At Home in the World: Human Nature, Ecological Thought, and Education after Darwin*. Albany NY: SUNY Press.

Schweid, Eliezer. 1970. *The World of A.D. Gordon*. In Hebrew. Tel Aviv: Am Oved.

Schweid, Eliezer. 1985. *The Land of Israel: National Home or Land of Destiny*, trans. Deborah Greiniman. Rutherford, Madison, and Teaneck: Fairleigh Dickinson University Press; London and Toronto: Associated University Presses.

Schweid, Eliezer. 2014. *The Foundation and Sources of A.D. Gordon's Philosophy* (in Hebrew). Jerusalem: Bialik Institute.

Segal, Miryam. 2010. *A New Sound in Hebrew Poetry: Poetics, Politics, Accent*. Bloomington, IN: Indiana University Press.

Shamir, Eilon. 2018. *For the Sake of Life: The Art of Living According to Aaron David Gordon* (in Hebrew). Tel Aviv: Hakibbutz Hameuchad.

Shapira, Avraham. 1996. *The Kabbalistic and Hasidic Sources of A.D. Gordon's Thought* (in Hebrew). Tel Aviv: Am Oved.

Shimoni, Gideon. 1995. *The Zionist Ideology*. Hanover and London: University Press of New England and Brandeis University Press.

Siach: An Environment and Social Justice Conversation. n.d. https://siachconversation.org/.

Sprinzak, Ehud. 1991 [1986]. *The Ascendance of Israel's Radical Right*. Oxford: Oxford University Press.

Strassberg-Dayan, Sara. 1995. *Individual, Nation and Mankind: The Conception of Man in the Teachings of A.D. Gordon and Rabbi Abraham I. Hacohen Kook* (in Hebrew). Tel Aviv: Hakibbutz Hameuchad.

Tal, Alon. 2002. *Pollution in a Promised Land: An Environmental History of Israel*. Berkeley, Los Angeles, and London: University of California Press.

Tal, Alon. 2008. "Enduring Technological Optimism: Zionism's Environmental Ethics and Its Influence on Israel's Environmental History." *Environmental History* 13 (2): 275–305.

Tal, Alon. 2011. "The Development of the Environmental Movement in Israel." *Israeli Center for Third Sector Research*, vol. 25.

Tal, Alon. 2013. *All the Trees of the Forest: Israel's Woodlands from the Bible to the Present*. New Haven: Yale University Press.

Tal, Alon, Shira Leon-Zchout, Itay Greenspan, Liat Oshri, and Shira Akov. 2013. "Israel's Environmental Movement: Strategic Challenges." *Environmental Politics* 22 (5): 779–791. DOI:doi:10.1080/o9644016.2013.825139.

Troen, Ilan S. 2003. *Imagining Zion: Dreams, Designs and Realities in a Century of Jewish Settlement*. New Haven and London: Yale University Press.

Troen, Ilan S. 1988. "The Transformation of Zionist Planning Policy: From Rural Settlements to an Urban Network." *Planning Perspectives* 3 (1): 3–23.

Turner, Yossi (Joseph). 2020. *Quest for Life: A Study in Aharon David Gordon's Philosophy of Man in Nature*. Boston: Academic Studies Press.

Werczberger, Rachel. 2016. *Jews in the Age of Authenticity: Jewish Spiritual Renewal in Israel*. New York, Bern, Frankfurt, Berlin, Brussels, Vienna, Oxford, and Warsaw: Peter Lang.

Yisraeli, Oded. 2006. "Eretz Israel: Between the Sanctity of Place and Purity of the Land: Trends in Medieval Jewish Mysticism." *Pathways through Aggada* 9: 117–142.

Zakim, Eric. 2006. *To Build and Be Built: Landscape, Literature, and the Construction of Zionist Identity*. Philadelphia: University of Pennsylvania Press.

Zertal, Idit and Akiva Eldar. 2007. *Lords of the Land: The War over Israel's Settlements in the Occupied Territories, 1967–2007*. New York: Nation Books.

Zerubavel, Yael. 1995. *Recovered Roots: Collective Memory and the Making of Israeli National Tradition*. Chicago and London: The University of Chicago Press.

Zipperstein, Steven. 2018. *Pogrom: Kishinev and the Tilt of History*. Princeton: Princeton University Press.

6 Environment, Science, and Secularization

In the "New Promised Land"

America: The "New Promised Land"

The modern Jewish experience was dominated by the experience of mass migrations. In the 19th century millions of Jews migrated to America, which they regarded as the "New Promised Land," settling it in droves. For the European Jewish immigrants who settled in America between the 1820s to 1920s, America was not exile, which Zionism decried, but a country full of promise and hope, where poor Jews could become economically secure, oppressed Jews could have access to all avenues of society and culture from which Jews had been previously excluded, and the Jewish community as a collective could celebrate its distinctive ethnic culture.

If America, rather than the Land of Israel, was the "Promised Land," what did the American natural landscape mean for the immigrant Jews? Did the Jewish immigrants feel at "at home" in a country that was still (theologically speaking) not their "homeland?" If so, how did American Jews reinterpret the Judaic messianic belief which nourished Jewish identity for millennia? What role did technoscience play in the secularization of the Jews during the first two thirds of the 20th century, including their attitudes toward the environment? And how did the embrace of religiosity toward the end of the 20th century gave rise not only to religiously inflected Jewish environmentalism but also to religiously inflected Jewish politics? By exploring the nexus of religion, science, and politics in relations to the environment we gain a deeper understanding of Jewish environmentalism in America.

America's promise for Jews lay in the distinctive features of American democracy: the legal separation of religion and state enshrined in the Constitution enabled Jews to be equal citizens without giving up their religion, take part in American society and culture without legal disabilities, and, when necessary, fight discrimination by invoking the Constitution itself. The separation of church and state, of course, did not mean that religion was missing from the American public sphere. In fact, the opposite was the case: from its inception, civic culture in America was framed in religious categories. The Puritans saw themselves as "New Israel," that is, as the newly chosen people of God, and that notion has remained pervasive in America,

DOI: 10.4324/9781351123945-7

accounting for America's sense of "mission" and "manifest destiny" to promote freedom and human rights around the world. It also gave rise to religious social movements that sought to shape American politics according to their interpretations of the biblical narrative, especially the Book of Exodus (Smith 1996; Gutterman 2005, 1–19).

The prominence of the Bible in American self-understanding enabled the Jews who settled in America to present their own ancient history as part of the American myth of origin and reinterpret the Jewish past in light of America democratic ideals. Because the nation's ideals were inspired by the Bible, Jews could promote the "Judeo-American symbiosis" and could "insert themselves as central players in the nation's origin" (Wenger 2010, 11). But in so doing, Jews in America, no less than their brethren in Europe's multi-ethnic empires or liberal nation-states had to address poignant questions: Who are the Jews collectively? Are the Jews a people, a nationality, a faith community, a religious minority, an ethnic culture, a religious civilization, or all the above? And what does it mean to be Jewish in America?

Throughout the 19th century and the first half of the 20th century Jews vehemently debated their collective identity in the context of thinking through modernity, nationalism, cosmopolitanism, and utopianism. The passionate debates yielded a variety of religious outlooks, secular ideologies, and social movements, of which Zionism was but just one minority position (Goren 1999; Rabinovitch 2012). Most Jews were not Zionists, adhering to either anti-Zionist or non-Zionist religious variants of modern Judaism or to anti-Zionist or non-Zionist secular ideologies that promoted Diaspora nationalism, cultural autonomism, territorialism, Jewish Socialism, or cultural pluralism. Those who understood Jewish collective identity religiously made the synagogue the center of Jewish collective identity; by construct, the secularists who focused on ethno-national culture centered Jewish collective identity on the social institutions of the Jewish community. But regardless of the terms used to express Jewish collective identity, it was imbued with the biblical imperative to pursue justice (Deut. 16:20). Articulated by the ancient prophets, the ideal of justice (*tzedek*) generated an ethos that framed the story of Jewish life in America, including Jewish attitudes toward nature. Religious Jews of whatever stripes viewed the Torah as the guide to the betterment of the world, whereas secularists interpretated the ameliorative imperative to engage a range of social, political, and technoscientific activities in the public sphere. Not unlike Christians in America, Jews have been engaged in "prophetic politics," the abiding commitment to mend the world, but the expression of the Jewish passion for justice would change over time: during the first three quarters of the 20th century, secularized Jews translated the religious ideals of their tradition into a secular idiom, marshalling science and getting involved in politics to accomplish societal ideals that included the protection of nature, because nature could not protect itself. Since the last quarter of the 20th century to the present, Jewish intellectuals who grew up in secular homes became more critical of modern scientism and the social ills it generated and called for the re-sacralization of nature as well as the re-sacralization of politics, contributing to post-secularism.

The dialectic of secularization and re-sacralization is quite complex because America is a deeply religious country and Environmentalism in America had strong religious roots. It emerged in the 19th century when far-sighted individuals such as John Muir (1838–1914), the founder of the Sierra Club (1889), became "the well-loved champion of Yosemite and the national parks" (Stoll 2015, 7). Following the Scottish Presbyterian path from religion to natural history, Muir's studies at the University of Wisconsin "undermined [his] biblical orthodoxy but fortified his natural theology" (Stoll 2015, 146) because he studied God's created world with the help of the sciences, botany and geology, and his own penchant for mechanics and engineering.

In the late 19th century Congress established forest reserves which President Theodore Roosevelt and his Chief Forester, Gifford Pinchot, put under the Forest Service in 1905. Congress created the National Park Service in 1916, and hunters, fisherman, and their advocates successfully achieved game preserves and game laws. In the 1930s, the Dust Bowl, an unprecedented environmental disaster, created wide-spread support for government conservation, so that protection of the environment was taken to be a governmental (i.e., political) responsibility. In the first half of the 20th century several environmental organizations were founded including the National Audubon Society, the Isaac Walton League, the Wilderness Society, the National Wilderness Federation, and Defenders of Wildlife. Two Jews—Louis Marshall (d. 1929) and his son Robert Marshall (d. 1939)—greatly contributed to the early phases of American environmentalism, but in general it is safe to say that until the 1950s preservation and conservation of nature did not concern the organized American Jewish community. The reason is quite simple: as the historian Debra Dash Moore puts it, "Jews cast their lot with American cities" because "Jews saw in urban space opportunities both to preserve traditions and to devise new patterns of living" (Moore 2014, 1–2). To understand how nature functions in the dialectical story of secularization and re-sacralization, the relevant information about the history of Jews in America is in order.

Jewish Presence in America: Inclusion and Exclusion

Jewish presence in America began in 1654 when twenty-three Jews from the Dutch colony of Recife, Brazil fled from the Portuguese and settled in New Amsterdam (Diner 2000, 1–19). The Jewish population in North America grew very slowly, from 200 in 1700, to 2500 in 1776, to 4,000 in 1820. Concentrated at first in the port cities of the Atlantic coast, and after 1770 in Richmond, New Orleans, Cincinnati, and Baltimore, Jews made their livelihood mostly as merchants who owned small stores. Comprised of Sephardic Jews and (after 1770) of Jews from Germany and Poland, American Jews observed Jewish law, and their small congregations were governed by the wealthier merchants.

In pre-Revolutionary America, Jews could not vote, hold elected office, or serve on juries. Nonetheless, in America of the 17th and 18th centuries, Jews enjoyed greater freedom and fewer social restrictions than anywhere else in the world. Jewish skills in trade and commerce were highly valued and Jews and non-Jews established business partnerships and socialized with each other in their leisure hours. Socialization led to intermarriages and most often Jews who married Christians joined the Christian majority.

The Revolution altered the course of Jewish life in America. Starting with New York, in 1777 states guaranteed religious freedom to all. The federal Constitution of 1789 made it possible for Jews to enjoy the full benefits of American life: "no religious test shall ever be required as a qualification to any office or public trust under the United States." With this declaration, religion could no longer disqualify Jews from holding public office and a Jew could serve in the Congress, the army, the diplomatic service, or even be elected to the Presidency. American Jews no longer had to see the synagogue as the place to which they must belong, and whose rules they must follow, because now the synagogues and the Jewish community were but voluntary associations. As voluntary associations, synagogues now had to respect the wishes of their members and members could worship in ways they deemed desirable. In the "New Promised Land," reforming the Jewish tradition was inevitable.

Mass emigration of Jews from Europe during the 19th century changed Jewish settlement in America, dispersing Jews across the entire continent. Jewish mass migrations came in two main waves: the first wave (1820–1880) consisted of about 250,000 Jews from Central Europe, and the second wave (1880–1920) consisted of over 2.8 million Jews mostly from Eastern Europe (Diner 2004, 78–92). America's promise of freedom, equality, economic prosperity, and social mobility made it irresistible, but Jewish life in the New Promised Land began very humbly. Prior to 1870s most Jewish immigrants began the new chapter of their life as peddlers and after 1870s as factory workers. The peddling occupation brought Jews to settle across the entire United States, establishing new communities and constructing new ways of being Jewish that will harmonize with the style and social mores of the Protestant majority (Diner 2015). In small and midsized cities Jews were involved in civic affairs, serving on juries, joining volunteer fire departments, working in law enforcement, and holding elected offices.

Reform Judaism and the Ideal of Justice

Between 1820 and 1880 the last laws that discriminated against Jews because of their religion were eliminated and equality was now the law. During these years, the "German Jews," as they became known, developed their communal institutions in cities across the United States that sprouted along river routes. Despite Constitutional protection of all religions and religious neutrality in government, missionizing to Jews did exist for example in the American Society for Meliorating the Condition of the Jews (ASMCJ) founded in 1820 by Elias Boudinot, a minor official in the George Washington's administration (Diner 2000, 63). German Jews responded to these missionizing efforts by establishing their own Jewish Sunday schools, nontraditional schools designed to inculcate Judaism in modern ways, but mainly they attempted to "develop a form of Judaism equivalent to Episcopalianism, Presbyterianism, and especially Unitarianism … that minimized Judaic ritual and emphasized ethics in a universalist context" (Kaplan 2005, 14). With the focus on ethics, observance of the rituals and laws of Judaism became secondary to universal ethics which Jews shared with all human beings.

Felix Adler (1851–1933), the son of the Reform Rabbi, Samuel Adler, took this position to its logical conclusion (Kraut 1979). The Adler family settled in the United States in 1856, when Samuel Adler was invited to serve as the rabbi of Temple Emanu-El in New York. After earning a BA at Columbia University (1870), Felix returned to Germany to study at the University of Berlin and earned a PhD in Philosophy from the University of Heidelberg (1873). Upon his return to the United States in 1874, he secured an appointment as Professor of Oriental Languages and Hebrew at Cornell University, but did not remain there long, because his atheistic views were considered "dangerous." Returning to New York, he was asked to deliver a sermon at Temple Emanu-El, with the expectation that he would succeed his father as the leader. The sermon he delivered on May 15, 1876, "The Judaism of the Future," shocked everyone because it did not refer to God even once. Instead, Adler called for establishing a universal rational religion that will unite all of humankind in moral social action to ameliorate the human condition. Adler's Ethical Culture, which fused Kantian moral philosophy, prophetic Judaism, and Socialism, would be attractive to certain types of Jews, for example, Mordecai M. Kaplan who was Adler's student at Columbia University, but it was the last sermon he delivered at Temple Emanu-El. In February 1877 he established the New York Society for Ethical Culture, as a movement that should further the principles of ethics among all adults and children through education, and members of the Society should express their religious consciences through moral and humane actions.

Adler's commitment to better the human condition generated a range of new and revolutionary social services. The Society for Ethical Culture established the District Nursing Service, a precursor of the Visiting Nurse Service, and founded the first free kindergarten in the United States. In 1880 Workingman's School was founded and in 1895 the school was reorganized, becoming the Ethical Culture School. Committed to social justice for all, the Society worked to improve conditions in tenements houses and members of the society staffed clubs, libraries, gymnasiums, job training programs, a kindergarten, a mother's club, educational classes and two employment bureaus. As Felix Adler put it, "Ethical Culture is religious to those who are religiously minded, and merely ethical to those who are not so minded." Adler's impactful social reforms were recognized academically. In 1902 he was appointed the Professor of Politics and Social Ethics at Columbia University, a position until his death in 1933, and served on the editorial board of the journal *Ethics*, in addition to serving on several national committees including the national Child Labor Committee and later the Civil Liberties Bureau. Adler's blend of religious humanism and social activism attracted many socially mobile and intellectually inclined German Jews, including the Marshall family, whose four children attended the Ethical Culture School.

The Society for Ethical Culture greatly contributed to the secularization and Americanization of Jews, but most German Jews remained loyal to the less radical Reform Judaism. The 1857 prayer book, *Minhag America*, was accepted by most reform congregations; in 1875 the seminary for Reform rabbis, Hebrew Union College, was established in Cincinnati; and in 1885 the Pittsburgh Platform systematized the beliefs of Reform Judaism. It reconceptualized the relationship between the Diaspora and the Land of Israel. Instead of dreaming about the coming messiah who will gather Israel to the Land of Israel, Reform rabbis preached that the messianic age will be brought about by human efforts that will bring world peace. Max Lilienthal, a Reform Rabbi in Cincinnati, put it starkly:

> We Israelites of the present age do not dream any longer about the restoration of Palestine and the Messiah crowned with a diadem of earthly power and glory. America is our Palestine; here is our Zion and Jerusalem: Washington and the signers of the glorious Declaration of Independence—of universal human rights, liberty, and happiness—are the deliverers, and the time when their doctrines will be recognized and carried into effect is the time so hopefully foretold by our great prophets.
> (cited in Kaplan 2005, 25)

Translating Jewish messianism into the idiom of American democracy, Reform Judaism no longer considered Halakha legally binding, allowing each Jew to choose how to express one's Jewishness and regarding humanity and God as partners in the unfolding process of continuing revelation whose core is a rational morality.

East European Immigrants and Jewish Secularism

The mass migration of Jews from Eastern Europe profoundly changed the future of American Jewry. Fleeing debilitating poverty and series of state sponsored pogroms in the 1880s, Yiddish-speaking, traditionally observant Jews from the Pale of Settlement did not take to the highly cerebral religiosity of American Reform. Settling in New York, Chicago, Philadelphia, Boston, Baltimore, and other industrializing urban centers, the new immigrants organized their own communities according to their place of origin in Europe. Each community had its kosher butcher shops, bakeries, schools where the basics of Jewish religious life were taught, and their own synagogue. A few of the immigrants derived their livelihood from peddling, but most worked in factories of the bustling industrialized urban centers of America. The living conditions of the immigrants were exceedingly limited, their paltry wages insufficient to feed themselves, but a determination to better their life through hard work, learning, and political organization would ensure their path to Americanization. If "Russian Jews" were to be Americanized, they had to be modernized and secularized, but not in the ways charted by Reform Judaism.

For East European Jewish immigrants, Conservative Judaism rather than Reform Judaism was the vehicle to Americanization which had to be done in English rather than Yiddish. In the public schools, Americanized Jews studied science in English rather than in Yiddish, and they interpreted their Jewish past through the scientific study of Judaism in English rather than in German. As for the relationship between the Land of Israel and the Diaspora, Conservative Judaism from its inception was pro-Zionist, encouraging settlement in the Land of Israel and appreciating the spiritual significance of the Land of Israel in Jewish religious life. The religious adaptation to life in America was decidedly bourgeois and it accepted American Capitalism with its promise of wealth and social mobility.

By contrast, the secular form of Americanization replaced ethnicity for religiosity, Yiddish for Hebrew, and revolutionary ideologies for prophetic pathos. Secular Yiddish culture in America was forged in the streets of urban centers, above all in New York, where Jews socialized in cafes, lecture halls, and public parades rather than in synagogues. Secular Yiddish culture was created in America by Russian Jewish immigrants whose families were already Russified and secularized in Russia, or by Jews whose families endorsed the more progressive Haskalah elements of otherwise traditional Jewish upbringing (Cassedy 1997, 10–11). With various levels of rootedness in Judaism, East European Jewish intellectuals created a vibrant Socialist, secular culture in Yiddish that fused Jewish Diaspora nationalism with radical ideologies (Michels 2005). Ironically, because the Jewish radicals chose Yiddish to forge the class consciousness of the Jewish immigrants, while secular Yiddish culture modernized and secularized East European Jews, it also created a distinctive ethno-culture for Jews only. Diaspora Jewish nationalism in Yiddish was particularistic notwithstanding its universal aspirations.

Jewish Socialists, by contrast, promised a kind of a messianic utopia and the end of exile. But instead of gathering the Jews in the Land of Israel, as Zionism envisioned, for the Jewish Socialists exile would become irrelevant because social differences among people would disappear. Socialism promised Jews a better, more just world, in which poverty will be eradicated, and "freedom, equality, and brotherhood" would prevail. Not surprisingly, in

Europe and in the United States Jews flocked to Socialist parties because its "language meant … that we were regarded as human beings with human rights" (Michels 2012, 72). Socialism promised universal redemption and emphasized self-improvement through nonreligious activities: reading newspapers, delivering speeches, or marching in parades, all of which were new secular ways through which young Jewish men and women, often poor and learned, could take part in public debates about important issues, build new institutions, and take a role in running things.

Socially inspired secular Yiddish culture helped fashioned the ideal new world while retaining Jewish cultural identity. The Yiddish press, Yiddish theater, Yiddish literature, Yiddish schools, and Yiddish summer camps manifested Diaspora nationalism as an alternative to the (allegedly "bourgeois") Zionist vision. For the promoters of secular Yiddish culture, the Jewish working poor, the folk, were the carriers of Diaspora nationalism and cultural autonomism. The ethnic sub-culture of secularized, Yiddish-speaking Jews strengthened the American Left but also stood in conflict with it, for example, about the transition from capitalism to socialism and the participation in manual labor. Jewish Leftists were also highly divided by fierce ideological debates among the Socialists, the Communities, the Trotskyists, the Labor-Zionists, the Bundists, and the Social Democrats (Liebman 1997). Beginning in Europe in the second half of the 19th century, these ideological debates were perpetuated in America, preventing unity of the Jewish Left.

Jewish Agricultural Settlements

Most East European Jewish immigrants made their living in factories, especially in the garment industry, often owned by wealthy Uptown Jews. Because they vowed to protect their rights and to provide for their needs, Jews joined labor unions, and they inevitably interpreted their experience in terms of class struggle. Jewish Socialism not only framed the class consciousness of Yiddish-speaking immigrants it also inspired efforts to transform the occupational structure of Jewish society by encouraging farming. The idea to establish farming communes in America was conceived in Russia in 1881 by the founders of the Am Olam (Eternal People) movement and small groups of idealist Jews organized themselves to settle in America where they would create agricultural communes that would realize the dream to normalize the Jews and make them into "productive" people. The Am Olam movement was possible due to the Homestead Act of 1862 and with the financial support of the Jewish Colonization Association of Baron Maurice de Hirsch, the Alliance Israélite Universelle, and the Jewish Agricultural Society.

In the 1880s and 1890s a total of twenty-four Jewish agricultural communes were founded in Washington, Oregon, Michigan, Iowa, Louisiana, Mississippi, the Dakota territory, Colorado, Arkansas, Kansas, and New Jersey, but most did not survive for more than a few years (Herscher 2018). They were doomed by lack of agricultural experience, lack of external funding, and internal divisions, but several communities in New Jersey had a measure of success and lasted well into the mid-20th century, combining farming, communalism, and light manufacturing (Brandes 1971; Weintraub 2007; Weinstein 2020). The Russian Jews in the colonies of South New Jersey—Woodbine, Alliance, Brotmanville, Norma, and Carmel—became naturalized Americans as a matter of course but expressing their Jewish identity in the relative isolation of rural environment was a constant struggle (Brandes 1971, 170–253). Although Jewish life in these communities was very rich and diverse, including Orthodoxy, Reform, Ethical Culture, and Zionism, as a project of national renewal they failed because "they were developed as a privately subsidized social experiment, with neither the strongly socialist nor the nationalist connotations of the settlement in Palestine" (Brandes, 1971, 7). We can go further to say that although America was regarded as a "New Promised Land" it could not generate the emotional attachment to the land that Zionist pioneers felt toward the Land of Israel.

Nonetheless, the memory of Jewish farming in America would inspire and inform the Jewish environmentalists in contemporary America who (as we noted in Chapter 2), promote farming by Jews as a path for Jewish spiritual renewal (Seidenberg 2005; Krone 2021). But at the beginning of the 20th century, Jewish farming had little to do with protecting the well-being of the environment or with the preservation of nature. Rather, it intended to transform the occupational structure of Jewish society through Jewish farming suffused with Socialist utopianism. Alas, in America the back-to-the-land movement of a few Jewish Socialists did not transform the occupational basis of Jewish society. American Jews overwhelmingly preferred the urban environment over the rural environment, and social mobility and economic prosperity depended on entering the professional middle class committed to Capitalism. Jewish Socialism in Yiddish faded in America after the 1950s because Jews successfully Americanized by endorsing Capitalism and benefiting from it, and because European Jewry, the major reservoir of native Yiddish speakers, was annihilated in the Holocaust. And yet, the ideals and values of the Jewish Left in the first half of the century did not disappear; they morphed into the Spiritual Progressives who have integrated environmentalism with a range of social reforms that will make American society more just, egalitarian, and environmentally attentive. Whether Judaism was experienced in the synagogue or in the streets, the experience was distinctly urban, relegating "nature" to the distant background. Periodically urban Jews retreated to natural settings where they could enjoy the outdoors either in their vacation homes, in summer retreats, or in youth camps (Lorge and Zola 2006; Sales and Saxe 2003; S. Fox 2023). The Catskill Mountains were the main region of Jewish summer camps and retreats, known as the Borscht Belt, where many American Jewish entertainers began their careers. These summer retreats could take place in the Catskills because its forests were protected by law.

Science and Environmental Politics: Influential Individuals

This chapter focuses on key people telling a story about larger cultural processes. We turn first to Louis Marshall (1856–1929) and his son, Robert Marshall (1901–1939), who reflected the early dominance of "German Jews" in American Jewish history. Their pioneering defence of nature was rooted in the prophetic ideal of justice that required protecting the weak and powerless. We then focus on two secular, East European Jews—Barry Commoner (1917–2012) and Murray Bookchin (1921–2006)—who transformed American environmentalism by warning about the dangers of technoscience, be it nuclear technology, pesticides, or chemicals in foods. For both thinkers/activists, the proper attitude toward technoscience was predicated on correct understanding of the interdependence of humanity and nature, namely, the development of ecological consciousness. Commoner articulated the Four Laws of Ecology and insisted on scientists' social responsibility to educate the citizenry, and Bookchin promoted Social Ecology, a unique form of humanistic eco-anarchism that exposed the connection between human oppression of other humans and human oppression of nature.

Commoner and Bookchin were transformed by and responded to WWII, in which technoscience made possible the genocide of the Jews, the loss of millions of human lives, and the possibility of thermonuclear war that could destroy all life on Earth. Bookchin and Commoner were deeply involved in the campaigns against nuclear testing in the late 1950s and the 1960s that eventually gave rise to environmental organizations such as the Environment Defense Fund (1967), the National Resource Defense Council (1970), and Green Peace (1971). These organizations demanded stricter national pollution control laws and the establishment of federal agencies with power to enforce them. The campaigns yielded several legislative milestones in the early 1970s, such as the National Environmental Policy Act, Clear Air Act, Clean Water Act, and Endangered Species Act. The ecological awareness that Commoner and Bookchin called for was marked by the establishment of Earth Day (March 22, 1970).

As environmentalism penetrated American public discourse, American Jewry was coming to terms with the meaning of the Holocaust and the establishment of the State of Israel. Since the Holocaust vindicated the Zionist analysis, American Jews became staunch supporters of the State of Israel. Zionism became the official position of the organized American Jewish community. In the 1980s and 1990s some American Zionists committed to environmentalism settled in Israel, where they could integrate Jewish and environmental sensibilities. However, most American Jews did not believe Zionism demands settlement in the Land of Israel. Over time, especially among Left-leaning Jews, criticism of Israel's mistreatment of its Arab population and its disregard for Palestinian claims to national self-determination gradually led to distancing from Zionism and the State of Israel's policies.

Liberty, Justice, and Protection of Nature

Louis Marshall: Champion of Liberty

In America, Jewish involvement in environmental protection began with Louis Marshall, a paragon of the successful integration of "German Jews" in the "New Promised Land" and their profound impact on the institutions of the organized Jewish community in America. Louis Marshall was born on December 14, 1856, in Syracuse. His father, Jacob Marshall, began life in America as a porter and peddler. The family retained an Orthodox way of life, observing the Sabbath and keeping kosher. In 1861 Jacob opened a leather and hide business and was often away from home. It was Louis's mother, Zilli Strauss of Württemberg Germany, who instilled the German-Jewish commitment to *Bildung* (self-cultivation) in her son. She inspired his prodigiousness, industriousness, love of learning, and discipline that made him a successful lawyer and communal leader.

Marshall attended Syracuse High School, where all morning sessions opened with Bible reading and prayer. In the afternoons (except Fridays), Louis attended German and Hebrew schools, and on Sunday he devoted his time to sports, especially baseball, another important path to Americanization (Silver 2013, 9). After graduating in 1874, he worked for two years in the law office of Nathaniel Smith, where he availed himself of law books when not performing menial tasks. This self-training enabled him to attend Columbia University Law School in 1876 and complete a two-year degree in one year. Louis passed the bar in the State of New York, and in 1878 became the only Jewish member of the firm Roger, Wallace, Brooks, and French. His photographic memory, erudition, industriousness, and frugality facilitated his success. By 1890, Marshall's firm was the most lucrative practice in Syracuse. Marshall devoted some of his growing income to purchasing property in upstate New York that he never intended to develop; thus began his commitment to protecting open spaces from the encroachment of industrialization, preserving pristine nature for future generations.

In Marshall's life, legal work, public advocacy, and institution building were closely intertwined (Reznikoff 1957; Rosenstock 1965; Silver 2013). We will only focus on his environmental work here. Matthew Silver argues that "the main thrust of Louis Marshall's orientation toward the environment colored his work on Jewish affairs" (Silver 2013, 407), but the reverse is equally true: Marshall's defense of Jewish constitutional rights for individuals and groups colored his environmentalism. For Marshall, nature was entitled to legal protection because it could not defend itself from destructive, illegal human activities. Extending legal protection to nature could curb harmful and illegal human action but knowing how to wisely protect nature requires environmentalism grounded in science. For that reason, Marshall helped to establish the New York State Forestry School at Syracuse University in 1911, where scientific knowledge could yield best practices. Marshall's son, Robert, would enter that school in 1920, devoting his career and public activism to wilderness protection and conservation.

Louis Marshall's first and long-lasting environmental legal action came in 1894, when he participated in the Constitutional Convention of the State of New York. In 1892 the New York legislature established the Adirondack Park. In 1893 the Legislature clarified that the park

> shall be forever reserved, maintained and cared for as ground open for the free use of all the people for their health and pleasure and as forest lands necessary to the preservation of the headwaters of the chief rivers of the State and a future timber supply, and shall remain part of the Forest preserve.

However, the law also contained Section 103, allowing the Forest Commission to sell timber. The Forest Commission promptly granted wood-cutting contracts. Timber was stolen from State lands, and speculators found ways to attack the state's title to land and even removed some 100,000 acres from the Forest Preserve. The ineffective, unlawful, and corrupt practices associated with the state's management of the Adirondack Park and Forest Preserve necessitated the amendment to the New York Constitution that was presented on August 1, 1894, by David McLure, with assistance from Louis Marshall and others. Section 7 of Article VII of the Constitution of New York State (later to be renumbered as Article XIV, Section 1 in 1938) inaugurated the concept of "wilderness" in American law.

The amendment endowed the State with the mandate of stewardship duties to promote "forever wild forest land." The Forest Preserve of New York was conceived as a place where the government shall act "to sustain and enhance a productive and enjoyable harmony between humans and nature." This was "a *holistic* approach to humans in nature, not one that divides humans from nature, or sacrifices natural integrity to human greed" (Robinson, 2007, 10).

The amendment was put to voters in a referendum in 1895 and was widely defeated, but when Theodore Roosevelt became Governor in 1900 the Forest Commission was reorganized and its biases against the Constitution confronted. The timber lobby gradually lost its influence over the patronage, but problems persisted with vast forest fires in 1908 and extensive thefts of timber from State lands. Marshall was aware of the unsuccessful implementation of the "forever wild" Forest Preserve mandate. In 1902 he joined other supporters of the Constitutional Forest Preserve to launch the Association for the Protection of the Adirondacks. The Association began its investigation and advocacy in support of Article XIV, but the attacks on the "forever wild" mandate persisted. In 1914 enforcement against timber theft became effective, and by 1915 the Commission was replaced by a single commissioner, and all appointments were to have civil service protection with expert qualification. Marshall participated in the debate about the "forever wild" amendment in the Constitutional Convention of 1915 and in following years continued to "vigilantly guard against any legal effort to trespass and violate the role of full protection of the Adirondack and Catskill forests." Further attempts to weaken the "forever wild" provision were rejected by the voters, but Marshall was continuously involved during the 1920s in defending the Adirondacks against "commercial developers who wanted to lower the environmental bar that has been raised in 1894" (Silver 2013, 410).

In the 1890s the science of ecology was in its infancy, but appreciation of nature was already promoted in America. Marshall's love of the Adirondacks reflected the Romantic naturalism promoted by writers such as Henry David Thoreau and his mentor, Ralph Waldo Emerson, and the Transcendental movement. Marshall regarded the Adirondacks as "the most perfect mountain region in the world," and there in the Saranac Lake area, despite initial objections to the purchase of land by Jews, the Marshall family built its country home in Knollwood in 1899. There Louis Marshall cultivated his love of the outdoors, and with the help of guide Herb Clark, the four children—Jim, Ruth, Robert, and George—became avid mountaineers who scaled America's peaks; Robert would have one named after him. In Knollwood, Louis Marshall expressed his romantic longing for preserving the fast-disappearing, pre-industrial America but also asserted American Jews' right to enjoy America's natural beauty. The country home also granted him respite from the demanding legal and public advocacy in New York, where he was involved in building the communal institutions of the American Jewry while defending the constitutional rights of Jews and other minorities (e.g., Catholics and Japanese).

Rightly called a "champion of liberty," Marshall cherished personal liberties as the greatest aspect of American democracy and believed in the power of the Constitution to protect them. As a Republican, he mistrusted big government, and as a German Jew who was educated in the ideals of *Bildung*, he emphasized the power of individuals to achieve their goals through education and moral self-cultivation. Although he was familiar with Felix Adler's social reforms, Marshall did not support Progressive legislation; especially after World War I, he worried that "our government is gradually being transformed into a Socialist state" (Sarna 2008, 9). Unlike Louis Brandeis, another highly influential Jewish lawyer, Marshall supported health regulations and rules limiting the number of working hours, but he did not support minimum-wage legislation because he "believed that government had no business intruding into the free marketplace of relationship between employers and employees" (Sarna 2008, 10). Individual liberty was, in his mind, American citizens' "most precious possession." Yet, as Sarna showed, precisely because Marshall, the conservative Republican, was a "champion of liberty," he staunchly advocated for persecuted individuals and groups—even more than Brandeis. The Progressive Democrat and Supreme Court Justice Brandeis, for whom labor unions' interests and government interests cohered, was much less involved in defending individual or group rights. Marshall, on the other hand, defended those whose rights were violated, regardless of their ethnicity, creed, or ideology. So too with nature: since it could not defend itself against human illegal activities, humans must extend it protection.

In 1913 Congress enacted a statute protecting migratory birds from hunters, but in Arkansas and Kansas, federal courts ruled the statute unconstitutional. The Justice Department recognized the issue's validity and was concerned that the Supreme Court would uphold the states. The US government negotiated a treaty with Great Britain arranging closed hunting seasons in the United States and Canada through reciprocal legislation. In 1918 Congress passed a statute to bring the treaty into effect, but the State of Missouri objected and brought action to prevent a federal game warden from enforcing it. When *Missouri v. Holland* reached the Supreme Court, Louis Marshall filed an *amicus curiae* brief on behalf of the Association for the Protection of the Adirondacks. His engagement on the Migratory Bird Treaty Act reflected his enthusiasm for nature and his lobbying for Jewish minority rights. Marshall drew an analogy between Jews and migratory birds: as politically powerless people, European Jews required special forms of legal protection because, like birds, Jews resided in different legal jurisdictions, each with its own local laws and traditions. For Jews and birds alike, "it was not clear whether a central governing power had the legal authority, or practical enforcement power to protect these defenseless objects" (Silver 2013, 403). Supreme Court Justice Oliver Wendell Holmes appreciated Marshall's moral concerns for the protection of the legally disempowered and ruled that the expansion of congressional power via a treaty for the protection of birds does not violate the Tenth Amendment states' rights standards.

Robert Marshall: Nature, Science, and Social Policy

With Robert Marshall, Louis's son, protecting nature was a matter of justice, concern for human rights, and governmental responsibility on a large scale. Robert inherited Louis's love of nature, joy of the outdoors, and commitment to conserving nature for future generations. Father and son shared a romantic view of nature that juxtaposed "nature" and "civilization" and appreciated the awesome beauty and majesty of American landscapes. But there were clear differences between them. Whereas Louis was deeply involved in local, national, and international Jewish politics, Robert was not engaged in Jewish communal affairs. Whereas Louis grew up Orthodox and remained committed to some level of religious observance throughout his life, Robert, forged by Felix Adler's Ethical Culture School, was not Jewishly observant at all. Robert illustrated the secularization and Americanization of Jews during the first three decades of the 20th century. He found his joy and happiness in nature and experienced spirituality on the High Peaks of the Adirondacks, which, echoing Emerson, he regarded as "sacred land." The most important difference between father and son was political. Louis was a Republican who believed in limited government. Robert was a Democrat who believed that only the government could properly protect and manage America's natural resources—and only the integration of environmental science and sound public policy, not the Constitution, could ensure the preservation of nature. Robert would use his scientific knowledge as a forester and plant biologist to frame the conservationist policies of the US Forest Service to "fight for the freedom of the wilderness" (Glover 1986, 3).

Born in New York on January 2, 1901, Robert was the third of four children. He and his siblings were educated in Felix Adler's Ethical Culture School, whose curriculum reflected Adler's educational philosophy and social reforms. Robert was an unexceptional student who devoted most of his energy to sports, an important venue for Americanization (Gurock 2005). Equally important to formal schooling in New York City were the vacations in Knollwood. For Robert, New York City and Knollwood represented "civilization" and "nature," respectively, with palpable tension between them. In contrast to the hustle and bustle of New York City, Knollwood days began with Bible study, followed by tennis or swimming and long afternoon walks or baseball games. The guide Herb Clark, more hunter and angler than trained mountaineer, introduced the family, especially Robert and his younger brother George, to the beauty, richness, and mystery of open spaces (Glover 1986, 27).

After graduating from the Ethical Culture School in 1919, Robert spent a year taking liberal arts courses at Columbia University but determined that his scientific interest demanded graduate training in the sciences. In 1920 he enrolled in the Forestry School of Syracuse that his father helped found a decade earlier, taking courses in chemistry, botany, engineering, forestry, geology, and zoology. For American Jews, who continued to encounter social and occupational discrimination, science was "a vocation that Jews *could* enter, and one on which they would be allowed to rise to the highest levels" (Efron 2014, 26–27).

During his college years (1920–1924), Robert remained involved in sports and hiking and began formulating his ideas about preserving the Adirondacks for the benefit of the people. In a paper written for one of his courses, Robert averred that an urban civilization "needed a wilderness to offset the crowding and commotion of urban conditions. Especially urgent was the preservation of the last vestiges of virgin forest in the state" (Glover 1986, 51). Robert matured during his college years, and his leadership skills began to emerge.

After graduating from Syracuse University, he worked as a field assistant at the Wind River Forest Experiment Station near Carson, Washington, and there launched his career as a scientific forester. He then enrolled at Harvard University, whose graduate forestry program was carried out at Harvard Forest, a 2,100-acre tract of rolling hills in north-central Massachusetts. After graduating with a Master of Forestry degree in 1925, Robert Marshall was qualified to enter the US Forest Service as an experimental scientist. Robert's entry into the US Forest Service launched a lifelong fusion of science, civil service, politics, and policy that lasted until his untimely death from cancer in 1939.

Robert Marshall emerged as an environmental activist who used his scientific knowledge to impact public policy and transform public consciousness about human relations with the natural world. Crucial to his understanding of human-nature relations were his two Alaska excursions, in 1929 and 1930–1931 (Marshall 1930; 1956; 2000 [1933]; 1991 [1933]). These experiences, during which he lived with and ethnographically studied the Koyukuk people, greatly contributed to his advocacy for protected wilderness.

Throughout the 1920s, wilderness preservation was controversial in the US Forest Service, which positioned itself opposite the National Park Service, established by Congress in 1916. The Forest Service wanted to prevent the National Park Service from gaining jurisdiction over several tracts of their lands, believing that national forests offered a "blue collar alternative to the national parks" as "places for the ordinary guy who could not afford the hotels and souvenir shops of the national parks. In the national forests there will be few developments but also few rules and few crowds" (Glover 1986, 95). These ideas were developed by Robert, who saw a connection between wilderness and the needs and desires of ordinary people. Robert noted that in the national forests, a small share of the American people

> have an overpowering longing to retire periodically from the encompassing clutch of a mechanistic civilization. To them the enjoyment of solitude, complete independence, and the beauty of undefiled panoramas are absolutely essential to happiness. In the wilderness they experienced the most worthwhile or perhaps the only worthy part of life.
>
> (cited in ibid.)

Marshall's biographer, James Glover, notes that the idea of wilderness as minority right was already articulated by Aldo Leopold, who in 1921 said that "foresters have a duty 'to meet the needs and desires of the minority' who like to use the wilderness" (Glover 1986, 93). But the framing of wilderness in the context of minority rights also echoes Louis Marshall's work on behalf of various minorities (e.g., Jews, Japanese, Indians, and Socialists). During Robert's years in Montana, he met workers destroyed by the exploitative economic system in Spokane's slums, Montana's copper mines, Appalachian sawmills, and "many other places where poor people worked for rich people's profits" (ibid., 72). Something was "radically wrong" in America.

In *The Social Management of American Forests* (Marshall 1930a), written for the League for Industrial Democracy, Marshall advocated comprehensively examining the need for public ownership and public management of the nation's forest resources. After Franklin Delano Roosevelt became president in 1932, Marshall's call for a "large scale public acquisition of forest lands and the use of the unemployed to safeguard these public forests" was easier to implement. This protected the wilderness for the future and offered access to the public. In 1932 a congressional investigation of forestry was launched by Senator Royal S. Copeland of New York. The resulting report, supervised by Earl Clapp, later the Chief Forester of the US Forest Service, appeared in April 1933 under the title *A National Plan of American Forestry*. Robert Marshall authored a section describing and evaluating all aspects of forestry, from timber, water, range, and wildlife to recreation, state aid, and fire protection. The report became the blueprint for forestry in Roosevelt's New Deal, which offered solutions to the intersecting social, scientific, and economic problems in forest management. Following his work on the Copeland Report, Marshall was appointed Chief of Forestry for the Bureau of Indian affairs, which brought him closer to the nation's policy makers. That same year Marshall also published *The People's Forests* in which he elaborated earlier arguments and supported them with data from the Copeland report (Marshall 2000 [1933]).

The book explained human and environmental interdependence, with forests both necessary for soil and water conversation and crucial for human happiness. "It is in nature," he asserted, that one can have "the most intimate and deepest spiritual experiences" (Marshall, 2000 [1933], 70). Marshall himself reported exhilarating experiences that were more spiritually rewarding than worship in the synagogue. After Louis died in 1929, Robert would regard himself as an agnostic. Sacred nature was more meaningful to him than communal worship. The forest was not only the source of spiritual experiences but also the context for scientific exploration. Indeed, "the forest is laboratory, unbound by the conventional four walls, floor and ceiling of the usual research center, but fully significant in the development of a knowledge of the laws of nature" (ibid., 70–71).

Robert's most lasting environmental achievement was establishing the Wilderness Society. Together with Robert S. Yard, Benton MacKaye, Harvey Broome, and Aldo Leopold, Robert co-founded the Wilderness Society on January 20, 1935 "to protect public lands, now and forever." Today, the Wilderness Society's membership is over one million people, and it protects 111 million acres of wilderness.

When Robert Marshall died unexpectedly in 1939, his brother George became the trustee of the Robert Marshall Wilderness Fund, which supported the Wilderness Society's conservation activities. George Marshall (1904–2000), who together with Robert scaled forty-six "High Peaks" of the Adirondacks, was a trained economist with an MA from Columbia University (1927) and PhD from the Brookings Institution (1930). George was even more attuned to the needs of the masses than his brother and father. He worked for the National Recovery Administration of President Roosevelt's New Deal and chaired the National Federation for Constitutional Liberties and the Civil Rights Congress, a Leftist organization that provided funding and leadership in the early civil rights movement. In the late 1940s and early 1950s, George collaborated with Paul Robeson, Dashiell Hammett, and

William L. Patterson on litigation protecting the rights of African Americans and American Communists. He was called before the House Committee on Un-American Activities, cited for contempt of Congress for refusing to submit National Federation records, and served three months in federal prison in 1950. George was committed to nature conservation, donating large sums and serving on the boards of the Wilderness Society and Sierra Club. In retrospect, the life of the Marshall family captures a certain arc within American Jewry for whom protecting nature and an environmental sensibility were linked to protecting minority rights and civil rights.

Environment, Science, and Politics among "Non-Jewish Jews"

Whereas the Marshalls protected the rights of nature because nature was powerless to defend itself, a different approach was articulated by children of East European Jewish immigrants who were imbued with Marxist ideology and who had faith in science, progress, and American democracy as pathways for the betterment of humanity. Two Leftist New York Jews—Barry Commoner and Murray Bookchin—greatly contributed to the emergence of environmental awareness and ecological consciousness in America. Their efforts shifted American environmentalism from human protection and conservation of nature to awareness of the interdependence of humanity and nature. Commoner and Bookchin exemplified what Isaac Deutscher (1968) labelled the "non-Jewish Jews," namely, those who did not observe the Jewish religion but whose ethnic identity, patterns of socialization, and cultural sensibilities were profoundly Jewish. The life and careers of Commoner and Bookchin developed independently of each other, but we discuss them together to shed light on science education's place in Americanizing East European Jews, in secularizing the prophetic ideal of justice (which they applied to protecting nature and protecting humanity from the threat of annihilation by technoscience), and in recognizing that humanity and nature's interdependence requires restructuring American politics. In their distinctive ways, Commoner and Bookchin illumine how the interplay of science and the environment functioned in American prophetic politics in the mid-20th century, paving the way for the Spiritual Progressives of the late 20th and early 21st centuries.

Barry Commoner: Technoscience, Environment, and Social Responsibility

In 1970 a drawing of Barry Commoner appeared on the cover of *Time Magazine*, who declared him the "Paul Revere of Ecology." On March 22 of that year America celebrated the first Earth Day, and *Time* correctly recognized Commoner's profound contribution to the emerging American ecological awareness. Ahead of others, Commoner, a cell biologist, understood the connection between the achievements of modern technology, represented by the atomic bomb, and the emerging environmental crisis. Convinced that scientists have a moral obligation to educate the public on issues of social relevance, Commoner warned Americans about the risks to human health posed by detergents, pesticides, herbicides, radioisotopes, and smog. Most distinctively, for Commoner the environmental crisis was directly related to problems of poverty, injustice, racism, public health, national security, and war (Drier 2017).

Commoner's contribution to American environmentalism is well known (Kriebel 2002) and appropriately he has been called the "the father of grass-roots environmentalism" (Montague 2002). Most relevant to this chapter is that Commoner has been depicted as a "new Jeremiah," by his biographer, Michael Egan (Egan 2007, 99). By labelling Commoner a "new Jeremiah" and linking him to the religious rhetoric of the Great Awakening, Egan assimilated Commoner into the paradigm of American "prophetic politics" in which religious motifs inform and transform public discourse and social policies.

Egan's portrayal, however, is somewhat misleading because it does not acknowledge the distinct Jewishness of Commoner's prophetic politics. Commoner came out of a distinctive Socialist, Yiddish culture of East European Jews in New York during the interwar years, which accounts for his views on science and science education's social responsibility. Moreover, even if Commoner was an "American Jeremiah," we should recall that the biblical prophet was not only a messenger of doom and gloom who called for repentance but also a prophet who counseled the Jewish exiles in Babylonia

> to build houses and dwell in them, plant gardens and eat the fruit they produce. Take wives and have sons and daughters; take wives for your sons, and give your daughters in marriage, so that they may bear sons and daughters; multiply there and do not decrease.
>
> (Jeremiah 29:5–6)

After the Fall of the Jerusalem Temple in 586 BCE, Jeremiah instructed Judean exiles in Babylonia to accept their political loss and accommodate themselves to life in Babylonia. If Commoner was like Jeremiah, his message was not like that of a Great Awakening preacher but of a secular, Yiddish-speaking, Socialist Jew concerned with the quality of human life in this world, which is shaped by human action.

Commoner was a product of the robust secular Jewish culture in New York that boasted not only literature, art, theater, and music, which the Russian-Jewish immigrants brought with them (Veidlinger 2009), but also science and science education (Cohen 2007; Maor 2013; Kadar 2017). Born on May 28, 1917, in Brooklyn, New York, Barry Commoner experienced the hardship of the Great Depression when his parents lost all their savings. Already in high school, he was fascinated by the natural world and spent a lot of time in Brooklyn's Prospect Park. In addition to economic hardship,

Jews in those days had to contend with anti-Semitism and social prejudice, barring them from many economic and intellectual pursuits. To cope, Barry followed the advice of his uncle, Avraham Yarmolinsky, a librarian in the Slavonic Department of the New York Public Library, to change his surname from "Comenar" to "Commoner." Americanizing Jewish surnames was one way for Jews to "pass" in America. It was Yarmolinsky who recognized Commoner's interest in biology and bought him his first microscope. Yarmolinsky and his wife, the poet Babette Deutsch, a lecturer at Columbia University, also encouraged and made it possible for Commoner to enroll in Columbia College in 1933, where he studied zoology. At Columbia, Commoner became involved in social activism, participated in Socialist and Communist meetings that promoted and defended the labor unions, and protested the lynching of African Americans in the South (Egan 2007, 19; Krier 1990). Commoner then enrolled in the graduate program in plant biology at Harvard University, where he began to develop his ideas about science's social function. His inspiration was the Irish-British scientist and Communist activist, John Desmond Bernal, whose book *The Social Function of Science* (1939) made a deep impression on Commoner. Commoner's fusion of science, science education, and radical politics was forged during the 1930s when other Left-leaning scientists were social activists (Kuznick 1987).

Commoner received his doctoral degree in 1941, just before the US entered WWII. He was drafted into the Navy, where his scientific expertise served the war effort. He worked on a device that would allow torpedo bombers to spray DDT on beachheads to reduce the incidence of insect-borne diseases among soldiers. While the device was effective in killing the flies, Commoner also noted that new swarms appeared, attracted by tons of decaying fish who were the accidental victims of DDT (Egan 2007, 21). The direct encounter with technoscience's unintended environmental consequences inspired his postwar environmental activism. This crystallized the atomic bombs that were dropped on Hiroshima and Nagasaki, which ended the war but birthed public debate about how to best manage nuclear power. The atomic bomb changed the perception and meaning of science: it was no longer just a tool for human progress and prosperity but also a threat to the future of humanity and indeed of life on Earth. In 1947 the Atomic Energy Act established the Atomic Energy Commission "to manage the postwar program to develop military, scientific and industrial potential of atomic and nuclear energy" (Egan 2007, 28). But how science should frame governmental policies on resource use was hotly contested as scientists took opposing positions in the debate. From 1947 to the publication of *Science and Survival* in 1966, Commoner stood at the center of these debates.

In 1947 Commoner took an academic position at Washington University in St. Louis and there began his public efforts to bring "science and science policy out of the laboratory and into public politic" (Egan 2007, 31). That effort was carried out through his involvement in The American Association for the Advancement of Science (AAAS), especially its Interim Committee on Social Aspects of Science, established in 1956 (Cittadino 2018, 723–24). Throughout the Cold War, the Korean War, and the pressure of the House Committee on Un-American Activities, the contested "social aspects of science" focused on nuclear weapon development and testing. Especially after March 1954, "when a U.S. Hydrogen bomb test in the Pacific went terribly awry, exposing Marshall Islanders and the crew of a Japanese fishing vessel to dangers radiation levels" (Cittadino 2018, 727), the connection between controlling nuclear power and protecting the environment emerged with greater clarity. Barry Commoner was convinced that determining what constituted acceptable risk was an inherently moral and public (rather than scientific) issue and that a free and democratic society requires citizen involvement in decision-making about risky technologies. He therefore encouraged the growth of grassroots information groups that facilitated public participation in critical questions of science and its application. Leading the Greater St. Louis Citizen's Committee for Nuclear Information, Commoner worked with other scientists and citizens making scientific information publicly available. In so doing, he positioned himself against the Atomic Energy Commission (AEC).

In 1957–1958 the AEC was engaged in an ambitious program to use underground nuclear explosions, mainly hydrogen bombs, for assorted large-scale engineering schemes (Cittadino 2018, 720). Barry Commoner vehemently opposed the program and lobbied the public to his side. In 1958 he published "The Fallout Problem" in *Science*, where he exposed the long-range effects of worldwide fallout from nuclear weapons testing, outlined the relationship between scientific knowledge and public policy, and left the decisions about fallout questions to informed citizens. Additionally, Commoner delivered informal talks in schools, women's clubs, and other local venues, leading to the creation of *Nuclear Information*, a newsletter published nine times a year. He thus put in place the type of adult education that thrived among the Jewish immigrants in New York in the first half of the 20th century. For Commoner, science was to serve society and could not be separated from society's guiding values; educated citizens, rather than the small scientific elite, should determine science policy (Egan 2007, 31–46). Commoner was by no means the only scientist who agitated for control of nuclear testing. In 1957 Linus Pauling, the Nobel Laureate in chemistry, issued the "Appeal by American Scientists" that garnered signatures of over 2,000 American scientists. In 1958 the appeal became international and signed by 11,038 scientists from forty-nine nations. Pauling's appeal was drafted by Commoner and his cohorts.

Commoner's efforts to educate the public about the danger of the atomic bomb were highly successful. In 1963 the Nuclear Test Ban Treaty was signed; Commoner hailed it as "the first victorious battle in the campaign to save the environment—and its human inhabitants—from blind assault of modern technology" (Egan 2007, 47). Protecting the public from the hazards of modern technology became his public mission. The dangers of modern chemicals were also recognized in the 1950s by Murray Bookchin, who published "The Problem of Chemicals in Food" in 1952 under the pseudonym Lewis Herber. In 1962, Bookchin (as "Lewis Herber") published *Our Synthetic Environment*—six months before Rachel Carson published *Silent Spring*, which challenged the use of pesticides by agricultural scientists and the government and called for a change in how humankind views the natural world. We discuss Bookchin in the following section; here, let us note that Commoner's and Bookchin's commitment to educating the public about modern technology's harms was forged during the 1930s and 1940s in the Left-leaning circles immersed in Marxism. Recent scholarship on Marx makes clear that Marx not only theorized about how human economic activities utilize natural resources but also correctly understood the impact of economic activities on air, water, and soil quality (Foster 2000; 2020; Foster and Burkett 2016). Marx's view of humankind's place in the environment was more holistic than most recognized. Bookchin and Commoner perpetuated that holistic approach by insisting on the interdependence of humanity and nature. These were not "anthropocentric" views but rather eco-Socialist views that offered a nuanced understanding that humans are part of nature as much as they have the capacity to transform natural ecosystems (Pepper 1993).

Commoner's *The Closing Circle: Nature, Man, and Technology* (1971) articulated "a framework for the modern environmental movement that sought to address many of the social and ecological problems that continue to plague a movement that has since suffered countless setbacks and false starts" (Egan 2002, 444). The straight line of modern technology breaks the circle of life, making collapse inevitable. To correct the problem, we need to "close the circle" by recognizing the interdependence of "nature, man, and technology," as Commoner's subtitle made clear. The closing of the circle is neither a mystical intuition nor an emotional attitude, but a rational process based on scientific information that any educated person can quickly grasp. Commoner distilled scientific information in his famous Four Laws of Ecology: (1) Everything is connected to everything else; (2) Everything must go somewhere; (3) Nature knows best; and (4) There is no such thing as a free lunch. The book became enormously successful because it was written in the nonacademic stye of adult education.

Commoner was not an ecologist by training, but he teased out the social significance of the science of ecology more effectively than professional ecologists. He explained that the conflict between technological innovations and environmental health was the most significant aspect of the environmental crisis. In the late 1960s the environmental movement focused on fighting pollution and protecting wild places, but Commoner argued that environmental issues are inseparable from social concerns including civil rights, women's rights, and consumer rights. Environmentalism, then, was about the pursuit of justice and correcting social, economic, and political inequalities.

This was also the key insight of Bookchin's Social Ecology. Commoner and Bookchin positioned themselves against Paul R. Ehrlich (b. 1932), whose book *The Population Bomb* (1968) blamed overpopulation for the environmental crisis and called for controlling population growth, even by means of coercion. Ehrlich, yet another Jewish scientist who shaped public perceptions of environmentalism, grew up in a middle-class, German-Jewish (Reform) family more similar to Louis Marshall rather than to the secularism of Commoner and Bookchin. Ehrlich was deeply moved by the misery and suffering caused by poverty, and his compassion for the poor and the socially marginal led him to insist on curbing population growth.

Ehrlich's Neo-Malthusian analysis led him and his co-author wife Anne Howland Ehrlich, to predict that "in the 1970s hundreds of millions of people will starve to death in spite of any crash programs embarked upon now. At this late date nothing can prevent a substantial increase in the world death rate [...]" (Ehrlich 1968, 1). Ehrlich was more of a Jeremiah than Commoner or Bookchin, and although his predictions did not come to pass, he accurately diagnosed the connection between overpopulation and poverty. Ehrlich persisted in his beliefs and in *The Population Explosion* (1991) connected population with greenhouse gases, global warming, and climate change. In the earlier debates between Ehrlich, Commoner, and Bookchin, Ehrlich considered overpopulation as the main environmental threat, whereas Commoner and Bookchin viewed modern technology, especially the proliferation of new polluting technologies, as causing environmental degradation.

Unlike the Marshalls and other early environmentalists, who wished to protect pristine wilderness from human intervention, Commoner's attention to air and water pollution, toxic chemicals, radioactive materials, ozone depletion, and carbon dioxide levels in the atmosphere was fueled by a desire to find healthier, more sustainable systems of production that can grow and develop in harmony with the environment. He thus prescribed altering wasteful economic and harmful technological practices: "If we are to survive economically as well as biologically, industry, agriculture, and transportation will have to meet the inescapable demands of the eco system" (Commoner 1972). Because Commoner viewed the environmental crisis as resulting from social and economic mismanagement, he argued for social and economic reorganization—not to eliminate technology but to change its orientation.

This comprehensive vision, of course, could become a reality only if translated into policies, which in turn required him to enter politics. In April 1980 Commoner became the presidential candidate of the Citizens' Party, which aggregated diverse Left-leaning constituencies that gave voice to the poor, laborers, and minorities who were alienated from American politics. The Citizens' Party was a political failure; Commoner himself admitted that it was "an insane idea, which utterly failed" (Egan 2007, 169), but he remained politically involved and advised the Reverend Jesse Jackson's Rainbow Coalition campaign in 1984. In the late 1980s Commoner's lifelong attention to poverty, social inequality, environmental health, and safety focused more directly on communities of color, since he understood that "there is a functional link between racism, poverty, and powerlessness, and the chemical industry's assault on the environment" (ibid., 170, n.11). In 1987 he bitterly noted that "in effect, the environmental hazard is regarded as smaller if the people it kills are poor—a standard that could be used to justify situating heavily polluting operations in poor neighborhoods" (ibid., n.13). In his later book (1990), Commoner argued that if we wish to prevent environmental pollution, we must redesign production systems to eliminate pollutants at their source.

The placement of power facilities, manufacturing plants, and waste incinerators unequally distributed risk exposure along class and color lines. The disempowered suffer from greater exposure to environmental pollutants and the resultant health hazards, and typically they have less access to health care. The notion that social injustice and ecological injustice are two sides of the same coin was articulated by secular Jews in the 1950s and 1960s, long before the environmental justice movement emerged in the 1980s and Jewish environmental spirituality emerged in the 1990s. The thinker who made the connection between social justice and the environment was Murray Bookchin.

Murray Bookchin: Social Ecology and the Ideal of Freedom

Like Commoner, Murray Bookchin was a child of Russian-Jewish immigrants in New York, but unlike Commoner, he encountered radical ideas in early childhood. Bookchin's maternal grandmother, Zeitel Carlat Kalutzky, was a Jewish revolutionary who disseminated Marxist ideas to the Russian masses to arouse their proletariat consciousness. After the failed revolution of 1905 and the death of her husband, she migrated to the United States in 1913 with her two children, Rose and Dan. Zeitel rented a room in the Lower East Side and found work for herself and her children in the needle trade. While attending a summer camp for Communist youth, Rose met Nathan Bookchin, a fellow Russian-Jewish immigrant, and despite her mother's disapproval, they married. Although the marriage was short-lived, the child born of it, Murray Bookchin, left a legacy.

In 1920 the entire family settled in the East Tremont neighborhood of the Bronx, a

> Jewish village laced with Old World traditions […] but also … a radical ambience—Workmen's Circle clubs, union locals, and social meeting halls were more numerous and influential in that neighborhood than synagogue. Shops were family owned; the kosher butcher, the greengrocer, the cigar maker, and more.
>
> (Biehl 2015, 4)

Murray was born on January 14, 1921, and in keeping with the family's secular orientation, he received no formal Jewish religious education. At home they spoke only Russian, dressed young Bookchin in Russian-style clothes, and introduced him to Russian revolutionary literature. As Bookchin later stated, "my family educated me in revolution" (Biehl 2015, 4). But growing up in East Tremont was a Jewish ethnic experience *par excellence*, and the neighborhood would inspire Bookchin's later vision of urban municipalities. Like Commoner, Bookchin was a "non-Jewish Jew" who translated a secularized Jewish way of life into radical secular politics.

Bookchin's family was hard hit by the Depression. What saved him was The Young Pioneers of America, the children's branch of the international Communist movement. Functioning like surrogate parents, the Young Pioneers gave him structure, purpose, education, stability, and validation (Biehl 2015, 7). In the early 1930s, when the Depression deepened and people lost their jobs and homes, the Communists offered hope and purpose. Facing multiple evictions and subsisting from church breadlines, Bookchin grew up fast and earned his living. The Communists gave young Murray work to sell the *Daily Worker* in East Tremont. But more importantly, they educated him. In the Workers School, the Cooperative Cafeteria, and the city streets, where Communists, Socialists, and Trotskyists argued politics, Bookchin absorbed political and economic ideas and the pugnacious rhetorical style that ridiculed and even demonized opponents. That style remained with him for life, long after he abandoned the Communist Party and left Trotskyism to endorse Anarchism, making him a fierce debater but not a beloved comrade.

To support himself and his mother, Bookchin worked in a foundry in Bayonne, New Jersey. The work was physically demanding, and it introduced Bookchin to both the plight of the workers and the organized labor movement (in which Jews were leaders) and to modern technology's toll on human bodies. Through labor in the auto industry, Bookchin acquired considerable technological knowledge, but he never glamorized or romanticized physical labor. In fact, he considered technology a means of liberating humanity from physical toil. Bookchin found respite from the harsh factory work in the headquarters of the Socialist Worker Party (SWP) at University Place in New York, his political home. While Bookchin, the labor organizer, sharpened his theoretical skills, turning the foundry workers into a revolutionary proletariat remained an elusive goal, notwithstanding his attempts to address their grievances. In frustration, Bookchin left the foundry in 1944 for the GM machine shop in Manhattan, where he joined the United Auto Workers (UAW), the country's most militant union (Biehl 2015, 41–42). The class conflict brewing at GM seemed more promising to the budding Leftist intellectual. In 1944 Bookchin met Joseph Weber, a Jewish intellectual who escaped Nazi Germany and became the inspiration to many New York Socialists. Weber became Bookchin's mentor, turning him from an industrial worker into a published intellectual, but the relationship soured before Weber died in 1959.

In 1946, Bookchin was drafted, stationed at Fort Knox, Kentucky, the army's center for mechanized cavalry. Trained as a tanker, he and his unit practiced maneuvers, but he mostly served as auto-parts clerk for the motor pool (Biehl 2015, 43). Bookchin's wartime draft ended in June 1947, and he returned to care for his ailing mother in New York while working at GM. Now more mature and realistic, Bookchin realized that "the proletariat was not revolutionary, Socialism was an illusion, and Marxism was based on a fallacy" (ibid., 44). He left GM and joined a group of former Trotskyists and Labor Zionists who created a new socialist movement that published *Contemporary Issues: A Magazine for a Democracy of Content*, whose goal was to offer "a democratic solution to the crisis of mankind" (ibid., 52). Led by Joseph Weber, the group saw itself as a shaper of public opinion through informal education. Most members were secular Jews, and one of them, mathematics teacher Beatrice Appelstein, became Bookchin's wife. The group met regularly, every Friday evening, no less, to talk politics and to debate current issues, secularizing the experience of the Sabbath meal. The group collectively edited *Contemporary Issues*, to which Bookchin contributed essays under various pseudonyms: M.S. Shiloh, Robert Keller, Harry Ludd, and Lewis Herber.

In 1952, Bookchin (as "Lewis Herber") published "The Problem of Chemicals in Food" in *Contemporary Issues.* While researching and writing for the magazine during the 1950s, Bookchin began articulating his worldview, fusing the critique of Capitalism, philosophy of organic development, opposition to the atomic bomb, advocacy of environmentalism, and promotion of municipal anarchism. The magazine finally folded in 1959, after the death of Joseph Weber, who toward the end of his life denounced Bookchin as a mere "journalist," "agitator," and "dilettante" (Biehl 2015, 77). Yet it was precisely Weber's death that most likely enabled Bookchin to emerge as a public intellectual in his own right, with a distinct environmental message honed while protesting nuclear testing and nuclear power plants in the late 1950s and early 1960s. In 1962, still under the assumed name of "Lewis Herber," Bookchin published *Our Synthetic Environment*, in which he exposed the hazards of food additives and noted the negative impact of pollution on urban living. Although the book was eclipsed by Rachel Carson's *Silent Spring* (1962), Bookchin began articulating an environmental message couched in Anarchist rather than Marxist terms. In the 1960s he co-founded and was the principal figure of the New York Federation of Anarchists, and in 1968 he founded another group that published *Anarchos Magazine.* Bookchin's Social Ecology uniquely blended environmentalism, counterculture creativity, anarchism, and radical politics.

In 1969–1970 Bookchin taught at Alternate U., a free, countercultural radical school in Greenwich Village, the seedling for what would later become the Institute of Social Ecology (ISE). In 1971 Bookchin and a group of friends moved to Burlington, Vermont, where they attempted to put eco-anarchism into practice. In 1973 he was hired by Goddard College in

Plainfield, Vermont, to lecture on technology. The lectures turned into a teaching position and the creation of the ISE in 1974. The Institute would move to its own campus in Cate Farm in Plainfield in 1981, where it established a community garden, art center, and geodesic dome, integrating communal living with environmentalism. In the following decades, ISE introduced numerous people, including secular, Left-leaning Jews, to the principles of Social Ecology, an ideology that intended to eliminate all forms of domination and oppression, including human domination of nature. With his reputation growing, Bookchin was hired to teach at Ramapo College in Mahwah, New Jersey, becoming an academic and rising quickly to the rank of Full Professor, even though he lacked formal academic credentials. In this regard, Bookchin was one of "the last intellectuals," as Russell Jacoby called them, namely, thinkers who integrated vast knowledge of Western thought with politics and social activism but who functioned on the margin of the academy (Jacoby 2000 [1987], 96–99).

Bookchin made seminal contributions to the early warning literature on environmental problems. In the late 1960s he allied with Commoner against Paul Ehrlich's concern for overpopulation, which distracted from the far greater environmental problems the US economy generated. For the Marxist Bookchin, economic inequality was rooted in social hierarchy, leading to domination, oppression, and the social and economic disparities driving the nature/society dichotomy. Social hierarchies harm nature because people "identify human progress with the idea of dominating nature" (Bookchin 1990, 202). Ending human domination of nature requires liberating humanity and creating a truly egalitarian society.

Associated with the New Left, Bookchin positioned Social Ecology (Bookchin 1980; 1982; 1986) as a critique of Deep Ecology, articulated in the early 1970s by Norwegian philosopher and mountaineer Arne Naess (d. 2009) and popularized by Warwick Fox (1986), George Sessions (1995), and Bill Devall (Devall and Sessions 2007). Bookchin (1988) sharply opposed Deep Ecology because it ignored the cognitive and ethical differences between humans and other forms of life; conversely, Deep Ecologists dismissed Bookchin's view as the very "anthropocentrism" that engendered the ecological crisis (Stoll 2001). The public debate between Bookchin and the Deep Ecology activist Dave Foreman, the founder of Earth First!, took place in November 1989 organized by the Learning Alliance, New York City's alternative education and action organization. The organizers planned a dialogue that featured "two articulate spokespeople from different wings of the relatively small but potentially powerful, radical ecology movement," at the time continually harassed by the Federal Bureau of Investigation (Levine 1991). Bookchin and Foreman managed to clarify their agreement and not merely dwell on their differences (Chase 1991), but Bookchin continued to critique Deep Ecology and similar spiritual trends in American environmentalism (Bookchin 1990; 1995; 1996). Bookchin secured Social Ecology's place within American environmental ethics, but his legacy remains contested. No one person has emerged as Bookchin's intellectual heir (Light 1998).

Bookchin's contribution to Jewish environmentalism has gone unrecognized, precisely because he was a "non-Jewish Jew." However, I argue that Social Ecology could not be understood without its Jewish roots, albeit secularized, and that Bookchin, more than Commoner or Robert Marshall, paved the way for a socially engaged Jewish environmental movement that secularized *Tikkun Olam* to address society's social, political, and environmental ills. Bookchin articulated "a truly natural spirituality centered on the ability of an awakened humanity to function as moral agents in diminishing needless suffering, engaging ecological restoration, and fostering an aesthetic appreciation of natural evolution in all its fecundity and diversity" (Clark 2009 [1998], 286).

Bookchin's naturalistic humanism insisted on humanity and nature's interdependence and human responsibility to care for nature by eliminating domination and exploitation. His Social Ecology was not anthropocentrism, as Deep Ecologists claimed, but the secularization of the Judaic commitment to making a just world by eradicating all forms of social injustice. Deeply inspired by Hans Jonas's philosophical biology, one of the few Jewish thinkers Bookchin cited (Bookchin 1982, 455), Bookchin denied the fact/value, is/ought dichotomies undergirding modern science and the culture of scientism. Instead, he argued that the natural world manifests values of mutualism, care, and responsibility. There is no conflict or contrast between ethics and nature, because ethics emerges out of nature; and since evolution is unidirectional, humans must care for nature responsibly to ensure its and their well-being. Social Ecology is not to "re-enchant nature," as Deep Ecologists sought, but to "re-enchant humanity," namely recognize humanity's continuity with the natural world and humanity's distinct capacities, achievements, and failures. Rejecting the natural sciences' materialist reductionism and Marxism's rigid historical materialism, Social Ecology argues for a dialectical relationship between "first nature" (i.e., the physical world) and "second nature" (i.e., human culture) without reducing one to the other.

Bookchin's dialectical, humanistic naturalism was suffused with secularized messianic passion. Joel Kovel, a fellow Jewish public intellectual who contributed to Social Ecology, saw in Bookchin "something of the prophet in the ecoanarchist. Or a fallen angel, for there is also something haunted, a sense of rancor and harsh discord" (Kovel 1998, 27). Kovel insightfully suggested that Bookchin's legacy was torn between two narratives. One narrative promoted Enlightenment humanism and universalism, in which harmonizing humanity and nature is a real possibility. In this narrative, justice means "corrective alterations in a basically irrational society," and freedom is "the participation in a fuller, more ethical life through the overcoming of hierarchy/domination according to the adoption of dialectical naturalism" (ibid., 29). The antithesis of freedom is hierarchy/domination: all hierarchies are dominative, and all dominations are hierarchical, and the ubiquity of hierarchy/domination requires a total revolutionary transformation, to which Bookchin was deeply committed. The other narrative secularized the Judeo-Christian tradition, in which the figure of the Anarchist is the

> prophetic instrument of higher power—the self-creativity of nature—on whose behalf he seeks to liberate humanity [...] The Anarchist is also fated to suffer and to endure persecution [...] The Anarchist, rather, is greatly burdened with false prophets from his own tribe.
>
> (ibid., 33)

Bookchin himself admitted that his *Ecology of Freedom* (1982) had "unabashed messianic character, a messianic character that is philosophical and ancestral." And by "ancestral," Bookchin referred to the Bible, without which Western culture could not be understood, as well as to his own Jewish ethnic and cultural roots.

Bookchin was a prolific writer who mastered several disciplines, including biology, philosophy, history, economics, sociology, anthropology, and politics. In the sweeping brushstrokes of a social theorist, he articulated Social Ecology as a grand meta-narrative about humanity and nature. Doing justice to Social Ecology and to the development of Bookchin's ideas requires a full-fledged monograph (e.g., White 2008; Eckersley 1989; 1992; Biehl 2015) which cannot be undertaken here. We only note that Bookchin's secular Social Ecology influenced the emergence of Jewish environmental spirituality in the 1970s. To anticipate: whereas Bookchin, the "non-Jewish Jew," articulated secular ethics of care and responsibility toward nature, secular Jews indebted to Social Ecology, such as Arthur Waskow, returned to more traditional forms of Judaism and gave Social Ecology a distinctive Jewish coloring. These Jewish Spiritual Progressives reappropriated the mystical impulses that Bookchin had sharply criticized, articulating a distinctly Judaic environmental spirituality. We will discuss them in the following chapter.

Bookchin's Social Ecology sought to heal the gap between man and society, between culture and nature, by showing that culture is rooted in nature. By integrating humanity and nature, or society and ecology, Bookchin rejected modern science's reductionism and the ideology of scientism as much as he dismissed the critics of the Enlightenment who were skeptical of human rationality's emancipatory possibilities.

Bookchin was above all a dialectical thinker who embraced developmental, organic thinking that recognized nature as the totality of evolutionary processes: existing entities are always A and not-A, both what they are and what they could be. Dialectical reason, Bookchin argues, captures the complexity of reality more accurately than the "conventional reason" that characterizes mathematical physics, which is too general, too simplistic, too reductionist, too arbitrary (Bookchin 1996, 11–12).

Dialectical thinking is especially appropriate for grasping evolution, in which inorganic materials evolved into organic forms of life that later developed into multicellular organisms that eventually developed intelligent neural apparatuses that allow for innovative choices (Bookchin 1990a). Bookchin labels this cumulative evolution of the natural world, especially the organic world, as "first nature." Out of "first nature" evolved lifeforms that can manipulate the natural environment as needed; adaptation is marked by growing degrees of subjectivity, self-awareness, and freedom. This emergent society is "second nature," which Bookchin defines as "the development of uniquely human culture, with a wide variety of institutionalized human communities, effective human technics, richly symbolic languages, and carefully managed sources of nutriments" (Bookchin 1996, 119). Insisting that "first nature exists in both continuity and discontinuity with second nature" (ibid., xi), Bookchin reminds us that humans are animals with animalistic emotional states (e.g., fear and desire) but are also unique for their "unprecedented capacities" for self-reflection, conceptual thought, and ability to consciously alter "first nature." Animals do have "communities" marked by functional differentiation, but these are not "societies," which require a level of rationality and abstract thinking that are absent in the animal kingdom. In humans alone, organic nature reaches the highest degree of self-consciousness.

According to Bookchin's dialectical naturalism, "nature belongs in humanity just as humanity belongs in nature." The relationship between them is not structured around subordination and domination but "highly reciprocal and complimentary." Neither "society" nor "nature" dissolves into the other (Bookchin 1996, 92). Because "second nature" or "culture" evolved out of "first nature," Bookchin denies that humans are either "alien" to nature (as many Christians believe) or "parasites" who destroy nature's inherent balance (as Deep Ecologists maintained). Similarly, Bookchin rejected romanticizing "wilderness," where one presumably experiences the oneness of nature, as Robert Marshall felt while hiking the Adirondacks. As a rationalist, Bookchin regarded human technology as "natural" because it was part of "second nature" and therefore, he reasoned, there is nothing "unnatural" about human intervention in nature. Bookchin emphatically states that "to denigrate this capacity is to deny the thrust of natural evolution itself toward organic complexity and subjectivity—the potentiality of first nature to actualize itself in self-conscious intellectuality" (ibid., 31).

Technology can either serve or impede nature's thrust toward greater complexity, diversity, and freedom. In other words, technology per se causes neither the rift between humanity and nature nor the human tendencies to dominate and exploit nature. Rather, the problem lies in social, uniquely human hierarchies. To ethically employ technology, humans must recover a genuine ethical sensibility that recognizes the prevalence of mutuality, care, and cooperation in the natural world. They must recognize, as Hans Jonas already had, that freedom inheres in nature, albeit only in incipient form.

In referring to Hans Jonas's *The Phenomenon of Life*, which Bookchin praised as "admirable" (Bookchin, 1982, 372), he explains that the evolution is neither "blind" nor totally random; it is neither inexorably predetermined nor directionless. Rather, it is dialectical, rational, historical, and ethical, propelled by the transition from potentiality to actuality, in which "things and phenomena are incomplete and unactualized in their development" (ibid., 19) but always moving toward greater differentiation, rationality, and freedom. In the transition from potentiality to actuality, what "is" is transformed by "what ought to be," and the "ought," is always more rational and more developed than the "is." The process is inherently cumulative: the old is always present in the new, and the past always shapes the future. It is ethical in the sense that "what should be" (i.e., the "ought") dictates the direction of the development and the emergence of "what is." Values, ideas, and ideals are not subjective constructions that humans invented and imposed on nature but expressions of the very process in which potentiality becomes actualized; values, ideas, and ideals are manifested in the way nature evolves toward greater freedom. As Bookchin puts it, "mutualism, freedom, and subjectivity are not strictly human values and concerns. They appear, however germinally, in larger cosmic and organic processes" (ibid., 365). Thus ethics "is not a matter of subjective taste; it is objectively determined" (Bookchin, 1996, 24).

Because value inheres in nature, the conventional binary dichotomy of "culture" vs. "nature" is misleading. They are different but influence each other. Human societies manifest a greater degree of freedom and participation that are already incipient in the natural world. But human societies also give rise to social hierarchies in which certain groups dominate others. In the *Ecology of Freedom* Bookchin explains how social hierarchies emerged and how they created the substrate of oppression out of which class relations were formed. Hierarchical social relations emerge from social and material factors (e.g., population growth, physical force) and changes in consciousness. In hierarchical sensibilities, people are deemed "superior" or "inferior" by a given standard and then use that ranking to justify domination. Mere differences in ability, experience, or knowledge are turned into hierarchical relations that presumably justify domination.

The goal of Social Ecology is to overcome hierarchies and patterns of domination. That means bringing an end to social exclusion, marginalization, and injustice as well as to the human domination and exploitation of nature. For Bookchin, the root cause of ecological degradation is the human domination of other humans, and the remedy to ecological degradation is to bring an end to all patterns of oppression. A truly ecological society lives in accordance with the thrust of evolution toward greater freedom, and such a society "takes on the responsibility of evoking, elaborating and giving ethical content to the natural core of society and humanity" (Bookchin 1996, 89). Social Ecology undergirds and promotes society's reorganization in a mutual, cooperative manner that enhances the freedom extant in nature. For humans, that means eco-anarchism is the only political arrangement that assures a genuinely emancipatory and egalitarian society

(Davidson 2009). For Bookchin eco-anarchism was not a mere theory but a lifelong commitment to social activism.

The history of American environmentalism could not be written without referring to Bookchin, but Bookchin, we should note, distinguished "ecology" and "environmentalism," promoting the former and critiquing the latter. He disdained "environmentalism" as

> environmental engineering [that] does not bring into question the underlying notion of the present society, notably that man must dominate nature. On the contrary, it seeks to facilitate that domination by developing techniques for diminishing the hazards caused by domination.
>
> (Bookchin, 1980, 77)

Environmentalism was inherently political because it "easily lends itself to the lure of statecraft, that is, to participation in electoral, parliamentary, and party-oriented activities" (Bookchin 1990, 160) and "uses the power of self-corrupting professionalized corps of legislators, bureaucrats, and military forces called the State" (ibid.). In contrast to environmentalism, ecology (specifically Social Ecology) "seeks to eliminate the concept of the domination of nature by humanity by eliminating domination of human by human" (Bookchin 1980, 77). Bookchin challenged the ecology movement "to focus its efforts on changing the basic structure of our anti-ecological society, not merely on providing more palatable techniques for perpetuating it or institutional cosmetics for concealing its irremediable diseases" (ibid., 79). A truly ecological society could emerge if we realize that "neither sexism, ageism, ethnic oppression, the 'energy crisis,' corporate power, conventional medicine, bureaucratic manipulation, conscription, militarism, urban devastation or political centralism can be separated from the ecological issues. All these issues turn round hierarchy and domination, the root conceptions of a radical social ecology" (ibid., 82–83). For Bookchin, all social problems are ecological problems, and all ecological problems are rooted in social problems of hierarchy. The ideal society is therefore the ecological society, in which hierarchy and domination are dealt with once and for all.

Bookchin addressed the ecological crisis through its social causes. Speaking in a secular, naturalistic, and worldly idiom, Bookchin infused Social Ecology with ethical passion that aspired to create a just, participatory, responsible, and rational democratic society that employs technology to harmonize humanity and nature. The ecological society is neither anti-technological nor based on false egalitarianism between humans and other forms of life. Rather, it exercises rational intervention in the natural world that actualizes a "deep-seated nisus in evolution toward self-consciousness and freedom" (Bookchin 1989, 203). Bookchin did not approve of "natural engineering," because he insisted that nature is too complex to be "controlled by human ingenuity, science, and technology" and although he celebrated human rationality, he insisted on the power of spontaneity, imagination, intuition, aesthetics and "a sense of wonder for the marvelous" as integral dimensions of human rationality.

Bookchin's Social Ecology was truly "an ecology of freedom" (Bookchin, 1989, 204). But was it Jewish? I answer in the affirmative. It reflected the secular Jewish experience in New York during the first third of the 20th century and manifested the secularization of the biblical commitment to justice: humans, not God, are responsible for the just treatment of other humans and of nature. Although Bookchin did not engage Jewish literary sources (except for the Bible), neither did he ignore the two hallmarks of Jewish life in the 20th century: the Holocaust and the State of Israel. He saw WWII and the Holocaust as human domination of other humans, which led to the destruction of nature. And Bookchin was a Zionist who spoke approvingly of the communalism of Israeli kibbutzim. Indeed, if any society came closest to actualizing Bookchin's vision, it was the Israeli kibbutzim.

Bookchin's Social Ecology dominated the environmental discourse during the second half of the 20th century, and since Jews since the 1930s were highly represented in the American Left, they could not have been oblivious to it. Yet, for some Left-leaning Jews, secularism, egalitarianism, and eco-anarchism failed to satisfy their spiritual longings, because Marxism erased their Jewish distinctiveness. Jewish Spiritual Progressives would reassert their Jewishness in religious terms, promoting Jewish environmental spirituality that integrated the insights of Social Ecology, care for the environment, and Jewish religious values and practices. *Tikkun Olam*—the basic messianic impulse of Judaism—became their overarching slogan. We illustrate that perspective by discussing the lives and works of Arthur Waskow, Michael Lerner, and Roger S. Gottlieb in the following chapter. Beyond articulating a Jewish environmental spirituality, they transformed the nexus of environment, religion, and politics in American public life, marking the transformation from secularism to post-secularism.

References

Biehl, Janet. 2015. *Ecology or Catastrophe: The Life of Murray Bookchin*. New York: Oxford University Press.

Bookchin, Murray. 1952 [under the pseudonym Lewis Herber]. "The Problem of Chemicals in Food." *Contemporary Issues*. Available at: Anarchy Archives: An Online Research Center on the History and Theory of Anarchism (www.dwardmac.pitzer.edu).

Bookchin, Murray. 1962 [under the pseudonym Lewis Herber]. 1962. *Our Synthetic Environment*. New York: Alfred A. Knopf. Reprint: 2018. Eastford, CTL: Martin Fine Books.

Bookchin, Murray. 1980. *Toward an Ecological Society*. Montreal: Black Rose Books.

Bookchin, Murray. 1982 [2005]. *The Ecology of Freedom: The Emergence and Dissolution of Hierarchy*. Oakland, CA, and Edinburgh, Scotland: AK Press.

Bookchin, Murray. 1986 [1971] *Post-Scarcity Anarchism*. San Francisco: Rampart Books. 2nd ed., Montreal: Black Rose Books, 1986. Oakland, CA, and Edinburgh, Scotland: AK Press, 2004.
Bookchin, Murray. 1988. "Social Ecology versus Deep Ecology: A Challenge for the Ecology Movement." *Socialist Review* 18 (3). Originally published in *Green Perspectives: Newsletter of the Green Program Project*, nos. 4–5 (summer 1987). Available at: www.theanarchistlibray.org.
Bookchin, Murray. 1989. *Death of a Smal Planet: It's Growth That's Killing Us.* Available at: http://dwardmac.pitzer.edu/Anarchist_Archives/bookchin/planet/planet.html.
Bookchin, Murray. 1990. *Remaking Society: Paths to a Green Future*. Boston: South End Press.
Bookchin, Murray. 1990a. "Recovering Evolution: A Reply to Eckersley and Fox." *Environmental Ethics* 12 (3).
Bookchin, Murray. 1995. *Re-Enchanting Humanity: A Defense of the Human Spirit against Anti-Humanism, Misanthropy, Mysticism and Primitivism*. London: Cassell.
Bookchin, Murray. 1996. *The Philosophy of Social Ecology: Essays on Dialectical Naturalism*. 2nd rev. edition. Montreal, New York, London: Black Rose Books.
Brandes, Joseph. 1971. *Immigrants to Freedom: Jewish Communities in Rural New Jersey since 1882*. Philadelphia: Jewish Publication Society.
Cassedy, Steven. 1997. *To the Other Shore: The Russian Jewish Intellectuals Who Came to America*. Princeton: Princeton University Press.
Chase, Steve (ed.). 1991. *Defending the Earth: A Dialogue between Murray Bookchin and Dave Foreman*. Boston: South End Press.
Cittadino, Eugene. 2018. "Barry Commoner and Paul Sears on Project Chariot, and the Atomic Energy Commission." *Isis* 109 (4): 720–743.
Clark, John. 2009 [1998]. "A Social Ecology." In *Environmental Philosophy: From Animal Rights to Radical Ecology*, ed. Michael Zimmerman, J. Bird Callicott, George Sessions, Karen Warren, and John Clark, 416–440. Upper Saddle River, NJ: Prentice Hall. Reprinted in *Earthcare: An Anthology in Environmental Ethics*, ed. David Clownmey and Patricia Mosto, 284–297.
Cohen, Stephen M. 2007. "Chemical Literature in Yiddish: A Bridge between the Shtetl and the Secular World." *Aleph* 7: 183–251.
Commoner, Barry. 1966. *Science and Survival*. New York: The Viking Press.
Commoner, Barry. 1971. *The Closing Circle: Nature, Man, and Technology*. New York: Alfred A. Knopf.
Commoner, Barry. 1972. "Ecology and Technology." *Resources Magazine*. Available at: www.resources.org.
Commoner, Barry. 1990. *Making Peace with the Planet*. New York: Pantheon Books.
Davidson, Stewart. 2009. "Ecoanarchism: A Critical Defence." *Journal of Political Ideologies*. 1 (1): 46–67.
Deval, Bill and George Sessions. 2007. *Deep Ecology: Living as If Nature Mattered*. Gibbs Smith Publisher.
Deutscher, Isaac. 1968. "The Non-Jewish Jew." In *The Non-Jewish Jew and Other Essays*, ed. Tamara Deutscher, 25–41. New York: Hill and Wang.
Diner, Hasia R. 2000. *A New Promised Land: A History of Jews in America*. Oxford: Oxford University Press.
Diner, Hasia R. 2004. *The Jews of the United States: 1654–2000*. Berkeley, Los Angeles, and London: University of California Press.

Diner, Hasia R. 2015. *Roads Taken: The Great Jewish Migrations to the New World and the Peddlers Who Forged the Way.* New Haven and London: Yale University Press.

Drier, Peter. 2017. "Celebrating Barry Commoner, Father of the Modern Environmental Movement," *Huffington Post* (posted May 28).

Eckersley, Robyn. 1989. "Divining Evolution: The Ecological Ethics of Murray Bookchin." *Environmental Ethics* 11 (2): 99–116.

Eckersley, Robyn. 1992. *Environmentalism and Political Theory.* London: University College London.

Efron, Noah J. 2014. *A Chosen Calling: Jews in Science in the Twentieth Century.* Baltimore: Johns Hopkins University Press.

Egan, Michael. 2002. "The Social Significance of the Environmental Crisis: Barry Commoner's *The Closing Circle.*" *Organization & Environment* 15 (4): 443–457.

Egan, Michael. 2007. *Barry Commoner and the Science of Survival: The Remaking of American Environmentalism.* Cambridge, MA: The MIT Press.

Ehrlich, Paul R. (with Ann Howland Ehrlich). 1968. *The Population Bomb.* New York: Ballantine Books.

Ehrlich, Paul R. (with Ann Howland Ehrlich). 1991. *The Population Explosion.* New York: Simon and Schuster.

Foster, John Bellamy. 2000. *Marx's Ecology: Materialism and Nature.* New York: Monthly Review Press.

Foster, John Bellamy. 2020. *Socialism and Ecology: The Return of Nature.* New York: Monthly Review Press.

Foster, John Bellamy and Paul Burkett. 2016. *Marx and the Earth: An Anti-Critique.* Chicago, IL: Haymarket Books.

Fox, Warwick. 1986. *Approaching Deep Ecology: A Response to Richard Sylvan's Critique of Deep Ecology.* University of Tasmania.

Fox, Sandra. 2023. *The Jews of Summer: Summer Camp and Jewish Culture in Postwar America.* Palo Alto: Stanford University Press.

Glover, James, M. 1986. *A Wilderness Original: The Life of Bob Marshall.* Seattle: The Mountaineers.

Goren, Arthur A. 1999. *The Politics and Public Culture of American Jews.* Bloomington: Indiana University Press.

Gurock, Jeffrey S. 2005. *Judaism's Encounter with American Sports.* Bloomington, IN: Indiana University Press.

Gutterman, David S. 2005. *Prophetic Politics: Christian Social Movements and American Democracy.* Ithaca and London: Cornell University Press.

Herscher, Uri D. 2018. *Jewish Agricultural Utopias in America, 1880–1910.* Detroit: Wayne State University Press.

Jacoby, Russell. 2000 [1987]. *The Last Intellectuals: American Culture in the Age of Academe.* New York: Basic Books.

Kadar, Naomi Prawer. 2017. *Raising Secular Jews: Yiddish Schools and their Periodicals for American Children, 1917–1950.* Waltham, MA: Brandies University Press.

Kaplan, Dana Evan. 2005. *American Reform Judaism: An Introduction.* New Brunswick: Rutgers University Press.

Kovel, Joel. 1998. "Negating Bookchin." In *Social Ecology after Bookchin,* ed. Andrew Light. New York: Gillford Press.

Kraut, Benny. 1979. *From Reform Judaism to Ethical Culture: The Religious Evolution of Felix Adler.* Cincinnati: Hebrew Union College.

Kriebel, David (ed.). 2002. *Barry Commoner's Contribution to the Environmental Movement: Science and Social Action*. Amityville, NY: Baywood Publishing Company.
Krier, James E. 1990. "The Political Economy of Barry Commoner." *Environmental Law* 20 (1): 11–23.
Krone, Addrienne. 2021. "Farming on the Front Lines: Jewish Environmentalism and Kinship in Chthulucene." *Worldviews: Global Religions, Culture, and Ecology* 26(1–2): 148–161. Available at: doi:10.1163/15685357020220203.
Kuznick, Peter J. 1987. *Beyond the Laboratory: Scientists as Political Activists in 1930s America*. Chicago: University of Chicago Press.
Levine, David. 1991. "Turning Debate into Dialogue." In *Defending the Earth: A Debate between Murray Bookchin and Dave Foreman*. Montreal and New York: Black Rose Books.
Liebman, Arthur. 1997. "The Ties that Bind: Jewish Support for the Left in the United States." In *Essential Papers on Jews and the Left*, ed. Ezra Mendelsohn, 322–357. New York: New York University Press.
Light, Andrew (ed.). 1998. *Social Ecology after Bookchin*. New York and London: The Guilford Press.
Lorge, Michael M. and Gary P. Zola (eds.). 2006. *A Place of Our Own: The Rise of Reform Jewish Camping*. University of Alabama Press.
Maor, Eli. 2013. "Science and Yiddish Don't Mix: Really?" *Journal of Scholarly Publishing* 44 (4): 340–354.
Marshall, Robert. 1930. *Alaska Wilderness: Exploring the Central Brooks Range*. Los Angeles and Berkeley: University of California Press.
Marshall, Robert. 1930a. *The Social Management of American Forests*. Madison, WI: League for Industrial Democracy.
Marshall, Robert. 1956. *Arctic Wilderness*. Los Angeles and Berkeley: University of California Press.
Marshall, Robert. 2000 [1933]. *The People's Forests*. Iowa City: University of Iowa Press.
Marshall, Robert. 1991 [1933]. *Arctic Village: A 1930s Portrait of Wiseman, Alaska*. University of Alaska Press.
Michels, Tony. 2005. *A Fire in Their Hearts: Yiddish Socialists in New York*. Cambridge, MA: Harvard University Press.
Michels, Tony. 2012. "The Jewish-Socialist Nexus." In *Jewish Radicals: A Documentary History*, ed. Tony Michels, 1–24. New York and London: New York University Press.
Montague, Peter. 2002. "Barry Commoner: The Father of Grass-Roots Environmentalism." In *Barry Commoner's Contribution to the Environmental Movement, Science and Social Action*, ed. David Kriebel, 5–4.
Moore, Deborah Dash. 2014. *Urban Origins of American Judaism*. Athens and London: The University of Georgia Press.
Rabinovitch, Simon (ed.). 2012. *Jews & Diaspora Nationalism: Writings on Jewish Peoplehood in Europe and the United States*. Waltham, MA: Brandeis University Press.
Pepper, David. 1993. *Eco-Socialism: From Deep Ecology to Social Justice*. London and New York: Routledge.
Reznikoff, Charles (ed.). 1957. *Louis Marshall, Champion of Liberty: Selected Papers and Addresses*. Philadelphia: Jewish Publication Society of America.

Robinson, Nicholas A. 2007. "*Louis Marshall—Champion of Civil Rights, Conservation and 'Forever Wild.'*" Center for the Forest Preserve in Niskayuna, NY.

Rosenstock, Morton. 1965. *Louis Marshall: Defender of Jewish Rights*. Detroit: Wayne State University Press.

Sales, Amy L. and Leonard Saxe. 2003. *How Goodly Are Thy Tents: Summer Camps as Jewish Socializing Experiences*. Newton, MA: Brandeis University Press.

Sarna, Jonathan. 2008. "Two Jewish Lawyers Named Louis." *American Jewish History* 94(1–2): 1–19.

Seidenberg, David. 2005. "Jewish Environmentalism in North America." In *Encyclopedia of Religion and Nature*, ed. Bron R. Taylor, 909–913. London: Continuum.

Silver, Matthew. M. 2013. *Louis Marshall and the Rise of Jewish Ethnicity in America: A Biography*. Syracuse, NY: Syracuse University Press.

Smith, Christian (ed.). 1996. *Disruptive Religion: The Force of Faith in Social Movement Activism*. New York: Routledge.

Stoll Mark, 2001. "Green versus Green: Religion, Ethics and the Bookchin-Foreman Dispute." *Environmental History* 6(3): 412–427.

Stoll Mark, 2015. *Inherit the Holy Mountain: Religion and the Rise of American Environmentalism*. Oxford and New York: Oxford University Press.

Veidlinger, Jeffrey. 2009. *Jewish Public Culture in the Late Russian Empire*. Bloomington: Indiana University Press.

Weinstein, Ruth. 2020. *Back to the Land: Alliance Colony to the Ozarks in Four Generations*. Galloway, NJ: South Jersey Culture & History Center.

Weintraub, Irving. 2007. *Jews in American Agriculture: An Annotated Bibliography*. Jefferson, NC: McFarland.

Wenger, Beth S. 2010. *History Lessons: The Creation of American Jewish Heritage*. Princeton and Oxford: Princeton University Press.

White, Damian F. 2008. *Bookchin: A Critical Appraisal*. Las Vegas, NV: Pluto Press.

7 Tikkun Olam

Resacralizing Nature and Politics

Public Theology: From Secularism to Post-Secularism

Whereas secularization characterized the American Jewish experience for most of the 20th century, since the 1970s there was a shift in the orientation of American Jews toward a greater public embrace of religiosity. For some, this revived religiosity is expressed in the return to Orthodoxy, which indeed is thriving today. For others, that meant a quest for spirituality inspired by New-Age movements, whether outside or within Judaism. And for still others, the affirmation of religious identity yielded the rise of Jewish environmentalism that seeks to protect God's created world from environmental degradation by recovering and reinterpreting the religious sources of Judaism. Focusing on the latter trend, this book situates Jewish environmentalism as one manifestation of contemporary post-secularism, namely, the conviction that religion must not be privatized because it has deep and relevant insights about the human condition and the working of the good society.

Emerging in the 1980s, post-secularism has critiqued the view that in modern societies religion must be relegated to the private sphere and with time religious practices and convictions will disappear. As a commentary on public affairs from a religious standpoint, public theology is a feature of this post-secularism (Barbieri 2022). The phrase 'public theology' was coined in 1974 by Professor Martin Marty, the sociologist of American religions. Public theology is first and foremost a theology, namely, a discourse about God, but unlike 'systematic theology,' or 'pastoral theology,' public theology engages social, economic, and political issues in the modern public sphere, which according to the Secularization Thesis was supposed to be devoid of religion.

Public theology is the practice of relating one's faith in the public sphere for the benefit of the common good (Hübenthal and Alpers 2022). Concerned with the moral quality of the society, public theology is committed to the ideal of justice; it speaks truth to power, critiques power relations, and combats abuses of power. On the level of *theory*, public theology consists of reasoned, critical analysis of categories and principles carried out within the academy within the contours of academic disciplines; on the level of *praxis*,

DOI: 10.4324/9781351123945-8

public theology takes a stance on all aspects of public life (social, economic, political, cultural, and environmental) and gets involved in all the issues that concerns the society at a given time. In a pluralistic, global society, public theology carries out an apologetic task, namely, it explicates the doctrines, symbols, and rituals of the religious traditions to people who do not share its faith perspective (Graham 2022). Since public theology is engaged in the education of the public, which may or may not accept its points of departure, public theology must expect some contestation. Whether communicative rationality can resolve conflicting viewpoints remains an open question, a persistent challenge for pluralistic, post-secular, democratic societies.

Judaism and Public Theology

How does public theology apply to Judaism? Not easily. Judaica scholars and Jewish theologians do not commonly employ this moniker to self-describe. Why do Jewish scholars avoid the phrase 'public theology'? Here are two possible reasons. First, as a theoretical discourse within the academy, Jewish theology is written by academic scholars who write for other academics, who are mostly but not exclusively Jewish. The modern research university sought to professionalize higher education through disciplinary specialization, which created rigid boundaries between itself and the rest of society and established strict rules regarding how to talk about religion without advocating religion. Public theology is therefore problematic because it undermines the assumptions and conventions of the modern research university.

Second, and more fundamentally, the Jewish discomfort about 'public theology' arises from the fact that in the 19th century Jews welcomed the differentiation of the public and the private spheres. Indeed, the religiously neutral public sphere enabled Jews to enter non-Jewish society and function in domains that were previously closed to them. This is why Jews welcomed the formal separation of church and state in the United States. In the 19th century, "Jews were concerned with freedom of religion and not freedom from religion" (Sarna 2002, 52), but in the post–Civil War period, many Jews, especially Reform leaders, joined "liberal Christians, religious free thinkers, and secularists" in their endorsement of the "separatist agenda." (ibid., 56). Reform Rabbis were adamant that America is not a Christian nation and that view became the prevailing Jewish view on church-state issues. In the late 19th century, the organized Jewish community applauded the efforts "to secularize the State completely," and although this effort was debated in the first decades of the 20th century, after WWII American Jewry largely endorsed the so-called "separatist agenda." Jews endorsed the so-called "high wall of separation" between church and state, believing that it could help bring about "liberty and justice for all."

It was precisely this "separatist agenda" and the secularization of the public sphere that Rabbi Michael Lerner rejected when he founded *Tikkun* magazine, the Tikkun Community, and the Network of Spiritual

Progressives. In this concluding chapter we focus on three Spiritual Progressives—Arthur Waskow (d. 2025), Michael Lerner (d. 2024), and Roger S. Gottlieb (b. 1946)—to illustrate the transition from secularism to post-secularism. All three have been associated with the secular Left but over their lifetime they embraced religious forms of Jewish identity associated with the Jewish Renewal movement; all three greatly contributed to religious environmentalism and to Jewish environmentalism in the United States, and all three demonstrate the discomfort of Jewish academics in the research university with public theology. Most importantly, these individuals have greatly contributed to the emergence of *Tikkun Olam* as the framing for social and environmental activism and they have greatly contributed to the re-sacralization of nature and politics. We end the chapter with a discussion of Rabbi Jonathan Sacks as the paradigmatic post-secular Jewish public theologian.

Arthur Waskow: Radical "Godwrestling"

Arthur Waskow was pivotal to the emergence of Jewish environmental spirituality in America. Between 1968 and 1972, he transitioned from being a "Jewish radical" to being a "radical Jew" (Waskow 1971, 11–46)—that is, from a Jew by birth who was involved in secular, radical politics to a religious Jew who sought to renew American Judaism by infusing it with a passionate commitment to social justice, egalitarianism, and pluralism. For Waskow, however, Judaism's spiritual renewal in the second half of the 20th century was predicated on both recognizing the bond between the well-being of humans (*adam*) and all other beings on Earth (*adamah*) and on the awareness that the ideal of justice encompasses human and nonhuman beings. Precisely because humanity and nature are interdependent, social justice always has an ecological dimension (i.e., eco-justice) and conversely, environmental issues always have a social dimension, since human labor transforms the physical environment. Social justice and eco-justice are two sides of the same coin, expressing the interdependence of humanity and nature.

Like his Leftist predecessors, Commoner and Bookchin, Waskow developed his environmental outlook through opposition to nuclear testing and nuclear energy during the 1950s and early 1960s. But he rediscovered and affirmed his Jewishness as a religious—not just ethnic or cultural—identity. Like Commoner and Bookchin, Waskow also recognized the technological achievements of WWII. For Waskow, wrestling with the meaning of Auschwitz and Hiroshima together—the former threatening the future of the Jewish people and the latter threatening the future of humanity on Earth—was necessary for Judaism's spiritual renewal. The quality of Jewish life and Earth's well-being were intertwined. In Chapter 2 we discussed the institutions Waskow built to promote an environmentally sensitive Judaism, eco-Judaism: the Shalom Center and Aleph: The Alliance of Jewish Renewal, associated with the Jewish Renewal Movement. Here we focus on his path from a secular social scientist and peace activist promoting disarmament to a religious Jew connecting Jewish genocide in the Holocaust to ecocide by thermonuclear war.

Born in 1933, Waskow grew up in Baltimore, Maryland, in a Jewish neighborhood like the New York City neighborhoods where Bookchin and Commoner grew up. Although his grandparents were not Russian revolutionaries like Bookchin's, they were thoroughly secularized. Unlike Bookchin, Waskow was Bar-Mitzvahed and received the rudiments of Jewish education, but the only religious ritual his family observed was the Passover Seder. It was no coincidence that Waskow's return to Jewish tradition after April 1968 was tied to the celebration of Passover, which he would reinterpret anew (Waskow 1971).

Waskow enrolled in Johns Hopkins University for his undergraduate degree (1954) and the University of Wisconsin-Madison for graduate studies in history and political science (1963). There he was exposed to Marxism as the path to humanity's (secular) redemption and matured intellectually into the New Left (Renaud 2021).

In 1959 Waskow moved to Washington, DC, to work for the Democratic congressman from Wisconsin, Robert L. Kastenmeier, and later became a senior staff writer at the Peace Research Institute. In 1963 he joined Richard Barnet and Marcus Raskin, two civil servants on the staff of the National Security Council, who founded the Institute for Policy Studies as an independent think tank in which Waskow served as a Resident Fellow until 1977. Waskow believed that public education about nuclear energy, the atomic bomb, and the arms race could further the cause of world peace. His early publications (Waskow 1962; 1963; 1965) gave no indication of his Jewishness or that Judaism was at all relevant to his advocacy of nuclear disarmament.

In 1963 Waskow published *The Worried Man's Guide to World Peace*, written for all the "people who are not only worried about the world but are ready to do something about their worry" (Waskow 1963, xv). For this audience, Waskow continued to evoke Gandhi's and Martin Luther King Jr.'s nonviolent resistance. Waskow linked opposition to nuclear power with concern for social justice and critique of capitalism throughout his career. The bomb, he explained, carries "productivity to its grotesque extreme, by enabling a few men to kill many, many millions in an extremely short time. In a sense, the bomb is hyper-wealth" (ibid., 4). To attain world peace, America had to confront its poverty and racism, which required grassroots activism. Peace activists needed to mobilize the public to discuss war and peace in study groups, social clubs, professional associations, church groups, labor unions, and business luncheons. Direct, face-to-face encounters would disseminate expert knowledge about the threat of annihilation. At this point in his public career, Waskow spoke about "moving" the world but not yet about "healing the world" or "repairing the world." After embracing Judaism, Waskow shifted his life's goal from "moving" the world toward peace to "healing" the world by restructuring social relations, including equality between men and women. Feminism would take central stage in his reconstruction of Judaism.

In 1966, Waskow published the book *From Race Riots to Sit-Ins: 1919 and the 1960s*, in which he analyzed the historical connection between the 1919 race riots and the sit-ins, freedom rides, and rent strikes of the early 1960s. He showed that achieving peace worldwide required transforming American social life: controlling violence at home and developing nonviolent methods to achieve political goals. Strongly in favor of racial equality, Waskow advocated "creative disorder." In 1967 Waskow co-authored with Marcus Raskin "A Call to Resist Illegitimate Authority" in support of those resisting the draft and the Vietnam War. Thousands signed, and Raskin and Waskow took part in turning in a thousand draft cards, for which they were tried but eventually exonerated. Until 1969 Waskow's political radicalism was not informed by Judaism. But "the earthquake of 1968" (Waskow 2020, xx) profoundly changed him. The murder of Martin Luther King Jr., the ensuing race riots across America that brought armed soldiers to American cities, and the Democratic National Convention in Chicago, where he was a delegate, reoriented his life.

In 1968, as Waskow left work to prepare for Passover, the only vestige of Judaism in his life, he had a revelation. He realized that contemporary struggles for liberation and justice evoked Jews' struggles against the biblical Pharoah. He thus composed the Freedom Haggadah, which interspersed the rabbinic text with passages from Martin Luther King Jr., Gandhi, Thomas Jefferson, Thoreau, Martin Buber, the Berrigan brothers, Emmanuel Ringelblum, and others (Waskow 1971, 19). In 1969, Waskow and Jews for Urban Justice used the Freedom Haggadah in the first Freedom Seder. The Seder, celebrated in a church basement in Washington, DC, with 800 people, called for the liberation of all people, not just Jews, from oppression. Thus began Waskow's transition from secularism to post-secularism, from being an ethnic Jew to a religiously engaged Jew who interpreted politics and secular life through Judaic rituals and beliefs.

Jews for Urban Justice (JUJ), a "radical Jewish movement [fusing] 'religious' celebration and 'political' radicalism" (Waskow 1971, 21), provided the social context for Waskow's spiritual transformation. From 1966 to 1971, JUJ opposed anti-Black racism within the white Jewish community. Within this short-lived organization Waskow engaged social, economic, and political issues, including the looming environmental crisis, from a Judaic perspective. JUJ illustrated "prophetic politics" in action: members staged public protests and mass marches against the war in Vietnam or for civil rights as religious services that featured readings from the Bible interspersed with speeches from political leaders, past and present. The command to seek justice shaped the content and form of protests in which "Black ministers and Black militants, Catholic priests and Left-wing atheists, rabbis and Bundists" (Waskow 1971, 21) joined hands to demand the end of the Vietnam War and liberation from oppression and injustice. In the massive anti-war protests in 1969,

Waskow and his JUJ comrades accentuated their Jewishness by marching behind a Torah and sounding the Shofar as they railed against "the Idols of Power, War, Wealth, and Technology." Ending the protest with the blowing of the Shofar signified the "echo of the coming Messianic Age of peace, justice, and the abandonment of all idolatry" (Waskow 1971, 23).

During the transformative period of 1968–1972, Waskow realized that the Jewish religious tradition had wisdom that modern radicalism lacked and that the Judaic experience is incomplete if it turns away from global struggles for liberation. The anti-bomb movement, countercultural movement, civil rights movement, women's movement, and anti-war movement coalesced as a struggle for liberation from oppression and a call for restructuring the social order. Waskow saw that the same yearning for freedom these struggles expressed is shared by the Passover Seder and many other Jewish rituals, especially the Sabbath, manifesting Judaism's messianic impulse. The Judaic messianic ideal entails that religion and politics cannot be bifurcated. This post-secular consciousness informed Waskow's book *The Bush is Burning: Radical Judaism Faces the Pharaohs of the Modern Superstate* (1971), which he dedicated to "my sisters and brothers of Jews for Urban Justice in Washington, D.C. (the Diaspora of the Diaspora) and of the radical Jewish movement across America—who agree with some of this and disagree with some of it, but who taught all of it."

By 1972, Waskow's transition to post-secular Jew was complete, but his public career within the Jewish community was only beginning. He was especially critical of Labor Zionism, which after WWII predominated in the organized Jewish community in America. Israel's resounding victory in the Six-Day War of 1967 generated euphoria but made Palestinians determined to oppose the Israeli occupation of the West Bank, Gaza, and the Golan Heights. Waskow ardently opposed Israel's control of these territories and called on the "Jewish Establishment" to recognize what he considered the Palestinian people's legitimate claims to self-determination. He criticized "Zionism imperialism" that aligned itself with the "American Empire" and its militaristic global capitalism. Critical of the leadership of American Jewry and of Israeli governments, Waskow argued that "radical Jews have an obligation to speak honestly and clearly to the Palestinian nationalist movement and their supporters in the American radical movement" (Waskow 1971, 54).

After the Yom Kippur War of 1973, Waskow joined other Jewish radicals to establish the organization Breira (Alternative) that argued for American Jews' right to criticize Israeli policies. Breira offered itself as the forum for open and critical discussion of all matters of concern to American Jews, especially their relation to Israel. Composed of Jewish veterans of the counterculture movement of the 1960s who were grouped mainly around *Response Magazine*, Breira received support from various Jewish intellectuals and from Reform and Conservative rabbis committed to the pursuit of justice.

Waskow and fellow radical Jews (e.g., Arthur Green and Michael Lerner, who is discussed below), called for Judaism's renewal from within by reexamining and reinterpreting its sacred sources. This is precisely what midrash (i.e., the traditional style of Jewish hermeneutics) does, but Waskow invested it with more radical meaning when he labelled this reexamination "Godwrestling." This is the translation of the name "Israel," which was given to the patriarch Jacob at Beth El after he wrestled with the angel of God (Gen. 35:9). To be a Jew, or "Israel," is to be a "Godwrestler" (Waskow 1996, 13). In *Godwrestling* (1978) and its revised version (1996), Waskow wrestled with complicated and contested issues such as the relationship of Diaspora Jews and the State of Israel, Zionism and Palestinian nationalism, Blacks and Jews, Auschwitz and Hiroshima, and environmentalism and capitalism. As his Jewish learning deepened until he eventually was ordained as a Rabbi by Zalman Schachter-Shalomi, Waskow wrestled with the meaning of Judaism by adopting midrash to reconstruct the ancient tradition to address contemporary challenges, chief among them the environmental crisis.

Auschwitz and Hiroshima

The Holocaust transformed American Judaism, but it was only in the mid-1960s that Jewish theologians (e.g., Richard Rubinstein, Yitzhak (Itz) Greenberg, Eliezer Berkovitz, Arthur A. Cohen, and Emil Fackenheim) addressed the trauma systematically (Katz 1985). The evil of the Holocaust challenged Jewish belief in God's omnipotence, omnipresence, and benevolence, the covenantal relationship between God and Israel, and the problem of Jewish powerlessness. Waskow cites Martin Buber, who in *Paths in Utopia* (1945) understood that after WWII "what is in question, therefore, is nothing less than man's whole existence in the world" (Waskow 1978, 164). Waskow was fully aware that mentioning "Auschwitz and Hiroshima in the same breath […] makes some people angry" but argued that "not to do so ignores them both" (ibid.,164). He goes on to explain that

> It is not that they are similar events, but each in its uniqueness is a foretaste of possible nothingness. The Jewish People was not destroyed at Auschwitz: the human race was not even nearly destroyed at Hiroshima. But the threats are there now.
>
> (ibid., 164)

Christian theologians had already begun to examine their tradition in the light of the horrors of Auschwitz; Waskow now urged Jewish theologians to examine not only Auschwitz but also Hiroshima. Understanding the Holocaust and Hiroshima together places the responsibility for the future of humanity in human hands and requires action in several domains, starting with "achieving harmony between humanity and nature" (Waskow 1978, 170). By 1978 the scope of the environmental crisis was well understood. The work of early environmentalists provided sufficient knowledge about the crisis, but at the same time, the new science of ecology—the "science of the web of life," as it is commonly referred to, can return us to Eden by restoring harmony with nature and ending human exploitative practices. This was precisely Bookchin's view of modern technology, although Waskow neither refers to Bookchin nor cites him. Bookchin's Social Ecology is also recognizable in Waskow's reflections on the interplay of technology and labor in which, "at the same moment, the liberation of all from toil and the subjugation of all to alienated work and to utter proletarianization are both made possible" (ibid., 171). Marxism, of course, had much to say about overcoming alienated labor and proletarianization. But Waskow was now critical of Marxism, deeming its secularity as grounded in idolatrous human self-veneration. Rather, Waskow dressed Bookchin's Social Ecology in a Judaic garb, teaching Jews that Judaism, the "Messianic science" (ibid., 182) of ecology, and social justice are inherently intertwined. During the 1980s the eco-justice perspective mobilized many Christian organizations. In the Jewish world, Waskow was eco-justice's main voice.

Universal Import of Jewish Environmental Spirituality

Linking Auschwitz and Hiroshima led Waskow to specify the Holocaust's universal significance, charting Jews' prophetic role. The Holocaust, he claimed,

> may have made the Jews into humanity's prophets, humanity's canary-people in the poisonous air of a poisoned mine-shift, humanity's early warning system of the danger of global holocaust. In the same way, the effort of the Jewish people to begin shaping a new version of Jewish peoplehood may be of use to other peoples and to the human race as a whole, as all of us try to move past modernism into a new shape of human society.
>
> (Waskow 1983, xi)

Waskow joined the Jewish Renewal Movement, studying under and collaborating with Zalman Schachter-Shalomi, its visionary founder, whose work was discussed in Chapters 2 and 3. As a religious Jew, Waskow committed himself to Judaism's renewal, to healing a world shuttered by secular modernity and its consequent traumas, Auschwitz and Hiroshima. Waskow placed his work within the theological framework of Lurianic Kabbalah, the 16th-century elaboration of medieval Kabbalah. The Lurianic myth provided the conceptual framework within which Waskow undertook the task of Jewish renewal. The myth is echoed in the title of Waskow's book, *These Holy Sparks: The Rebirth of the Jewish People* (1983). There, Waskow promoted the movement for Jewish renewal and outlined a program consisting of restoring life's wholeness, rhythm, and meaning; building grassroots religious communities and "reconnecting, revitalizing, and reaffirming" Jewish communities around the world; infusing politics with the Jewish tradition of inquiry; and tapping into the joy of Jewishness and of "the universal God of universal justice" (Waskow 1983, 6).

How to accomplish Judaism's renewal? Mainly through "small, intimate, face-to-face congregations without hired functionaries," that is, "fellowships" (*havurot*) in which Jews renew their own individual Jewish lives, encourage amateur participation, and integrate the physical, emotional, intellectual, and spiritual dimensions of their lives. The Havurah movement, which we mentioned in Chapters 2 and 4, indeed revitalized American Judaism within established synagogues, on college campuses through Hillel Foundation, and in Jewish community centers, through bureaus of Jewish education. In the context of *havurot*, Jews recovered traditional Jewish learning, expressed their passion for peace and social justice, experimented with egalitarian forms of Jewish rituals, and promoted inclusive Jewish leadership. In many ways, the *havurot* were a Judaic variant of Bookchin's eco-anarchism. To this audience of young, socially progressive, and religiously interested and committed Jews, Waskow promoted his reconstructed Judaism through teaching, writing, and social activism that revolved around establishing a healthier relationship between humanity and nature. Waskow (appropriately, in my view) maintains that Judaism itself charts the paths for healing human-nature relations.

In a series of books, some written with his second wife, Rabbi Phyllis Berman (Waskow 1982; Waskow 1983; Waskow and Berman 1996; Waskow and Berman 2002; Waskow and Berman 2011), Waskow showed how "to renew Judaism by wrestling with its sources" (Waskow 2012 [1982], xiii). In *Seasons of Our Joy* (1982), Waskow systematically explicated the Jewish calendar, in which festivals and holy days are tied to nature, history, community, and individual spiritual growth and "shape time into spirals—spirals of months, years, cycles" (Waskow 1982, 237). Understanding Judaism as a spiral means that Jews reconnect with the remote past (spiraling back in time) to move toward the Messianic Age (spiraling forward in time). Inspired by Martin Buber, Waskow's project of Jewish renewal was indeed designed as a "path in utopia," to use Buber's phrase.

Two decades later, Waskow and Berman published *A Time for Every Purpose under Heaven* (2002), in which they explicated the Jewish rhythm of life as a spiritual path for the 21st century. The exposition of the key moments in one's life—birth, maturation, marriage, death, illness, divorce, adoption, emotional upheaval, separation, unemployment, and changing residence—are all reinterpreted in the context of a Holistic Judaism that is simultaneously rooted in and different from rabbinic Judaism, as much as rabbinic Judaism was different from but also rooted in biblical Judaism. Judaism's symbols, narratives, and rituals provide spiritual meaning that makes it possible for Jews today to live in a new way: "trying to spit out Modernity and return to a richer Judaism of the past [...] and trying to digest Modernity and move forward to a richer Judaism of the future" (Waskow and Berman 2002, xxv). Without using the term "post-secular" or "post-secularism," this sentence captures the very essence of being post-secular, of moving beyond secularism and secularization to reconnect with the Judaic religious past. Waskow's process of giving Jewish meaning to the cycles of the earth and of human life is presented as a spiral that metaphorically captures the midrashic way of thinking, in which "Jews gaze into an old text and see in it something utterly new;" going back to the ancient past makes possible going forward into the future.

Social Justice as Eco-Justice

The Sabbatical Year (*Shemitah*) best exemplifies looking to the remote past for the sake of present and future Jewish renewal. In Chapter 5 we mentioned the Sabbatical Year when discussing how early 20th-century Zionism struggled with the ancient practice when Jews settled the Land of Israel. Since the laws of the Sabbatical year do not apply outside the Land of Israel, Waskow, the American Jew, could not experience the Sabbatical Year. Yet the laws of the Sabbatical Year appealed to his passion for social justice because they consisted of releasing Hebrew slaves, forgiving debts, and supporting the poor. After the seventh Sabbatical Year comes the Jubilee (*Yovel*), during which all Sabbatical Year regulations were to be observed, but all landed property were to be reverted to the original owners or their legal heirs, and all Hebrew slaves whose six-year term was unexpired or who refused to leave their master upon their terms' completion were to be emancipated.

For Waskow, the biblical Jubilee offered a framework for social justice in America. In 1976, just a few years after he "reclaimed" Judaism, America celebrated its bicentennial. Since mathematically "a bi-centennial is also a Jubilee," it followed that "1976 should be an American Jubilee Year" (Waskow 1978, 113). Waskow sees in the Jubilee a chance to renew a ravaged America by honoring the cycles of life—plant, animal, human—and the cycles of seasons and generations. Jewish renewal is both a renewal of Judaism and of the secular, urban world that has long abandoned the ancient rhythm of rest, roots, and nature. Waskow translated Jubilee ideals into concrete programs for social change. Tax reform, decentralization, self-empowerment, food self-sufficiency, and rest all factor into "a plan how to achieve social justice and community in a society where there is no economic growth" (ibid., 127).

Waskow's programs gave Bookchin's eco-anarchism a "Judaic twist," so to speak. Under the banner of eco-Kashrut (or eco-kosher), Waskow emphasized that Jewish spirituality is always embodied, socially embedded, and communally oriented. Food can be ritually suitable for consumption (i.e., kosher) if the labor that produced and distributed it was just and dignified, and not based on exploitation. And since all forms of production, including food, require energy, the eco-justice perspective (and environmentalism in general) had to consider energy production and its economic, social, and political implications. Energy based on extraction of natural resources is inherently problematic because it has led to excessive corporate wealth rooted in workers' exploitation. Capitalism, therefore, is both inherently immoral and a major contributor to the environmental crisis, especially to climate change.

Evoking Tractate Avot 1:14: "If not now, when?" Waskow relentlessly demanded that American Jews regard climate change as *the* Jewish issue of our time. Waskow's unremitting climate advocacy resulted in numerous Reform, Conservative, Reconstructionist, and Orthodox rabbis following his lead and signing (October 29, 2015), the Rabbinic Letter on the Climate Crisis, inspired by the Papal Encyclical *Laudato Si'*. Addressed "to the Jewish people, to all community of spirit, and the world," the Rabbinic Letter calls for a "a new sense of eco-social justice—a *tikkun olam*—that includes *tikkun tevel*, the healing of our planet." The 425 signatories "urge those who have been focusing on social justice to address the climate crisis and those who have been focusing on climate crisis to address social justice."

Waskow's climate advocacy also gave rise to Jewish organizations that focus exclusively on climate advocacy as social justice. The Jewish Climate Action Network works through education activism and organizing to add an urgent and visionary Jewish voice to the climate crisis. Chapters throughout the United States build community focused on climate change issues. The organization promotes sustainability, clean energy, and partnering with other environmental organizations, footprint reduction through changes in food choices and adding biodiversity to grounds. And Dayyenu: A Jewish Call to Climate Action "mobilizes Jewish support for climate solutions, builds our collective power, and raises up a spiritual, religion and moral voice in the national and global movements confronting the climate crisis." These organizations are indebted to the work that Arthur Waskow was doing since the 1970s. His engagement in "redemptive activism" transformed American Judaism by making it more attentive, caring, and involved in environmentalism, including climate advocacy.

Even in his nineties, Waskow continued to inspire, to be socially and politically engaged, and to hone his message. *Dancing in God's Earthquake: The Coming Transformation of Religion* (2020) offers his most explicit explanation of Jewish environmental spirituality. In it, he framed the tale about the Garden of Delight (Eden) as a narrative of growth in which humankind learns that "what we sow is what we reap. Bending the Earth to our will forces the Earth to bend us to its will" (Waskow 2020, 7). Writing during the COVID-19 pandemic, Waskow added the "Coronavirus Plague" to "the various plagues of global scorching" (e.g., undrinkable water, intrusive plastics, droughts, famines, floods, and fires) that make up our contemporary ecological crisis, which was caused by the human predilection for "Domination and Hierarchy," the same perennial sins that Bookchin identified as fueling the environmental crisis. As he reflected on his life, Waskow related how he had begun "to connect to the broken fragments of the Old Tablets of Certainty as the fragment began to glow and glimmer new thoughts, new patterns" (Waskow

2020, xx). Evoking the Lurianic cosmogonic myth, Waskow affirmed that "it is up to us to invent the tool we need to struggle to repair the broken vessel" (ibid., xxii). In devoting his life to Judaism's renewal, Waskow has uncovered the

> unexpected truth that I owe a great deal to the culture of ancient Earth-oriented Judaism—the Hebrew Scriptures. They were the spiritual expression of what we would now call an indigenous people—shepherds and farmers in close touch with the Earth.
>
> (ibid., xxv)

Repairing the broken world was Waskow's life's mission, shaping Jewish environmental spirituality and Jewish prophetic politics in America.

Michael Lerner: Post-Secular Spiritual Politics

Addressing climate change has become the overriding concern of Jewish environmentalism today, although the ecological crisis is much broader than climate change. But addressing climate change meaningfully requires political action. Can Jewish environmental spirituality impact American politics? Waskow is an example of a Jew who "reclaimed" his Jewish religious identity later in life after he departed from American politics. Michael Lerner, a close friend of Waskow, had a different trajectory: his Jewish identity was forged earlier in life through his parents' commitment to Zionism and their involvement in Democratic politics. A disciple of Abraham Joshua Heschel, Lerner emulated his mentor's prophetic politics not only by seeking to renew Judaism from within but also by aspiring to transform American society and politics based on the insights of Judaism. Active in national politics as a religious Jew, Rabbi Michael Lerner advocated for the remaking of American politics into "politics of the image of God" (Lerner 1996, 4). Lerner was a powerful and influential expression of post-secular Jewish earthly spirituality.

Born in 1943 in Newark, New Jersey, Michael Lerner first studied in a private school committed to "interdenominational Christianity," while attending afternoon Hebrew school three times a week. Unhappy with his private school experience where he stood out as a Jew, when Lerner reached seventh grade, he asked to switch to a public school which was predominantly Jewish. More important than the supplementary Hebrew school, summer camps were the primary social context for the cultivation of positive Jewish identity among American youth. Religious Jews of all stripes, Yiddish-speaking secular Jews, and American Zionists all sent their children to summer camps whose ideologies differed, but their educational practices were the same, all geared to ensure Jewish survival in the open society (Fox 2023). In Lerner's case it was Camp Ramah, a Hebrew-speaking summer camp associated with the Conservative movement (Lerner 1994, 16), even though Lerner was Bar Mitzvahed at Temple Bnai Abraham of Newark, a nondenominational synagogue whose spiritual leader from 1939 to 1976 was Rabbi Joachim Prinz, a vocal Civil Rights leader associated with Reform Judaism. Temple Bnai Abraham was a very large synagogue that served an affluent Jewish community of Jewish professionals (ibid., 17) and the perceptive youngster would quickly sense the tension between the preached prophetic teachings and the actual practice of Judaism.

Lerner graduated high school in 1960 and enrolled at Columbia University, majoring in philosophy while also taking courses at the Jewish Theological Seminary (JTS). The faculty of JTS, as we noted already, included two major thinkers that framed the way American Jews interpreted contemporary life: Abraham Joshua Heschel and Mordecai Kaplan. As Lerner relates, "it was under Heschel's tutelage that I first read Hegel, Heidegger, Buber, Levinas and most importantly, the prophets" (Lerner 1994, 18), and it is from Heschel that Lerner learned the concept of God as "Being which transcends all our categories and defies all our expectations, because its essence is radical freedom and necessarily beyond all our categories" (ibid., 50). However, Lerner's reflections about God as the "Force that makes for salvation" echoed the ideas of Mordecai Kaplan, the other influential teacher at the Jewish Theological Seminary, even though Heschel rather than Kaplan served as Lerner's role model. During the turbulent decade of the 1960s until his death in 1972, Heschel was the most prominent traditional Jew involved in the Civil Rights movement and the Anti–Vietnam War movement. In his religiously saturated political activism, Heschel (like Martin Luther King and other leaders of the Civil Rights movement) was engaged in prophetic politics *par excellence*, and Lerner emulated him.

Unlike Commoner and Bookchin, who were either trained in or informed about the natural sciences and immersed in technoscience, Lerner was trained in the humanities and the social sciences, and he made a living as a psychotherapist rather than a researcher in a lab. As a therapist, Lerner sought to diagnose what was ailing American society and culture, identifying its pathologies, which he thought were manifested most acutely in the American Jewish community. Based on his social scientific research in the Institute for Labor and Mental Health, Lerner launched the social movement to heal and repair American Judaism first and later American society, politics, and culture. In 1986 he founded *Tikkun* magazine as a community of like-minded Jews and non-Jews who were committed to the "healing and transformation … and to generating ideas and understandings that can be useful to all of humanity, not just Jews" (Lerner 1994, xviii). He spelled out his vision for the healing of Judaism in *Jewish Renewal: A Path to the Healing and Transformation* (1994), a book directed mostly but not exclusively to the Jewish community in America. In 1996, after receiving a private ordination from Rabbi Zalman Schachter-Shalomi, Lerner established his own synagogue, Beyt Tikkun, in Berkeley as an interfaith community associated with the Jewish Renewal Movement, for the purpose of integrating "spiritual and prophetic consciousness with spiritual activism to create a more loving and just world" (Tikkun 2025). Beyt Tikkun appealed to Jewish liberals, secularists, agnostics, as well as to non-Jews, especially the "spiritual-but-not-religious."

The Tikkun community was to be a "meaning oriented movement" whose "spiritual politics" would be rooted in prophetic morality replacing the ethos of greed, materialism, competitiveness, and selfishness (Lerner and West 1996). In 2005 this was extended to the Network of Spiritual Progressives that called on Americans "not to maximize money and power," but to "maximize love and caring, ethical and ecological sensitivity and behavior, kindness and generosity, non-violence and peace." Rejecting capitalism, meritocracy, and social hierarchies, NSP outlined a "prophetic agenda" that domestically demanded a living wage for all, an anti-poverty program, quality childcare, family support network, government-run healthcare domestically and internationally called for peace by treating all with caring and respect. Emulating Heschel, Lerner functioned as a contemporary prophet who exposes the ills of society while offering a religious vision to heal them.

Secularism and the Loss of Meaning

For Rabbi Lerner, secularism is the underlying cause of the contemporary spiritual crisis, manifested most acutely in American Jewry. Since the 19th century, Jews have enthusiastically embraced "the secular temperament of the modern age" (Lerner 1994, 153) which has caused many to make "a valiant effort to eliminate everything that distinguishes them as Jews" (ibid., 160). The result of this assimilation has been "the absorption of the dominant values of the lager society: materialism, selfishness, acceptance of the economic and political status quo … pessimism about fundamental social change, and individualism" (ibid., 161).

According to Lerner, Jewish secularism is problematic not only because it "has failed to develop Jewish culture that is distinguishable from the increasingly homogenized international culture imposed on the world by Western standards," but because it embraced the "materialist ontology that [has] contributed to the 'disenchantment of the world' and the denial of the validity of ethical truth" (Lerner 1994, 164). Philosophically speaking, so Lerner charges, "the secularist embrace of empiricism eliminated the foundation for hope that transformation is really possible" and Jewish secularists who have embraced the empiricist mentality have been "deaf to the crisis of meaning." Moreover, Jewish secularists enthusiastically embraced a radical interpretation of the separation of Church and State in the United States, consequently creating a public space that is "emptied of values and spirituality," a public space that is "merely technocratic and easily manipulated without restraint by the most powerful forces in the society" (ibid., 166). By "de-meaning," that is, eliminating meaning from public discourse, secularism created the void that made so many thirsty and hungry for meaning, which could be found only through the revival, or recovery, of spirituality in public life.

The spiritual malaise was environmentally consequential because it has led to "selfishness and me-firstism" that "has contributed to the destruction of the ecological support systems for the planet and the homogenization of thought, emotion and art" (Lerner 1994, 167). In contrast to Lynn White Jr., Lerner blames secularism and its ideology of scientism, which detached science from a moral worldview and preached "value-neutrality," as the cause of environmental degradation (ibid., 185). The separation of modern science from values has "allowed science and technology to be harnessed to the largest intentional genocide of world history" (ibid., 186), namely, the Holocaust, a horrific tragedy in which many members of Lerner's family lost their lives (ibid., 193). Bookchin, Waskow, and Roger Gottlieb, to whom we shall turn below, also considered modern scientific rationality to be the underlying cause of both the Holocaust and the ecological crisis, but, in contrast to Bookchin, Waskow, Lerner, and Gottlieb framed the recovery of values in religious/spiritual terms.

According to Lerner, the only way to heal the spiritual crisis of our time is to recover biblical religiosity. Why? Because the Bible articulated a conception of God as "radical freedom and transcendence." In the Torah one finds a revolutionary message of "healing, repair, and transformation" (Lerner 1994, 33) that can transform all aspects of American life: labor, economics, health care, foreign policy, race and ethnic relations, sexuality and family relations, education and culture. Echoing Mordecai Kaplan, Lerner speaks about God as the

> Force toward health, toward goodness, toward holiness, that does not obey the rules, does not act in a lawlike manner but rather enables us to break the expectations, to transcend the patterns of the past toward what would, could, and should be.
>
> (ibid., 50)

To believe in the God of Judaism means to believe in the possibility of change, the possibility of freedom, the possibility of hope, in other words, in the "possibility of possibility." Precisely because "radical freedom" belongs to God alone, the Jewish belief in God can challenge the existing social order and can bring about meaningful change. Therefore, if we are to heal the spiritual crisis our day and fill our life with meaning, we should bring "the God of freedom and transcendence" to bear on every aspect of our earthly, embodied, social, ethical, and political life. Focusing on the betterment of society, a biblically inspired earthly spirituality would reject "any form of oppression, or stunting or misdirecting of human capacity for love, caring, freedom and creativity [as] contradictory to the highest spiritual truth of the universe" (Lerner 1994, 67). Lerner thus envisions

> a religious revival that specifically seeks to reclaim those aspects of spirituality that are consistent with a response for human rights, the

> dignity of self-determination of the individual, and the tolerance for difference and diversity that were absent in some religious traditions in the past.
>
> (ibid., xxvi)

Such religious revival would come by turning to the God of the Bible and "recalling Judaism's message of transcendence and compassion, by reaffirming the revolutionary integration of spiritual and ethical truths that are central to Judaism."

Judaism: The Ecological Religion

Environmentalism is a central component of Lerner's program for Jewish and societal renewal. Reminding his readers that "the earth does not belong to us" (Lerner 1994, 329), he characterizes Judaism as an "ecological religion," namely, a religion that can best face the "potential extinction in the next two hundred years" because it teaches us "to reorient our attitude toward the earth, learn to respect its rhythms, respond to its needs, and forge a relationship with a planet more in keeping with our obligations as Jewish stewards" (ibid.). This reorientation will come if we realize that "human beings have rights only as stewards, and only to the extent that they are actually fulfilling God's commandments, living in accord with their capacity to actualize the divine" (ibid.). As an ecological religion, Judaism "offers concrete ways to do that, based on the Torah tradition" and Lerner moves on to summarize key Jewish precepts, values, and rituals that enables Judaism "to save the planet" because it can serve as a basis for a "worldwide plan for how to use the world's resources and how to distribute the world's wealth" (ibid., 333).

Acknowledging the pioneering work of Arthur Waskow, Lerner seeks to revive the ancient practice of the Sabbatical Year (Ex. 23:11): "one year out of seven we should stop working the earth, dedicating this year to replenishment of the earth" (Lerner 1994, 330). By reviving the ancient practice, we cultivate a commitment "to working on the ecology of the planet together" and we remind ourselves that "the earth needs a rest from all our pollutants and from all our complicated efforts to master and dominate it" (ibid., 330). The Sabbatical Year could function as

> an important instruction to the human race: you don't run things, you are stewards, and you can have power to run your lives only if you do it in accord with the higher purpose that you are here to serve.
>
> (ibid., 331)

No less important to ecological Judaism is the Sabbath, which Lerner presents as a "vehicle for introducing a different kind of consciousness, one that reconnects us with the earth's rhythm's and challenges the priority of making, conquering, subduing, shaping the world to our immediate needs" (Lerner 1994, 335). The Sabbath, as Heschel already famously taught, "is a standpoint of reverence and wonder that we can begin to develop attitudes that would make us put the survival of the planet above other desires" (ibid., 335). Similarly, Lerner echoes Waskow and Schachter-Shalomi in promoting the practice of eco-Kashrut that changes our attitude toward money, market economy, food consumption, labor, and family life. In language that brings to mind Bookchin, Lerner explains that if we follow the practice of eco-Kashrut, we

> imagine the world as one huge organic whole, a body suffused with consciousness but also having developed parts of it, namely the human species, that are invested with special consciousness and hence with special responsibility for figuring out the needs of the whole.
>
> (ibid., 341–42)

Earthly spirituality transforms the world to what it should be.

Lerner's post-secular program for Jewish renewal coheres with the efforts of other Jewish environmentalists discussed in this book. But if other Heschel-inspired American Jews were convinced that the fusion of Judaism and environmentalism can be fully actualized only in the Land of Israel, the Holy Land, Lerner diminished the centrality of the Land of Israel. His Jewish renewal is decidedly post-Zionist (Lerner 1994, 219–64). Recognizing the right of the State of Israel to exist and offering "fervent support" to Israel, Lerner (like Waskow and Seidenberg) voiced scathing criticism of the governments of Israel for their oppression of Palestinians, especially in the territories conquered in the Six-Day War of 1967. For the rest of his life, Lerner advocated for a post-Zionist vision for Jewish renewal.

The Politics of Meaning and Caring Society

In the ancient wisdom of the Bible, Lerner finds an ethics of care and responsibility that can overcome "false pride and egotism that come from conceptualizing ourselves as the center of the universe and acting from the arrogant assumption that the world was made to serve our individual needs" (Lerner 1996, 19). To heal the wounded earth and our wounded selves, Lerner envisions the "reorganization of our society as well as our personal lives to make them less an impediment and more congruent with our highest ideas" (ibid.). The goal of his "politics of meaning" is "to create a society that encourages and supports love and intimacy, friendship and community, ethical sensibility and spiritual awareness among people" (ibid.). Translating that lofty goal into politics was easier said than done.

Lerner observed that the political Right was more successful in fusing meaning into politics. To counter their success Lerner co-founded Network of Spiritual Progressivism in 2005 to transform American politics by resacralizing it, or more precisely by reinterpreting Leftist politics in spiritual terms. In *The Left Hand of God: Healing America's Political and Spiritual Crisis* (2006), Lerner spelled out "how a progressive spiritual politics can help rescue the country from its current self-destructive course" (Lerner 2006, 13). Railing against "technocratic rationality that has replaced an awareness of the spirit, filling our life with selfishness, materialism, and disconnection" (ibid., 14) while conceding that the political Right in the United States identified the spiritual crisis of America correctly as the disconnect from the sacred dimension of life, Lerner rejects the solution offered by the political Right, which he named "The Right Hand of God," because it is based on power, hierarchy, and exclusivism. He offers the Spiritual Progressives' counter-vision, which he names, "The Left Hand of God" that "embraces compassion, love, generosity of spirits, kindness, peace, social justice, environmental sanity, and nonviolence" (ibid., 17).

Lerner's spiritual agenda for American politics consists of

> greening the planet, developing sustainable forms of agriculture and production of goods, creating co-ops and neighborhood associations, transforming unions and civic associations, developing responsible investments and encouraging corporations to embrace a greater degree of accountability, transparency and ecological sensibility.
>
> (Lerner 2006, 220)

In a more overly religious language, Lerner, like Waskow, restates Bookchin's Social Ecology, when he articulates a Spiritual Covenant with America, as a "vision of hope" (ibid., 224). The Spiritual Covenant pertains to American families, personal responsibility, social responsibility, values-based education, health care; building a safer world; separation of church, state, and science; and most relevant for our discussion, environmental stewardship (ibid., 227–240).

Although Lerner was not an environmental activist, he succinctly articulated the principles of Jewish environmental spirituality, and, through the *Tikkun* magazine, he provided Jewish environmentalists a forum to disseminate their views, have a receptive audience, and workout theoretical issues. Crossing denominational divisions, the community of Jewish Spiritual Progressives articulates a distinct voice within American Judaism as evident in Lerner's edited anthology *Tikkun Reader: Twentieth Anniversary* (Lerner 2007), the special issues of *Tikkun* devoted to environmentalism (*Tikkun* Spring 2015; *Tikkun* Winter 2015; *Tikkun* Fall 2017), the collection *Righteous Indignation: A Jewish Call for Justice* (Rose, Kaiser, and Klein, 2007), and the anthology of essays, *Judaism and Justice: The Jewish Passion to Repair the World* (Schwarz 2006). Consisting of academics, Judaica scholars, rabbis, social activists, and educators these anthologies make clear how Jewish environmental spirituality fits within a larger vision of "Jewish justice agenda" that recognizes "the world is in need of repair," and that Jews are called "to do justice work."

Roger S. Gottlieb is our third example of a Jewish Spiritual Progressive whose life and works exemplify the transition from secularization to re-sacralization not merely within the American Jewish community but also within the American academy.

Roger S. Gottlieb: Religion, Politics, and Spirituality in the Academy

Roger S. Gottlieb is the most influential Jewish environmental philosopher today, although he may not be a household name among American Jews. We began this study by referencing Gottlieb's understanding of spirituality, and we now return to examine his work more closely. Associated with the Network of Spiritual Progressives and Michael Lerner's *Tikkun* magazine, Gottlieb, like Arthur Waskow before him, illustrates how a Jewish Marxist becomes a religious environmentalist. Like Waskow, Gottlieb was associated with the New Left: he was immersed in the intricacies of Marxist theory (Gottlieb 1992) and participated in the civil rights and women's movements of the 1960s and early 1970s. The experience of social and political activism led Gottlieb to affirm his own Jewishness and to appreciate Judaism's distinctive contribution to the nexus of politics and spirituality. Since the 1990s, Gottlieb has focused on religious environmentalism and has greatly contributed to the interdisciplinary and interfaith academic discourse of religion and ecology. If Waskow created the vision and practice of Jewish environmentalism and Lerner promoted the fusion of spirituality and progressive politics, Gottlieb articulated the theory of a pluralistic, politically engaged, and socially transformative environmental spirituality within which Judaism can enter into dialogue with other religious traditions. Gottlieb's life and work help us understand how and why environmental spirituality illustrates the post-secular moment in which religion functions within public, political life.

Gottlieb was born on October 20, 1946, in White Plains, New York. The postwar years brought suburbanization to American Jewry as Jews joined the middle class and enjoyed social mobility with the decline of anti-Semitism (Sklare and Greenblum 1979; Kranson 2017). Gottlieb grew up "in a moderately assimilated middle class suburban family, [and] it was not until my early thirties that I realized the deep connection I had to my Jewishness" (Gottlieb 2015, xix). His religious background was a "watered-down version of Reform Judaism," and he "reached adulthood untouched by most of traditional Jewish liturgy, theology, or spirituality" (ibid.). He attended public school in White Plains, and although he grew up in a nonobservant Jewish family, he enrolled in Brandeis University in 1964.

Gottlieb intended to major in psychology but after two years switched to philosophy, earning his BA in 1968 and PhD in 1975. While he was at Brandeis, many American colleges and universities witnessed the growing politization and radicalization of students. Organizations such as Students for Democratic Society (SDS) organized sit-ins, strikes, hostage-taking, and building seizures. At Brandeis, on January 8, 1969, 70 Black students gained control of Ford Hall (a student center at the time) and demanded hiring more Black faculty members, increasing Black students' enrollment from 4 percent to 10 percent, and establishing an independent department of African American studies. The standoff was resolved after two weeks, but Brandeis was rife with 1960s counterculture and politics of protest, molding young Gottlieb.

Having earned a PhD in philosophy with a dissertation on "Kierkegaard and Nietzsche's Theories of the Transition from One Value System to Another," Gottlieb launched his career as an academic philosopher. He first held a visiting assistant professorship at the University of Connecticut (1974–1977), another temporary position at Tufts University (1978–1980), and, after a yearlong National Endowment for the Humanities Fellowship (1980–1981), was hired as professor of philosophy at Worcester Polytechnic Institute, where he was granted tenure in 1985 and where he still teaches. Early on, Gottlieb was a typical secular Jewish intellectual for whom "Jewish observance meant little" and who "spent years trying to figure out what kind of Jew I wanted to be and what kind of Judaism might fit my particular blend of New Age and New Left sensibilities" (Gottlieb 1999, 2).

Gottlieb's first book, *History and Subjectivity: The Transformation of Marxist Theory* (1987), reflected these tensions within the New Left. With a command of Marxist theory, Gottlieb aspired to save Marxism from its own limitations. He summarizes the book's thesis thus: "a transformed Marxist theory must reject Marx's claim that the mode of production possesses sole social primacy [...] Economic class, processes of socialization, and politics, I argue, are now co-primary in determining social life" (Gottlieb 1987, xi–xii). Gottlieb sees social life as the product of both "History" (namely, structural economic changes) and "Subjectivity" (how people construct their personalities, especially in patriarchal societies), arguing that traditional Marxists overemphasized the former and underemphasized the latter. Influenced by the Western Marxists Karl Korsch, George Lukacs, and Antony Gramsci, Gottlieb argues that human beliefs and attitudes are critical in constituting social reality and that these are not mere economic by-products but help determine the mode of production and political activity within society. With leaders of the Frankfurt School, Max Horkheimer, Theodor Adorno, and Wilhelm Reich, the book shared the deep concern for subjectivity as revealed in their various applications of psychoanalysis and cultural criticism of bourgeois society. And from Herbert Marcuse and Jürgen Habermas, Gottlieb derives the conviction that social theory, unlike natural science, aims to liberate rather than control people. Because social life is subjective, Marxism must treat moral norms as essential to social theory.

The main inspiration for the proposed transformation of Marxism came from socialist-feminist theorists (e.g., Sheila Rowbotham, Juliette Michelle, Heidi Hartmann, Barbara Ehrenreich, and Ann Ferguson among others), who have expanded the subjective strain of Western Marxism under the slogan "the personal is political." Feminist theorists showed that family life, sexuality, child rearing, and emotional attachment are crucial areas of political concern and taught Gottlieb how to

> transcend the sexist ideology that separates feeling and thought, the personal and the objective, female and male. What is required is a consciousness that is both Historical and Subjective, both an accurate understanding of the reality of social life and a personal understanding and self-transformation of one's habits, needs, hopes, and fears.
>
> (Gottlieb 1987, 151)

In the feminist model, "subjective knowledge and the transformation of the self are combined with a structural account of social life" (ibid.). That insight would guide Gottlieb, who defines himself as a "feminist" (later, an "ecofeminist") to attend to subjectivity, which, later in life, he called "spirituality." We will say more about Gottlieb's understanding of spirituality below; for now, let us note that Gottlieb's engagement with feminist analysis of women's oppression and the intersection of gender, class, and labor led him to identify the limitations of both Marxist theory and Western Marxism. Socialist feminists overcame these limitations because "they have presented a model in which subjective knowledge and the transformation of the self are combined with a structural account of social life" (ibid., 151). Feminist theory gave rise to "a consciousness that is both Historical and Subjective," explaining how "gender-based socialization processes … shape that consciousness" (ibid.).

And what about Jews? How did they figure in Marxist analysis and New Left politics? In a nutshell, Marxist theory could not accommodate Jewish particularity (i.e., "social differentiation"), since it focused on the general laws of historical materialism, claiming to offer a universal solution to the problems of human existence by focusing on class rather than religion or nationhood. In the 1980s, Gottlieb (like Lerner) became increasingly aware of "the latent anti-Semitism of some of my political comrades—who embraced the cause of every oppressed group but to whom anti-Semitism remained invisible" (Gottlieb 2015, xix). Aware of Marxism's limitations, Gottlieb asserts that "when Marxism fails to understand the importance of social differentiation, leftist movements will not only fail politically, but also manifest racial, ethnic, or gender oppression" (Gottlieb 1987, 204). Left-wing anti-Semitism is evident in "the fact way the left has ignored Jewish oppression, denied its existence, or condemned as 'unprogressive' Jewish responses to it" (ibid.).

This is most evident in the case of the Holocaust and Zionism. The New Left either ignores the Holocaust or denies the Jewish success in creating a state in which Jews are not oppressed. Instead, Israel is presented as "a colonial-settler state, an agent of imperialism, or a case of 'genocide' against Palestinians comparable to the Nazi genocide against Jews" (Gottlieb 1987, 208). In addition to explaining the connection between Zionism and the

Holocaust, Gottlieb schools his Leftist colleagues about Jewish diversity. Social differentiation takes many forms, and Jewish culture is not homogenous. Moreover, while Jews have been historically oppressed, they can also be oppressors, which for Gottlieb is the tragedy of Zionism. Like Waskow, Gottlieb admits that "while Israeli Jews are free from immediate anti-Semitism, they still face racism, sexism and class domination. Arabs in Israel and the occupied territories suffer legal and extralegal oppression" (ibid., 213). Defending Jewish social differentiation and endorsing Zionism has not prevented Gottlieb from criticizing injustice committed by Jews, but he has not elaborated on the Israel-Palestinian conflict as much as his fellow Jewish Spiritual Progressives, Waskow and Lerner.

Spiritual Resistance to Genocide and to Ecocide

Confronting anti-Semitism within the New Left became personally important to Gottlieb after marrying Miriam Greenspan, a feminist, a therapist, and a daughter of Holocaust survivors. Greenspan influenced Gottlieb's thinking and writing as much as Phyllis Berman influenced Waskow's. In the 1980s Gottlieb studied the Holocaust on the assumption that "the Holocaust possesses terribly important meaning for us, meanings which may shape or alter our belief about human nature, morality, politics, and spiritual life" (Gottlieb 1990, 1).

In *Thinking the Unthinkable* (1990), Gottlieb connected the Holocaust and "the threat of nuclear war and ecological destruction" (Gottlieb 1990, 7) and articulated the "spirituality of resistance" as the response to both genocide and ecocide. In his own essay in the anthology, "The Concept of Resistance: Jewish Resistance during the Holocaust" (ibid., 327–44), Gottlieb returned to the History/Subjectivity nexus. He recounted how Jews responded to Nazi threats to their very religious identity in ways that "preserved their attachment to the principles of holiness, religiosity and faith," noting that "the most dangerous threat was not physical annihilation by the Nazis but their own betrayal of these principles" (ibid., 332). And in "Remembrance and Resistance" (ibid., 434–46), Gottlieb argued that

> the greatness of Jewish resistors lies not only in their courage, but in their capacity to call forth out of themselves that which is most holy: a love of life, and of other people, and the capacity to manifest that love in a way which is both gloriously selfish and selfless at the same time.
> (ibid., 444)

Jewish resistance in the Holocaust was mostly spiritual rather than physical, although the latter was not absent.

Valuing Jewish spiritual resistance was not new in Holocaust studies, but Gottlieb presents it as the posture all must adopt in response to the ecological crisis. In *A Spirituality of Resistance: Finding a Peaceful Heart and*

Protecting the Earth (1999), Gottlieb presents "spirituality of resistance" as a call to move beyond inner work such as Gottlieb the spiritual seeker experienced in the 1960s counterculture as he explored psychedelics, yoga, and Eastern traditions. His call, then, is "to live on this earth: fully conscious of what is happening on it, actively resisting that which we know to be evil or destructively ignorant" (Gottlieb 1999, 13). Echoing Waskow, Gottlieb observes that "what the Nazis did to their victims can serve as a kind of warning about the potential consequences of our own ecological madness" (ibid., 28).

It is easier to ignore or deny the truth of climate change, just as European Jews disregarded Nazi plans to annihilate them. Spirituality of resistance begins with telling the truth of the environmental crisis to mobilize "a kind of global despair" (Gottlieb 1999, 30). Through spiritual resistance,

> we encounter the face of God, awaken to the call of the Goddess, and realize our deepest connection to the mysteries of human life. The precious moment in which we face the truth and resist evil are the culmination of our spiritual journey.
>
> (ibid., 32)

Gottlieb's understanding of the need to confront truth, grief, and anger arises from the death of his infant son. The pain of this "wake-up call," as C.S. Lewis defined it, launched him on the spiritual path in which he reembraced Jewish religious life to deal with the Holocaust and the environmental crisis. Like Waskow, Gottlieb reconnected with the Jewish tradition whose values, symbols, and rituals now guide and enrich his political activism.

Religious Environmentalism: Spirituality and Political Activism

In the mid-1990s Gottlieb emerged as a leading contributor to the field of religion and ecology, which emerged in the 1970s. His anthology *This Sacred Earth: Religion, Nature, Environment* (1996) asserts that there are many ways to be "religious." Unlike those who are "spiritual but not religious" (see Chapter 4), Gottlieb sees "spirituality" as an aspect or dimension of "religion" and sees no tension between spirituality and social activism. Similarly, "religious environmentalism" is not a contradiction in terms. Indeed, the limited, modern technoscientific worldview dictates a spiritual response to the ecological crisis (Gottlieb 1996, 11). This rich selection of writings by theologians and activists from all world religions features the breadth, diversity, and complexity of the discourse of religion and ecology.

In another anthology, *The Ecological Community: Environmental Challenges for Philosophy, Politics, and Morality* (1997), Gottlieb asks academics ignoring the crisis "to acknowledge the severity of the environmental crisis and to alter our research and classroom conduct accordingly" (xiv). The interdisciplinary writings show how philosophers, political theorists, ethicists, and scholars of religious studies grapple with the meaning and implications of the environmental crisis, wrestling with various theoretical challenges. As much as Gottlieb had to convince secular academics of religion's relevance to environmentalism, he also had to show to religionists that awareness of the environmental crisis makes religion a liberating (namely, progressive) social force. His anthology *Liberating Faith: Religious Voices for Justice, Peace and Ecological Wisdom* (2003) demonstrates religion's profound role in public life and argues that a liberating faith offers spiritual resistance to the crises of our time, chief among them the environmental crisis, and that resistance is inter-religious, public, and thoroughly political.

Gottlieb's understanding of the relationship between religion and politics makes him a post-secular thinker, although he does not refer to himself as such. In *Joining Hands: Politics and Religion Together for Social Change* (2002), he dismisses the "religious"/"secular" binary as "not so enlightening" (xxvii) and admits the failure of the Secularization Thesis. In the chapter "Saving the World: Religion and Politics in the Environmental Movement," Gottlieb shows how "world-making politics and emancipatory religion have joined in environmental politics and ecological spirituality [...] at times virtually fusing to form a historically unprecedented phenomenon" (Gottlieb 2002, 153). The emergence of religious environmentalism led to identifying the connection between environmental racism (e.g., placing waste disposal plants or landfills in poor, mostly Black, neighborhoods) and environmental justice, giving rise to the ecojustice perspective we discussed above in reference to Arthur Waskow. That perspective breaks the barriers between religion and politics, theology, and social activism. The pursuit of eco-justice cuts across religious boundaries as people of various traditions engage the environmental challenges. Religious environmentalism has turned traditional religions "green" as much as it infused secular green politics with spirituality.

Gottlieb has made an indelible contribution to the academic discourse on religion and ecology, especially in his edited volume *Oxford Handbook of Religion and Ecology* (2006) and the monograph *A Greener Faith: Religious Environmentalism and Our Planet's Future* (2006a). The former consists of essays by leading contributors to the field, representing various faith traditions across the world. It reflects on the main themes that engage religious environmentalists, and it wrestles with contested issues. The latter showcases the accomplishments of religious environmentalism, "now a leading voice telling us to respect the earth, love our nonhuman as well as our human neighbors, and think deeply about our social policies and economic priorities" (Gottlieb 2006a, 9). By looking at environmental devastation from a religious perspective, we can identify our plight in spiritual as well as political, economic, and technological terms. After all, the spiritual crisis was caused by the very assumptions of modernity, such as "the social authority of science, the unceasing technological innovation, detached professional, unlimited scientific research, and stiff unfeeling doctors who practice medicine as if patients were collections of body parts" (Gottlieb 1999, 30–31). Against contemporary technoscience and hedonistic consumerism, Gottlieb promotes spirituality as a source of intellectual depth, psychological sensitivity, emotional openness, and love and care for others, all of which constitute what he terms "spirituality."

In *Spirituality: What It Is and Why It Matters* (2013), Gottlieb explains how the spiritual quality of nature offers the deepest healing to our struggles with "isolation, endless rushing, the fear of death, facing the failures of the isolated nuclear family, consumerism, and technology" (Gottlieb 2013, 147). Nature can address our spiritual hunger, but, again, the tragedy of the environmental crisis is that "we have pretty much eliminated nature" (ibid., 148). When we register the natural world's sacredness, we not only re-enchant nature, but we also "integrate ecological and social concern, conservation of endangered landscapes, and protection of minorities suffering disproportionately from pollution" (ibid., 149). The re-enchantment of nature can be explained in various intellectual frameworks; all express the idea that "life is sacred" and that all living things are our kin (ibid., 150). Every religious tradition encompasses some form of nature spirituality, which understands the reciprocal balance of humanity and nature as part of an entire way of life. Such a way of life demands balance, restraint, and listening. Gottlieb shows us that nature spirituality is not abstract or theoretical but emotional, embodied, and physical, very aware that all living beings age, decay, and die and that life is sacred even though it is not eternal. Re-sacralizing the world, Gottlieb tells us, requires fusing spirituality and politics. Politics is not antagonistic to spirituality: all political issues require us to take a stand on what a just society is.

Religious environmentalism offers spiritual politics that address the grief and despair we feel due to massive environmental degradation and moral quandaries it poses. Gottlieb engages these in his more recent *Morality and the Environmental Crisis* (2019). Gottlieb lays out how our relationship with the natural world has been undermined largely by a modern science bent on controlling the natural world and in thrall to the military-industrial complex (Gottlieb 2019, 17–18). With no moral grounding and rapidly developing technology, modern science has yielded "two catastrophic world wars [...] Mass death in concentration camps [...] and belief systems (Nazism, for example) whose lack of reason is palpable" (ibid., 18). Science's presumed rationality resulted in the madness of the massive environmental crisis, which science and technology cannot fix. The modern conception of rationality has disenchanted the world by debunking our notion of the sacred. The response of religious environmentalism and spiritual resistance is nonviolent environmental activism, which restores meaning to living on a vulnerable planet.

Gottlieb writes for a broad and culturally, religiously, and ethnically diverse audience. As a Jew who is proud of his Jewishness and who appreciates the depth of the Jewish religious tradition, Gottlieb writes as a scholar of comparative religious studies who is familiar with many expressions of religious environmentalism. He has shown how to integrate Judaism into a global academic discourse and how Jews could participate in interfaith environmental activism. Jewish environmental activists today may not undergo the same secularism-to-post-secularism life trajectory as Gottlieb, Lerner, and Waskow, but most would agree that addressing the environmental crisis and its manifold social and economic aspects demands spiritual politics or political spirituality that engages in *Tikkun Olam*. Spiritual progressives, Jewish or non-Jewish, confront environmental challenges in their own distinctive ways, but they share a critical stance toward contemporary consumerism; modern technoscience; greedy, exploitative corporations; and unjust social policies. Spiritual progressives may use diverse intellectual resources to articulate their fusion of spirituality and politics, but they all agree that the environmental crisis demands a spiritual response.

Tikkun Olam: A Popular and Contested Trope

The Spiritual Progressives have clearly made their impact on American Judaism and on American society and politics. Because of *Tikkun* magazine, the Tikkun Community, and the Network of Spiritual Progressives, the phrase *Tikkun Olam* became ubiquitous not only in Jewish discourse but also in American public discourse. Reform Judaism and other forms of Liberal Judaism employed the term as a shorthand for a very wide array of social action.

To Jewish Liberal activists, *Tikkun Olam* is synonymous with social justice, referring to acts of civic responsibility that are meant to address the political and social ills of the world. *Tikkun Olam* has been applied to support of marginalized communities, to adopting progressive politics, and, of course, to a range of ecological strategies such as eco-Kashrut and animal rights. In the name of *Tikkun Olam* Jews have supported Black Lives Matter, pushed for Peace in the Middle East, critiqued the Israeli government's policy toward the Palestinians, especially the way the government conducted the War in Gaza, and, most relevant to this book, justified environmental action and climate advocacy. Jonathan Krasner, a Professor of Jewish Education Research at Brandies University correctly summarized the ubiquity of *Tikkun Olam* by saying that "many American Jews consider it a cornerstone of their Jewish identity, a key reason they're committed to make the world a better place" (Krasner 2023; Cf. Rosenthal 2005; Cooper 2013).

The very popularity of *Tikkun Olam* has generated an internal debate among Jews.

Is there Jewish religious obligation to work for the moral, spiritual, and/or pragmatic progress of the world? If so, what are the sources of this obligation, how should Jews carry out the obligation, and what specific message does Judaism bring to the world in which we live? The Orthodox community reflected on these questions, and its views were divided between the "outward" and "inward" orientations (Shatz, Waxman, and Diament 1997), and across the denominational spectrum there were conflicting views on the place of Jews and Judaism in the American public sphere (Mittleman, Licht, and Sarna 2002; Mittleman 2003). Why is *Tikkun Olam* so popular a phrase? My answer is that *Tikkun Olam* captures the hybridity of the post-secular age, expressed in the seemingly paradoxical phrase "earthly spiritualty," the title of this book. For secular, non-observant, non-affiliated Jews, who seek to find meaning in the mundane word, the phrase *Tikkun Olam* allows them to wrap their social activism in a Jewish garb, thereby affirming their Jewish identity, kinship, and solidarity with other Jews. For religiously affiliated Jews, regardless of levels of observance or knowledge, invoking *Tikkun Olam* is a way to sacralize the secular, public sphere investing all aspects of life, including politics, with religious/spiritual meaning. Put differently, for secular Jews, the trope of *Tikkun Olam* functions as a marker of ethnic or cultural identity, whereas for religiously affiliated Jews, the popular trope performs the task of public theology.

Tikkun Olam has been invoked to justify environmental activism, but several theologians noted that in rabbinic Judaism "the world" that one must mend is not nature but the human social world (Dorff 2005, 7). Moreover, as Rabbi Dorff makes clear, in rabbinic Judaism the ultimate agent of mending the world is God rather than humans. In traditional Judaism the ideal of *Tikkun Olam* expresses Jewish social ethics whose practice ultimately results in a peaceful society (Dorff 2008, 181–96). Critical of the "promiscuous use" of *Tikkun Olam* to justify all forms of social activism, the late Rabbi Byron Sherwin (2013) showed that for Jewish philosophers *Tikkun Olam* meant the "repair" of the human person, body and soul, and that for the kabbalists, *Tikkun Olam* ultimately pertained to human theurgic impact on the (broken) Deity. Neither philosophers nor kabbalists were concerned about human intervention in the natural environment. Less critical than Sherwin, Rabbi Dorff explains that the human responsibility to care for the natural is derived from the Judaic ethics of responsibility and care rooted in the doctrine of creation and the sovereignty of God.

The debate about the meaning and proper application of *Tikkun Olam* captures the inherent ambiguity of "earthly spirituality." Those who maintain that Jews have an obligation to the society at large have no qualm about employing *Tikkun Olam*. When theologians do so, they engage in public theology, whether they declare so or not. The most influential Jewish public theologian is the late Rabbi Jonathan Sacks (d. 2020), a Modern Orthodox Jew who became a moral voice to humanity. I end this chapter and this book by looking at his legacy because he was a paradigmatic post-secular public theologian, because he addressed the ecological crisis head on, and because his legacy wrestled with the problematics of post-secularity. Although Rabbi Sacks was an Orthodox Jew, his project was not so different from that of Conservative and Liberal rabbis, such as Elliot Dorff, Arthur Waskow, Michael Lerner, and a host of others who have addressed a range of social, political, and economic issues from a Judaic perspective (e.g., Schwarz 2006; Rose et al. 2007; Cosgrove 2010).

Rabbi Jonathan Sacks: Post-Secular Public Theology

From the Secular to the Post-Secular

Rabbi Sacks was the paradigmatic Jewish post-secular public theologian. He was born in 1948 into a Jewish family that observed traditional Judaism, but he was educated in non-Jewish schools: Saint Mary's Primary School, Christ College, Oxford, and then Cambridge University. His father, Louis Sacks (d. 1996) was a Polish Jew who came to England at the age of six, after the family fled pogroms in Poland and settled in London's East End. Rabbi Sacks described his father as a pious Jew "who went to synagogue every day. He did not understand much Hebrew, but this he knew: he was a Jew, he believed in God, and his fate was in God's hands. That was enough for him. Louis walked tall" (Sacks 2011, 183–84; Cf. Sacks 2000, 62–64). As the official biography of Rabbi Sacks states, the parents instilled in Rabbi Sacks and his three siblings "enormous devotion to education, Judaism and the wider society, a blend of secularism and religion that would become the template of his lifelong pursuit" (Jonathan Sacks 1981–2025). This is what the 19th century ideal of *Torah im derekh Eretz*, which was articulated by Rabbi Samson Raphael Hirsch, meant in 20th-century England: openness to secular culture along with proud affirmation of one's Jewish religio-ethic identity.

At Cambridge University, young Jonathan Sacks read philosophy with the atheist moral philosopher Bernard Williams (d. 2004), who challenged Sacks to probe into his unexamined religious beliefs. This was in the mid-1960s, when the British academy was undergoing major transformation as Logical Positivism reached what Sacks called "a dead end" and as the dominant theories of moral philosophy—Deontology and Consequentialism—were challenged by the analytically trained religious philosopher G.E.M. Anscombe, who launched the recovery of ancient virtue ethics. Jonathan Sacks, the young philosophy student, was most attuned to these academic debates as well as to the cultural upheaval worldwide brought about by the countercultural youth movements that challenged authority, traditional and sexual mores, and that gave the world the New Age Spiritualty discussed in previous chapters. To grapple with these tumultuous times and clarify to himself what they believed, Sacks travelled to the United States and Canada in 1968 to meet with public intellectuals and religious leaders. This was a transformative experience. By his own admission, the encounter with Rabbi Joseph Soloveitchik introduced him to "a new kind of Jewish thought, based not on philosophical categories but on Halakhah, Jewish Law" (Sacks 2011, 90), and the encounter with Rabbi Menachem Mendel Schneerson, the seventh Lubavitcher Rebbe, led him to see the predicament of the Jewish people. Summarizing the meaning of these experiences, Rabbi Sacks famously quipped: "Rabbi Soloveitchik had challenged me to think. Rabbi Schneerson had challenged me to lead. In both—though neither spoke of it—I sensed the extent to what Jewish life had lost in the Holocaust" (Sacks 2011, 91).

Sacks returned to Cambridge to start his graduate training in philosophy, first continuing with Bernard Williams and later switching to Oxford University to study with Philippa Foot, a major contributor to the revival of virtue ethics. In 1973, after a few years of studying and teaching philosophy at these elite British universities, Sacks renounced the academic path and embarked on a serious study of Judaism. Although he eventually completed his PhD from University College London (1981), Sacks quenched his thirst for meaning not in the halls of the secular university but in Jewish religious institutions. In 1976 he received rabbinic ordination from Jews' College London and from Yeshiva Etz Chayim, London, and thereafter rose quickly to leadership of Modern Orthodoxy in England. From 1991 to 2013 he served as the Chief Rabbi of the United Hebrew Congregations of the Commonwealth and although his authority was challenged by the Ultra-Orthodox communities, on the right, and by Liberal and Progressive congregations, on the left, he emerged as the voice of moral wisdom not only in the English public sphere but also in the United States, in Israel, and indeed in the global community.

With over 40 books, many programs on television and radio, and numerous essays in the press, Rabbi Sacks became the most influential interpreter of Judaism in the global, pluralistic, post-secular age. Speaking simultaneously to Jews and non-Jews he captured the post-secular moment, while promoting Jewish earthly spirituality that validated religious pluralism. In 2005 he was knighted and made a Life Peer, taking his seat in the House of Lords in October 2009. In 2014 he was named as "the Becket Fund's 2014 Canterbury Medalist for his role in the defence of religious liberty in the public square." In 2016 he won the Bradley Prize in recognition of being as "leading moral voice in today's world." In the same year he also won the Templeton Prize, which honors a "living person who has made exceptional contributions to affirming life's spiritual dimension whether through insight, discovery, or practical works." Rabbi Sacks was awarded the Templeton Prize for promoting tolerance and understanding among religiously and ethnically diverse communities. The prestigious award was bestowed on him for the recent publication of *Not in God's Name: Confronting Religious Violence* (2015) in which he defended religious pluralism and challenged bigotry, hatred, and violence, especially when promoted in the name of religion. But the award committee also cited Rabbi Sacks's earlier book, *The Great Partnership: Science, Religion, and the Search for Meaning* (2011) stating that Rabbi Sacks "boldly defends the compatibility of religion and science, a response to those who consider them necessarily separate and distinct." Sacks's defense of religious pluralism was related to his view of the "great partnership" of science and religion (Tirosh-Samuelson 2023).

Religious Pluralism and the Partnership of Science and Religion

Rabbi Sacks was a critic of modern secularity, which he considered the source of many social ills, including the breakdown of the nuclear family, the corrosion of community, the prevalence of loneliness, hedonism, and consumerism, and the pervasive loss of meaning. To these social problems, Rabbi Sacks offered a well-reasoned, eloquent, and action-oriented Judaic solution that emphasizes "family, education, philanthropy and the sanctification of everyday life" (Sacks 1990, 13). Philosophically speaking, the contemporary malaise arises from the relativism characteristic of secular Liberalism. Contrary to Liberals (Jews or non-Jews) who are skeptical about the objective status of certain beliefs, values, and norms, Rabbi Sacks asserted categorically:

> I do not believe that the sanctity of human life and the inalienable freedoms of a just society are relative. They are religious absolutes. They flow directly from the proposition that it was not we who created God in our image but God who made us in His image.
>
> (Sacks 2002, 19)

These religious absolutes, Rabbi Sacks asserts, are "the very tradition that Jews, Christians, and Muslims … share" (ibid.), and therefore members of these religious traditions can and should communicate, collaborate, and cooperate with each other, despite their theological differences. The solution to our social ills will come not from the abstract universalism of the Enlightenment and its Liberal principles but from the recognition and respect we pay to religious particularism. As Sacks put it, "Universalism must be balanced with a new respect for the local, the particular, the unique" (ibid., 20). Instead of talking about "Religion" (with capital "R"), as if there is one universal religion for all of humanity, we should recognize the plurality of religious traditions that encompass individuals, communities, sacred texts, rituals, and habits of heart and mind with all their specificity. But if religious traditions are to thrive in the public sphere, how can they co-exist peacefully without seeking domination, without resorting to violence, and without denying the freedom to believe? Rabbi Sacks answered the question by coining the phrase: "the dignity of difference," the title of his book in which he defended religious pluralism, but without endorsing multiculturalism, and protected particularism from being flattened into generic universalism.

Rabbi Sacks was a post-Liberal thinker who acknowledged the shortcomings of Liberalism and critiqued the way Liberal theories were used to undergird multiculturalism. In *The Dignity of Difference,* he declared from the outset: "I am not a liberal Jew. My faith is Orthodox" (Sacks 2002, 18). Therefore, Rabbi Sacks did not have to show that Judaism is compatible with Liberalism, as did Jewish theologians associated with Liberal and Progressive Judaism. Instead, he had to show that particularistic Judaism has a universal appeal to all human beings. In his book, *The Home We Build Together: Recreating Society* (2007), Rabbi Sacks acknowledged the positive attempt to make ethnic and religious minorities in Britain more appreciated and respected. But he also charged that multiculturalism went too far, giving rise to identity politics, as minorities and aggrieved groups jockeyed first for rights, and then for special treatments. The process, which began first with Jews, and was later taken on by Blacks, women, and gays, gave rise to "a culture of victimhood" that "sets groups against group, each claiming that its pain, injury, oppression, humiliation is greater than the other" (Sacks 2007, 61). Multiculturalism, he insisted, is "inexorably divisive," because it stifles free speech and threatens liberal democracies. To restore the sense of common good and common purpose, multiculturalism must be rethought and reconfigured.

The same critique was reinterred in *Morality: Restoring the Common Good in Divided Times* (2020), published just before he passed away. The proponents of multiculturalism were correct to recognize historicity and cultural diversity, thereby correcting the Enlightenment determination "to abolish identity in the name of the universal" (Sacks 2020, 130). However, they went too far, giving rise to identity politics that over time corroded the common good, because it recognized "the individual as the bearer of rights, and of autonomy as the supreme value of the social order" (ibid., 116). In his judgment, "identity politics deepened the fragmentation caused by multiculturalism, adding to it not just culture and ethnicity but also other forms of identity based on gender and secular orientation" (ibid., 137). Mincing no words and with a nod to Habermas, Rabbi Sacks warned that "there is real danger here of the splitting of society into self-segregating, non-communicative ghettoes" (ibid., 137). Multiculturalism was mistaken because instead of respecting difference, it:

> turned difference into exclusion and suspicion. It builds walls, not bridges. It abdicated the hard world of understanding, respecting and working with and for the people not like us. It encourages a mindset of victimhood and oppression. It abandons ideas of the common ground and the common good.
>
> (ibid.)

If unbridled multiculturalism turned Western societies into fragmented, conflictual, and dysfunctional societies, because they had lost the vision of the common good based on religious absolutes, how can this vision be restored? By gaining the correct view on the relationship between the particular and the universal, between human reason and divine revelation, between science and religion, between the secular and the sacred. As an Orthodox Jew, Rabbi Sacks adamantly protected "the right to be different," a right that belongs not to the "liberal democratic state" but to the "compassionate society" (Sacks 2020, 139). Put differently, to protect the "dignity of difference" but without corroding the common good, we need to have a proper understanding of the two main frameworks that provide meaning to people today: science and religion. In the post-secular society, religious pluralism should go hand in hand with the *partnership* between religion and science, a partnership that respects and protects differences while establishing collaboration, cooperation, and mutual support.

A master communicator, Rabbi Sacks used what he called "drama of ideas" as a heuristic device to capture intellectual tensions and cultural processes. Writing in short, pithy sentences he wrote in a decidedly non-academic style that made complex ideas easily accessible to the general reader. Here is how he presents the relationship between "science" and "religion":

> Science is about explanation. Religion is about meaning. Science analyses, religion integrates. Science breaks things down to their component parts. Religion binds people together in relationships of trust. Science tells us what is. Religion tells us what ought to be. Science describes. Religion beckons, summon, calls. Science sees objects. Religion speaks to us as a subject. Science practices detachment. Religion is the art of attachment, self to self, soul to soul. Science sees the underlying order of the physical world. Religion hears the music beneath the noise. Science is the conquest of ignorance. Religion is the redemption of solitude.
>
> (Sacks 2011, 6)

Clarifying the differences between "science" and "religion" was not intended as an argument that they are necessarily in conflict with each other but rather that they are integral to the human condition, expressing different features of being human. Modernity indeed posited "science" and "religion" as a binary dichotomy of "either/or." If "science" is the path to truth, then "religion" must be excluded. Modern science thus contributed to the secularization of modern society, but with negative outcomes. Dominated by scientism, secular culture generated reductionist materialism, hedonistic consumerism, and technological determinism that flattened human existence. Rabbi Sacks rejects all of these 'isms' because they either ignore or deny freedom: the freedom to believe, the freedom to be different; the freedom to choose; the freedom to love; the freedom to share; the freedom to care, and even the freedom to err. Instead, he offers an integrationist or a complementary model for the relationship of science and religion, a "both/and" relationship rather than one of "either/or." His integrationist approach accommodates not only science *and* religion, but also transcendence *and* immanence, universalism *and* particularism, necessity *and* freedom, truth *and* meaning, facts *and* values, eternity *and* historicity, mind *and* body, reason *and* emotion, theory *and* praxis, spirituality *and* worldliness, humaneness *and* Jewishness.

The "great partnership" thus captures his non-dualistic approach to reality, the human condition, society and politics. In Jewish parlance, the "great partnership" is expressed in the fusion of Wisdom (*Chokhmah*) and Torah. As Rabbi Sacks explains:

> Chokhmah is the truth we discover; Torah is the truth we inherit. Chokhmah is the universal language of humankind; Torah is the specific heritage of Israel. Chokhmah is what we attain by being in the image of God; Torah is what guides Jews as the people of God. Chokhmah is acquired by seeing and reasoning; Torah is received by listening and responding. Chokhmah tells us what is; Torah tells us what ought to be. (Sacks 2009, 221)

Rabbi Sacks's integrationist approach thus combined the inward-looking and the outward-looking perspectives within Judaism, showing how particularity is universalized and vice versa, how the universal is particularized. Rabbi Sacks became such an influential global thinker because he invited world faiths "to understand themselves on their own terms with eye on how those terms can nonetheless form part of a global tapestry" (Harris, Rynhold, and Wright 2012, xvii).

Environmental Ethics of Responsibility

Writing for Jews and non-Jews, Rabbi Sacks expressed his ideas in two main literary genres: semi-academic treatises that reflected on the human condition and engaged contemporary social, political, and economic issues from a Judaic perspective, and exegetical works that explicated Jewish canonical texts, especially the Hebrew Bible, through conversation with philosophers, political theorists, social critics, psychologists, scientists, and literary authors, non-Jewish and Jewish. His five-volume *Covenant and Conversation: A Weekly Reading of the Jewish Bible* (2009–2019), as well as *Essays on Ethics: A Weekly Reading of the Jewish Bible* (2016) and *Studies in Spirituality: A Weekly Reading of the Hebrew Bible* (2021), are arranged according to the order of the weekly Torah portions, but instead of a linear, verse-by-verse commentary, Rabbi Sacks composed short essays, each with a pithy title, that highlight key ideas in the Torah portion. In this manner, he familiarized the (observant) Jewish reader with contemporary secular culture, while enabling nonobservant Jewish readers and non-Jews to appreciate the depth and relevance of the Hebrew Bible.

Ecological Approach to Politics

In the books written for the public at large, Rabbi Sacks acknowledged the severity of the environmental crisis. Although he did not cite the rich literature of environmental studies, his writings indicate some familiarity with it. For example, in *To Heal the Fractured World* (2005), whose title echoes the concept of *Tikkun Olam*, Rabbi Sacks referred to the path-breaking work of Hans Jonas (d. 1993), the Jewish philosopher who understood the connection between modern technologically and ecological crisis and who insisted on human responsibility for the well-being of future generations (Levy 2002, 77–94; Morris 2013).

An ordinary notion of responsibility, Rabbi Sacks argues, is not adequate to handle the ethical challenges of environmental crisis. Only a religious notion of responsibility—"responsibility to the infinite in terms of space, eternal in terms of time, can sometimes be more cogent than secular alternative" (Sacks 2005, 8). By this, he does not mean to say that "religious individuals are more environmentally active than their secular counterparts; we all know the problem and we all try to help."

The Judaic ethics of responsibility is anchored in the doctrine of creation: because humans are created by God in the image of God, humans are *obligated* to observe divine commands, including the command to preserve the world that God had created. Rabbi Sacks's understanding of the ethics of responsibility follows the footsteps of other Modern Orthodox thinkers such as Rabbis Joseph Soloveithcik, Norman Lamm, and Walter Wurzburger. For example, Wurzburger summarizes the environmental ethics of responsibility as follows:

> The Earth is not ours. We are responsible to God for the preservation of His World. When conquest of nature for the enhancement of human welfare is viewed not as an act of self-assertion but as a religious obligation, conservation of non-replenishable resources and protection of the environment cease to be mere matters of prudence but acquire the status of ethico-religious imperatives.
>
> (Wurzburger 1994, 107)

Rabbi Sacks shared that understanding but went beyond it by applying ecological thinking to social and political issues, or as he put it, by taking "an ecological approach to politics" (Sacks 1997, 224).

Rabbi Sacks's "ecological approach to politics" does not denigrate the significance of the individual or the relevance of the state, but adds "the families, communities and voluntary networks that make up so much of our lives" (Sacks 1997, 224). The approach is "ecological" in the sense that it views the sociopolitical world holistically; it recognizes the inherent dynamism of social systems; it focuses on the health, proper functioning, and growth of the sociopolitical system; and it recognizes the interdependence of all parts of the system. To think ecologically is to think in terms of life processes, that is, to be committed to "celebrating life," to use the title of one of Sacks's books (2000). Based on his "ecological approach to politics," Rabbi Sacks was able to speak about "the politics of hope," the title of the book from which the above quote is taken. Why did he remain hopeful despite the severity of the ecological crisis? The answer, I believe, lies in the similarity he saw between ecology, language, and morality.

The argument of *The Politics of Hope* (1997) is that "the social domain … is in essence a moral domain, and to reclaim it means that we must recover a shared moral language" (Sacks 1997, 210). And morality, Rabbi Sacks explains, is like a language,

> subtle, complex, but with its own distinct coherence, its rules of syntax and semantics, its grammar and vocabulary. With a language there is an open-ended variety of things we can say, and each of us makes sentences that have never been said before. But as individuals we do not make the language. We learn it. We inherit it. We enter its unique yet universal configuration of virtues.
>
> (ibid., 219)

Because both morality and language are acquired in the same manner, Rabbi Sacks rejects John Locke's theory that we invent language through reasoned reflection on private experience, as much as he takes issue with Alasdair MacIntyre's pessimistic assessment that "the possibility of a shared moral life has disappeared" (Sacks 1997, 219). Whereas MacIntyre concluded that we live in an "'after-virtue' age" (MacIntyre 1981, 49–54), Rabbi Sacks argued that

> language is never fragmented beyond recovery … We develop languages because we seek communication. We develop moralities because we seek community. The task of restoring community and morality is one and the same and derives from the same need: to rescue the self from solitude, so that in finding the "We" we can learn to say "I".
>
> (Sacks 1997, 222)

As a public theologian, he took upon himself the task of recovering the shared moral language of the West by focusing on the Hebrew Bible, the foundation of "Judaeo-Christian ethics." In his exegetical works, Rabbi Sacks explicated the "ecological morality" of the Bible, namely the principles of biblical environmental ethics, in order "to heal the fractured world," to use the title of one his books (Sacks 2005) and ensure the future of humanity on the wounded planet Earth.

Biblical Ecological Morality

The main principles of biblical environmental ethics are well known: the responsibility of humanity "to serve and conserve the world" the prohibition against wanton destruction (*bal tashchit*); the environmental significance of Sabbath, the Sabbatical Year, and the Jubilee; the connection between social justice and eco-justices, and the obligation to prevent unnecessary suffering of animals (Tirosh-Samuelson 2017). These principles have been identified already in the early 1970s when Jewish theologians (e.g., Lamm 1971; 1974) defended the Bible against the charges of Lynn White Jr. that the Bible is the cause of the environmental crisis. This book has shown how Jewish eco-theologians and environmental activists interpreted these elements, giving Jewish environmentalism its distinctive identity. Rabbi Sacks's contribution to Jewish environmental discourse lies not in identifying new principles or in finding previously unnoticed textual sources relevant to contemporary environmentalism but in making the moral ecology of the Bible accessible as public theology. When he explicated "the ecological imperative" and "environmental responsibility" in his biblical commentaries, he made these ideas accessible and even compelling for Orthodox Jews, who commonly associate environmentalism with secular culture or with Liberal and Progressive forms of contemporary Judaism. And when he teased out the universal message of the "Jewish Bible," he taught non-Jews, especially Christians, to read the biblical text through the lens of Judaic hermeneutics. In other words, both inwardly and outwardly public theology is an apologetic project as Elain Graham or Ted Peters use the term.

The doctrine of creation anchors biblical moral ecology. In his commentary on Bereshit, the first Torah portion of the Book of Genesis, Rabbi Sacks's essay, "Three Stages of Creation," explains "the most revolutionary as well as the most influential account of creation in the history of the human spirit" (Sacks 2009, 23). The first stage is "the stage of saying 'Let there be.' The second is the stage of 'and there was.' The third is the stage of seeing 'that it is good'" (ibid., 25). According to Rabbi Sacks,

> implicit in the first chapter of Genesis is thus a momentous challenge: Just as God is creative, so you be creative. In making man, God endowed one creature—the only one thus far known to science—with the capacity not merely to adapt to his environment, but to adapt his environment to him; to shape the world; to be active, not merely passive, in relation to the influence and circumstances that surround him.
>
> (ibid., 24)

The doctrine of creation thus

> foreshadowed the Jewish doctrine of revelation: that God reveals Himself to humanity not in the sun, the stars, the wind or the storm but in and through words—sacred words that make us co-partners with God in the work of redemption.
>
> (Sacks 2009, 25)

If the "first stage of creation" has to do with imagination, the second stage of creation is *will*. The biblical creation narrative teaches that "we are not pre-programmed machines; we are persons, endowed with will. Just as God is free, so we are free; and the entire Torah is a call to humanity to exercise responsible freedom in creating a social world which honours the freedom of others. Will is the bridge from 'Let there be' to 'and there was' (ibid., 26). And the third stage of creation—and "God saw that it was good"—teaches us that the created world "is the arena of blessing and good," and that means

> that the religious life is not to be sought in retreat from the world and its conflicts into mystic rapture or nirvana. God wants us to be part of the

> world, fighting its battles, tasting its joy, celebrating is splendour. But there is more ... When we recognize the goodness in someone, we more than create it, we help it to become creative. This is what God does for us, and what He calls us to do for others.
>
> (ibid., 27)

Judaic earthly spirituality and its ethics of responsibility could not have been expressed better.

The doctrine of creation in "the image of God" has been interpreted in many ways, but it remains problematic for environmentalists, because it appears to accentuate human exceptionalism. The biblical narrative famously and controversially stated: "God blessed them and God said to them, 'Be fertile and increase, fill the earth and master it; and rule the fish and the sea, the birds of the sky, and all the living things that creep on earth'" (Gen. 1:28). Did the Bible give humanity the license "to master" or "have dominion" over every living being, as Lynn White Jr. charged? Following other Jewish environmentalists and eco-theologians, Rabbi Sacks insists that this verse should not be read in isolation but together with the rest of the Torah portion. The second biblical creation narrative teaches that "The Lord God took the man and placed him in the Garden of Eden, to till it and tend it" (Gen. 2:15). For Sacks, that is clear evidence that the Bible did not interpret 'mastery' or 'dominion' to mean domination and exploitation. Humanity was given the right to use the natural world and benefit from its goodness but not to abuse it. In his essay "Environmental Responsibility" on Parashat Shofetim in Deuteronomy, Rabbi Sacks explained Genesis 2:15 as follows:

> Man was set in the Garden of Eden "to work it and take care of it" (Gen. 2:15). The two Hebrew verbs are significant. The first—*le'avda*— literary means "to serve it." Man is not just a master but also a servant of nature. The second—*le-shomrah*—means "to guard it." This is the verb used in later Torah legislation to describe the responsibilities of a guardian of property that does not belong to him. He must exercise vigilance in his protection and is liable for loss through negligence. This is perhaps the best short definition of man's responsibility for nature as the Bible conceives it. *Man's dominion over nature is thus limited by the requirement to serve and conserve.* The famous story of Genesis 2–3— eating the forbidden fruit and the subsequent exile from Eden—makes just this point. *Not everything that we can do, may we do. Transgress the limit, and disaster follows.*
>
> (Sacks 2016, 303, emphasis added)

According to the biblical text, then, the task of humanity is to care for and protect the world God had created.

In his essays—"Beyond Nature" on Parashat Bereshit (Sacks 2016, 9–14) and "Environmental Responsibility" on Parashat Shofetim (ibid., 299–304),

"Animal Welfare" on Parashat Ki Tetze (Sacks 2019, 195–200), and "The Ecological Imperative" on Parashat Shofetim (ibid., 175–80)—Rabbi Sacks explained the difference between scientific laws and moral laws and explored how the obligation to care for the natural world is derived from the covenantal relationship with God. Endorsing the insights of Rabbi Samson Hirsch, Rabbi Sacks imparted to his reader the following message: "'The earth is the Lord's and all that is in it' (Ps. 24:1). We are its guardians on behalf of its Creator, for the sake of future generations" (Sacks 2019, 180). This is Jewish environmental responsibility in a nutshell.

Published posthumously in 2021, Rabbi Sacks's *Studies in Spirituality* offers a nice coda to this chapter and to this study of Jewish earthly spirituality. Rabbi Sacks wrote the book because "so many people I meet are searching for it, and because they say they do not always find it in Judaism today" (Sacks 2021, xviii). And what is 'spirituality'? Rabbi Sacks defines it as follows: "Spirituality is what tells us that 'man does not live by bread alone but by everything that comes from the mouth of God' (Deut. 8:13)." His explanation of 'spirituality' in the Introduction to the volume can serve as the coda to our book:

> We are not just physical beings with biological drives. We have hopes, dreams, fears, loves. We strive for connection. We make sacrifices for the sake of others. We search for meaning. We experience transcendence … We are spiritual beings. We are not just random mutations of genes, blindly replicating themselves into the future. Nor are we merely profit-maximisers or pleasure-seekers. We seek meaning in our lives and we are lifted when we are able to endow ordinary acts with the charisma of holiness. In this, we have the precedent of the great lives of those who came before us, whose stories recorded in our sacred texts still speak to us, showing how ordinary people can be raised to extraordinary heights when they open their minds and moods to the One who is beyond us yet within us … We will need spiritual strength even more than military, economic or technological strength in the years ahead. For it is spirituality that teaches us that life is sacred, that there is more to happiness than the pursuit of wealth, power, success, or fame, and that though life is short we can, at blessed moments, experience the transformative power of joy giving us a sense of eternity in the midst of time.
>
> (Sacks 2021, xviii; xxi; xx).

Speaking in the public sphere and to the public from a Judaic perspective may rattle Jews who prefer to keep Judaism privatized, and it undoubtedly disrupts the conventions of the research university. However, only if Jewish environmental spirituality is promoted as public theology does it have a chance to make a difference. Public theology not only motivates people to act about those things they care most about but also critique public officials and civil servants who fail to abide by the values of their tradition. If we care about the future of the created world, we should welcome the work of Jewish public theologians who wish to protect the created world for the sake of the Creator God and for the sake of future generations, humans and more-than-humans.

References

Barbieri, William. 2022. "Post-Secularity and the Pluralization of Public Theology." In *T & T Clark Handbook of Public Theology*, ed. Hubenthal and Alpers, 69–85. London, New York, Oxford, New Delhi, and Sydney: T & T Clark.

Cooper, Levi. 2013. "The Tikkun Olam Catch All." *Jewish Educational Leadership* 11 (1) (Winter 2013): 46–53.

Cosgrove, Elliot J. (ed.). 2010. *Jewish Theology in Our Time: A New Generation Explores the Foundations and Future of Jewish Belief.* Woodstock, VT: Jewish Lights Publishing.

Dorff, Elliot. 2005. *The Way into Tikkun Olam (Repairing the World)*. Woodstock, VT: Jewish Lights Publishing.

Dorff, Elliot. 2008. *The Jewish Approach to Repairing the World (Tikkun Olam): A Brief Introduction for Christians.* Woodstock, VT: Jewish Lights Publishing.

Fox, Sandra. 2023. *The Jews of Summer: Summer Camp and Jewish Culture in Post-war America.* Palo Alto: Stanford University Press.

Gottlieb, Roger S. 1987. *History and Subjectivity: The Transformation of Marxist Theory.* Philadelphia: Temple University Press.

Gottlieb, Roger S. 1997. *The Ecological Community: Environmental Challenges for Philosophy, Politics, and Morality.* London: Routledge.

Gottlieb, Roger S. 1999. *A Spirituality of Resistance: Finding a Peaceful Heart and Protecting the Earth.* New York: Crossroad.

Gottlieb, Roger S. 2002. *Joining Hands: Religion and Politics Together for Social Change.* Boulder, CO: Westview Press.

Gottlieb, Roger S. 2006a [2009]. *A Greener Faith: Religious Environmentalism and Our Planet's Future.* Oxford: Oxford University Press.

Gottlieb, Roger S. 2013. *Spirituality: What It Is and Why It Matters.* Oxford University Press.

Gottlieb, Roger S. 2015. *Political and Spiritual Essays on Religion, Environmental, Disability and Justice.* Lanham, MD: Rowman and Littlefield.

Gottlieb, Roger S. 2019. *Morality and the Environmental Crisis.* Cambridge: Cambridge University Press.

Gottlieb, Roger S. (ed.). 1990. *Thinking the Unthinkable: Meanings of the Holocaust.* New York/Mahwah: Paulist Press.

Gottlieb, Roger S. (ed.). *Marxism, 1844–1990: Origins, Betrayal, Rebirth.* New York and London: Routledge.

Gottlieb, Roger S. (ed.). 1996. *This Sacred Earth: Religion, Nature, Environment.* New York: Routledge.

Gottlieb, Roger S. (ed.). 2003. *Liberating Faith: Religious Voices for Justice, Peace and Ecological Wisdom.* Lanham, MD: Rowman and Littlefield.

Gottlieb, Roger S. (ed.). 2006. *The Oxford Handbook of Religion and Ecology.* Oxford: Oxford University Press.

Graham, Elaine. 2022. "Public Theology as Apologetics." *T & T Clark Handbook of Public Theology*, ed. Christoph Hubenthal & Christiane Alpers, 109–124. London, New York, Oxford, New Delhi, and Sydney: T & T Clark.

Harris, Michael J., Daniel Rynhold, and Tamra Wright (eds.). 2012. *Radical Responsibility: Celebrating the Thought of Chief Rabbi Lord Jonathan Sacks.* London: London School of Jewish Studies; New York: The Michael Scharf Publication Trust, and YU Press; Jerusalem Maggid Books.

Hübenthal, Christoph and Alpers, Christiane (eds.). 2022. *T & T Clark Handbook of Public Theology.* London, New York, Oxford, New Delhi, and Sydney: T & T Clark.

Jonathan Sacks: *The Rabbi Sacks Legacy.* 1981–2025. "The Life of Rabbi Lord Jonathan Sacks." https://rabbisacks.org/life-of-rabbi-jonathan-sacks/timeline/.

Katz, Steve. 1985. *Post-Holocaust Dialogues: Critical Studies in Modern Jewish Thought*. New York: New York University Press.

Krasner, Jonathan. 2023. "*The World is Broken, So Humans Must Repair It: The History and Evolution of Tikkun Olam.*" Available at: https://www.brandeis.edu/jewish-experience/history-culture/2023/may/tikkun-olam-history.html.

Kranson, Rachel. 2017. *Ambivalent Embrace: Jewish Upward Mobility in Postwar America*. Chapel Hill: University of North Carolina Press.

Lamm, Norman. 1971. "Ecology in Jewish Law and Theology." In his *Faith and Doubt: Studies in Traditional Jewish Thought*, 162–186. New York: Ktav Publishing House.

Lamm, Norman. 1974. "Nature." In his *The Good Society: Jewish Ethics in Action*, 213–222. New York: Viking Press.

Lerner, Michael. 1994. *Jewish Renewal: A Path to Healing and Transformation*. New York: A Grosset/Putnam Book.

Lerner, Michael. 1996. *The Politics of Meaning: Restoring Hope and Possibility in an Age of Cynicism*. New York: Addison-Wesley Publishing Company.

Lerner, Michael. 2006. *The Left Hand of God: Healing America's Political and Spiritual Crisis.* San Francisco: HarperSanFrancisco.

Lerner, Michael (ed.). 2007. *Tikkun Reader: Twentieth Anniversary.* Lanham, MD: Rowman & Littlefield.

Lerner, Michael and Cornell West. 1996. *Jews and Blacks: A Dialogue on Race, Religion, and Culture in America*. New York: A Plume Book.

Levy, David J. 2002. *Hans Jonas: The Integrity of Thinking*. Columbia and London: University of Missouri Press.

MacIntyre, Alisdair. 1981. *After Virtue: A Study in Moral Theory.* Notre Dame, IN: University of Notre Dame Press.

Mittleman, Alan (ed.). 2003. *Religion as a Public Good: Jews and Other Americans on Religion in the Public Square*. Lanham, MD: Rowman and Littlefield Publishers.

Mittleman, Alan, Robert Licht, and Jonathan D. Sarna (eds.). 2002. *Jews and the American Public Square: Debating Religion and Republic*. Lanham, MD: Roman & Littlefield Publishers.

Morris, Theresa. 2013. *Hans Jonas's Ethics of Responsibility: From Ontology to Ecology.* Albany, NY: SUNY Press.

Renaud, Terence. 2021. *New Lefts: The Making of a Radical Tradition*. Princeton and Oxford: Princeton University Press.

Rose, Or N., Jo Ellen Green Kaiser, and Margie Klein (eds.). 2007. *Righteous Indignation: A Jewish Call for Justice*. Woodstock, VT: Jewish Light Publishing.

Rosenthal, Gilbert S. 2005. "Tikkun ha-Olam: The Metamorphosis of a Concept." *Journal of Religion* 85 (2): 214–240.

Sacks, Jonathan, 1990. *The Persistence of Faith: Religion, Morality and Society in a Secular Age*. London and New York: Continuum.

Sacks, Jonathan, 1997. *The Politics of Hope*. London: Vintage.
Sacks, Jonathan, 2000. *Celebrating Life*. London: Bloomsbury Continuum.
Sacks, Jonathan, 2002. *The Dignity of Difference: How to Avoid the Clash of Civilizations*. London: Bloomsbury Continuum.
Sacks, Jonathan, 2005. *To Heal a Fractured World: The Ethics of Responsibility*. New York: Schocken Books.
Sacks, Jonathan, 2007. *The Home We Build Together: Recreating Society*. London: Bloomsbury Continuum.
Sacks, Jonathan, 2009. *Genesis: The Book of Beginnings. Covenant and Conversation: A Weekly Reading of the Jewish Bible*. Jerusalem: Maggid Books.
Sacks, Jonathan, 2011. *The Great Partnership: God, Science, and the Search for Meaning*. London: Hodder & Stoughton.
Sacks, Jonathan, 2015. *Not in God's Name: Confronting Religious Violence*. New York: Schocken.
Sacks, Jonathan, 2016. *Essays on Ethics: A Weekly Reading of the Jewish Bible*. Jerusalem: Maggid Books.
Sacks, Jonathan, 2019. *Deuteronomy: Renewal of the Sinai Covenant. Covenant and Conversation: A Weekly Reading of the Jewish Bible*. Jerusalem: Maggid Books & the Orthodox Union.
Sacks, Jonathan, 2020. *Morality: Restoring the Common Good in Divided Times*. New York: Basic Books.
Sacks, Jonathan, 2021. *Studies in Spirituality: A Weekly Reading of the Jewish Bible*. Jerusalem: Maggid Books & OU Press.
Sarna, Jonathan D. 2002. "Church-State Dilemmas of American Jews." In *Jews and the American Public Square: Debating Religion and Republic*, ed. Alan Mittleman, Robert Licht, and Jonathan D. Sarna. Lanham, MD: Rowman and Littlefield Publishers.
Schwarz, Sidney. 2006. *Judaism and Justice: The Jewish Passion to Repair the World*. Woodstock, VT: Jewish Lights Publishing.
Shatz, David, Chaim I. Waxman, and Nathan J. Diament (eds.). 1997. *Tikkun Olam: Social Responsibility in Jewish Thought and Law*. Northvale, NJ and London: Jason Aronson.
Sherwin, Byron. 2013. "Tikkun Olam: A Case of Semantic Displacement." *Jewish Political Studies Review* 25 (3/4): 43–58.
Tikkun (Spring 2015). "The Place of Home in the Age of Climate Disaster."
Tikkun (Winter 2015). "Jubilee and Debt Abolition."
Tikkun (Fall 2017). "Ecological Civilization and the New Enlightenment."
Tikkun. 2025. www.tikkun.org.
Tirosh-Samuelson, Hava. 2017. "Jewish Environmental Ethics: The Imperative of Responsibility." In *The Wiley Blackwell Companion to Religion and Ecology*, ed. John Hart, 179–194. Hoboken, NU; Chichester, West Sussex, England.
Tirosh-Samuelson, Hava. 2023. "Rabbi Jonathan Sacks: Religious Pluralism and the Partnership of Religion and Science." *Kulturni Studia/Cultural Studies* 20 (1): 1–25. Available at: doi:10.7160/KS.2023.200101.
Waskow, Arthur I. 1962. *The Limits of Defense*. Garden City, NY: Doubleday & Company.
Waskow, Arthur I. 1963. *The Worried Man's Guide to World Peace: A Peace Research Institute Handbook*. Garden City, NY: Anchor Books.
Waskow, Arthur I. 1965. *The Debate over Thermonuclear Strategy*. Boston: D.C. Heath and Company.

Waskow, Arthur I. 1966. *From Race Riots to Sit-In, 1919 and the 1960s: A Study in the Connection between Conflict and Violence.* Garden City, NY: Doubleday & Company.

Waskow, Arthur I. 1971. *The Bush Is Burning!: Radical Judaism Faces the Pharaohs of the Modern Superstate.* New York: The Macmillan Company.

Waskow, Arthur I. 1978. *God-Wrestling*. New York: Schocken Books.

Waskow, Arthur I. 1982. *Seasons of Our Joy.* New York: Bantam Books.

Waskow, Arthur I. 1983. *These Holy Sparks: The Rebirth of the Jewish People.* San Francisco: Harper & Row.

Waskow, Arthur I. 1996. *Godwrestling Round 2*. Woodstock, VT: Jewish Light Publishing.

Waskow, Arthur O. 2020. *Dancing in God's Earthquake: The Coming Transformation of Religion*. Maryknoll, NY: Orbis Books.

Waskow, Arthur Ocean and Phyllis Ocean Berman. 1996. *Tales of Tikkun: New Jewish Stories to Heal the Wounded World.* Northvale, NJ: Jason Aronson.

Waskow, Arthur Ocean and Phyllis Ocean Berman. 2002. *A Time for Every Purpose under Heaven: The Jewish Life-Spiral as a Spiritual Path.* New York: Farrar, Straus and Giroux.

Waskow, Arthur Ocean and Phyllis Ocean Berman. 2011. *Freedom Journeys: The Tales of Exodus and Wilderness across Millenia*. Woodstock, VT: Jewish Light Publishing.

Wurzburger, Walter. 1994. *Ethics of Responsibility.* Philadelphia and Jerusalem: Jewish Publication Society.

Conclusion

One book cannot do justice to the complexity, richness, and importance of contemporary Jewish environmentalism. Intended as a contribution to several academic fields—Jewish Studies, religion and ecology, religion and science, women and gender studies, and environmental humanities—this book discusses Jewish environmentalism to shed light on the dynamics of post-secularity. Looking at Jewish environmentalism historically, the book situates it in the Jewish experience in the 20th century as the world became aware that the future of human life on Earth hangs in the balance and as the Jewish people experienced the horrors of the Holocaust, on the one hand, and the revival of Jewish political sovereignty, on the one hand. Examining Jewish environmentalism in the Diaspora and in Israel, the book recognizes the crosscurrents and internal tensions within the Jewish experience but also the relevance of Judaism to our current contemporary crisis.

Jewish environmental activists refer to their brand of Judaism as Eco-Judaism and to themselves as EcoJews to distinguish themselves within Judaism. However, this book does not use these monikers, nor does it refer to Jewish environmentalism as a 'movement,' because Jewish environmentalism is too diffused and decentralized. Preferring to talk about Jewish environmentalism as a sensibility and using 'environmental spirituality,' 'ecological spirituality,' and 'earthly spirituality' interchangeably, this book shows that it cuts across all strands of contemporary Judaism.

Even though Judaism today does not speak in one voice, on environmental matters there is considerable agreement among the various strands of contemporary Judaism, because they all take the Hebrew Bible as their point of departure for reflecting about the natural world. Hence, all strands of contemporary Judaism affirm that the natural world belongs to the Creator rather than to human beings; that God revealed the Torah to the People of Israel; and that the ideal future for which Jews hope depends on what Jews do in the present. The doctrines of creation, revelation, and redemption articulate an ethics of responsibility and care for the natural world, ethics expressed in certain rituals and practices. What a modern Jew chooses to believe or to do remains a matter of subjective decision, but the modern Jew has many resources that inspire care and concern for the created world.

DOI: 10.4324/9781351123945-9

All strands of contemporary Judaism maintain that humanity's task is to protect the world God had created because the world belongs to God. However, the denominational structure of contemporary Judaism causes fragmentation and duplication of endeavors that only weaken the environmental vision. Fortunately, there are national Jewish environmental organizations that transcend denominationalism, offering environmental education to Jews of diverse religious affiliations and levels of religious observance. Growing environmental awareness has brought about the "greening" of Jewish communal institutions—synagogues, schools, community centers, and youth camps—by introducing environmental practices such as recycling, implementing LEED building codes, installing energy-saving devices, or planting communal gardens and engaging in composing and encouraging environmental education. These earthly activities help deepen the spiritual life of Jews who now see themselves as part of the divinely created world which they are commanded to protect.

The book began with an overview of the ecological crisis brought about by human activities during the past 250 years or so. Contrary to scientists and technology experts who believe that humanity can engineer itself out of the ecological crisis, this study suggests that technoscience is not the panacea, although any discussion of environmental challenges must consider technoscience. Instead, like all forms of religious environmentalism, Jewish environmentalism is concerned with changing human attitudes, perceptions, and mindsets to generate the right conduct toward the natural world of which humans are a part. And Jewish environmentalism frames those attitudes in terms of the relationship between God and humanity.

By thinking about nature in religious terms, Jewish environmentalism negates that desacralization of the world that began with the Enlightenment and the triumph of modern science. Claiming to understand the laws of nature, humanity harnessed its forces for the sake of human "progress" and the "betterment of society." Indeed, the spectacular achievements of modern technoscience improved human standards of living, extended human life span, and facilitated more leisure time in which to pursue life's pleasures. But the bright story of modern progress also had a dark side. In the 20th century, the two World Wars, the industrialization of death in Auschwitz, and the specter of nuclear annihilation in Hiroshima exposed the dark side of modern technoscience. Toward the end of the 20th century and more so in the first two decades of the 21st century, rapid advances in technoscience profoundly changed human relations to the physical environment causing over-exploitation of natural resources, pollution and toxicity of air, water, and soil, and most dangerously, irreversible global warming and climate change. And if these were not bad enough, in the early 21st century, converging technologies, especially biotechnology, digital technology, and artificial intelligence further empowered the human species to transform not only the biological bodies of humans, including even their own brains, but also the bodies of their offsprings in future generations.

In the Anthropocene humans reign supreme, but human mastery also ruins the web of life. Does the Anthropocene spell "progress" or "peril"? Will the Anthropocene be the final triumph of humanity or the beginning of humanity's own demise? How should the human species relate to the non-human world, or rather the "more-than-human world"? Environmentalism responds to these questions by insisting that humans are not the masters of nature but integral parts of nature, interconnected with and interdependent on all other ecosystems and forms of life. Without ignoring or denying human distinctiveness, environmentalism also maintains that because human beings have destroyed natural systems, they have a moral responsibility to repair what they have destroyed, heal what they have wounded. Jewish environmentalism frames that obligation in religious terms: God created the world, and humanity was given the task to protect and preserve the world and all its inhabitants that ultimately belongs to God. In so doing, Jewish environmentalism is post-secular.

Rejecting the privatization of religion, post-secularism brought religion back to the public sphere, insisting on its relevance to social, economic, and political life. Post-secularism also critically reexamined the philosophical assumption of the Enlightenment Project and the social outcomes of technoscience. Post-secularism does not mean the end of the secular epoch, or the rise of a non-secular civilization. Post-secularism only qualified, critiqued, and corrected the disenchantment of the world brought about by the so-called Great Disentangling of the Enlightenment Project. Challenging binary dichotomies characteristic of modern secularity, post-secularism proposed new perceptions, narratives, and rituals, all under the rubric of "spirituality." In the post-secular age, we find spiritual forms that hybridized the "secular" and the "religious," but we also witness the open clash of conflicting crosscurrents in an increasingly polarized society, as the relationship between faith and politics is reformulated or renegotiated. In the post-secular age, a plurality of spiritual formations re-enchants what was previously disenchanted or re-entangles what was previously disentangled. And precisely because religion is now very present in the public sphere, taking a stand about a range of social, economic, and political issues, polarization, fragmentation, and debate now abound. The post-secular age is inevitably contentious.

When theologians attend to societal issues, they engage in public theology. Public theology speaks from the beliefs and practices of the religious community the societal issues of the nation and even the global community. In the Christian world, public theology promotes its vision through the Global Network for Public Theology (GNPT), the *International Journal of Public Theology* (IJPT), and research academic centers in universities in England, Europe, South Africa, and Australia. Among Jews, public theology is less prevalent because processes of secularization evolved differently than for Christians, because the differentiation of the public and private spheres enabled Jews to be part of predominantly non-Jewish society, and because

Jewish academics welcomed and endorsed the conventions of the modern research university. Nonetheless, Jewish public theology does exist today: Jewish theologians do reflect on the role of Judaism in American public life and the relationship between Judaism and the liberal democratic state; Jewish ethicists engage biomedical and bioethical issues, and Jewish political theorists ponder the desired relationship between religion and politics in the State of Israel.

Jewish environmentalism is an example of Jewish public theology. The Jews who articulated and promoted environmental spirituality are rabbis, theologians, and religious philosophers for whom the Jewish religion itself offers the most meaningful guide for the challenges of the eco-crisis. When these thinkers address "down-to-earth" issues of pollution, hazardous wastes, extracting industries, labor, energy policy, global warming, and climate change by invoking Jewish canonic texts, beliefs, values, and principles, they engage in public theology. When they instruct us to behave in pro-environmental ways to promote the public good as part of the covenantal relationship with God, they engage in public theology. When they reinterpret biblical, rabbinic, philosophic, kabbalistic, or Hasidic sources and apply them to environmental problems, they engage in public theology; and when they set *Tikkun Olam* as the religious goal of civic engagement that will heal the environmentally damaged world, they engage in public theology. Thus, Jewish environmental spirituality has constructive and prescriptive dimensions: the constructive aspect refers to the exegetical/hermeneutical work of Jewish environmentalism, for example, in the discourse on the Sabbath, the Sabbatical Year, and the Jubilee, the principles of *Bal Tashchit*, and *Tza`ar Ba`aley Hayyim* and the prescriptive aspect refers to the activities that constitute the environmentally mindful life (e.g., adopting plant-based diet, avoiding wastefulness, refraining from fracking and minimizing extraction of minerals, buying electric cars, or purchasing solar panels).

Like other forms of public theology, Jewish earthly spirituality is unavoidably political since dealing effectively with the ecological crisis requires legislation, regulation, and deliberative action. But even if environmentalism is necessarily political, it is not necessarily "secular" in the sense of being nonreligious. Indeed, as this book demonstrates, promoting the well-being of the environment does not belong to this or that side of the political divide, or put it differently, environmentalism, politics, religion, and secularity do not neatly align. Religious people on the political Right can and do advocate for the protection of the environment because they see it as divine creation; conversely, secular and secularized people who are affiliated with the political Left can and do frame their environmental activism in religious or spiritual categories. Jewish environmental spirituality, I contend, can transcend political divisions and that may create space for people of different political affiliations and persuasions to cooperate and collaborate to address environmental degradation.

Jewish environmental spirituality has much in common with other faith traditions who care about the well-being of the physical world, especially

Christianity and Islam. Judaism frames human obligations toward the natural world in terms of an ethics of responsibility and care that posits a causal connection between the moral quality of human life and the vitality of God's creation. Conversely, the corruption of society is seen as closely linked to the corruption of Nature. In both cases, injustice arises from human greed and failure to protect the original order of Creation. The just allocation of nature's resources is therefore a religious matter of the highest order, and all three Abrahamic traditions insist on the moral obligation to care for the marginalized in society to ensure that their needs are protected. Indeed, the poor and the marginalized disproportionately suffer the harm of environmental degradation and economic plight causes the poor to degrade their environment, which, in turn, further exacerbates their economic plight. Since the three religious traditions see social justice and eco-justice as intertwined, there is a broad foundation for interfaith cooperation and collaboration. In the age of climate change and ecological collapse, religious environmentalism can serve as the platform for constructive relationships among world religions.

The message of the book, then, is quite optimistic: despite the massive damage to the environment, world religions, and Judaism in particular, have articulated meaningful responses. The Jewish tradition undoubtedly supports an "environmentally correct" life, namely, a way of life that is deeply concerned about the natural environment because it was created by God. In the Judaic environmental spirituality, environmental mindfulness and practices deepen the spiritual life of Jews. Alas, the environmental mindset conflicts with the capitalist ethos of contemporary society wherein self-centered greed, craving for power and domination, and disregard to suffering of human beings and other forms of life are rampant. To take Jewish earthly spirituality seriously leads one to caring, protecting, and preserving God's created world, a world which ultimately belongs to God rather than to humans. Jewish environmental spirituality also critiques the ethos of the Anthropocene in which technological know-how reached unprecedented heights and technoscience is even invested with salvific powers. Against this idolatrous veneration of technology, Judaism teaches that created humans have the moral responsibility for the well-being of the physical world to ensure the well-being of future generations. Finding ways to halt and perhaps even reverse environmental decline and degradation is the moral task of our generations, and it could be accomplished only if we adopt the ethics care and responsibility taught by Judaism and shared by other religious traditions. The Judaic ethos of care and responsibility can be endorsed by secularized Jews, by observant Jews, and by Jews who are "spiritual but not religious," as well as by people of other religious and spiritual traditions. If we affirm that "The Earth is the Lord's and the fullness thereof" (Ps. 24:1), Jewish earthly spirituality instructs us to protect the world and ourselves from destruction. This moral task is ours and we are not free to desist from it.

Index

For Product Safety Concerns and Information please contact our EU
representative GPSR@taylorandfrancis.com
Taylor & Francis Verlag GmbH, Kaufingerstraße 24, 80331 München, Germany

www.ingramcontent.com/pod-product-compliance
Ingram Content Group UK Ltd.
Pitfield, Milton Keynes, MK11 3LW, UK
UKHW020144200926
475702UK00005B/913

* 9 7 8 0 8 1 5 3 5 7 7 3 5 *